Environmental Health and Toxicology

Environmental Health and Toxicology

Edited by **Raven Brennan**

SYRAWOOD
PUBLISHING HOUSE

New York

Published by Syrawood Publishing House,
750 Third Avenue, 9th Floor,
New York, NY 10017, USA
www.syrawoodpublishinghouse.com

Environmental Health and Toxicology
Edited by Raven Brennan

Contents

Permissions

List of Contributors

Preface

In my initial years as a student, I used to run to the library at every possible instance to grab a book and learn something new. Books were my primary source of knowledge and I would not have come such a long way without all that I learnt from them. Thus, when I was approached to edit this book; I became understandably nostalgic. It was an absolute honor to be considered worthy of guiding the current generation as well as those to come. I put all my knowledge and hard work into making this book most beneficial for its readers.

Toxins and hazardous compounds affect the environment in various ways, from gradual deterioration of ecosystems to severe chronic diseases. Toxicologists carry out various safety evaluations and risk assessments to analyse the damage caused to environmental health. This book compiles the recent studies in the field of nanotoxicology, treating toxic waste, forensic toxicology, and assays for toxicity assessment. The aim of this book is to present researches that have transformed this discipline and aided its advancement. With state-of-the-art inputs by acclaimed experts of this field, this book targets students and professionals alike.

I wish to thank my publisher for supporting me at every step. I would also like to thank all the authors who have contributed their researches in this book. I hope this book will be a valuable contribution to the progress of the field.

Editor

Effect of *Hibiscus sabdariffa* anthocyanins on 2, 4-dinitrophenylhydrazine-induced tissue damage in rabbits

A. Ologundudu[1]*, A. O. Ologundudu[1], O. M. Oluba[2], I. O. Omotuyi[1] and F. O. Obi[2]

[1]Department of Biochemistry, Adekunle Ajasin University, Akungba Akoko, Ondo State, Nigeria.
[2]Department of Biochemistry, Faculty of Life Sciences, University of Benin, Benin-City, Nigeria.

This study examines the effects of anthocyanin extract of the dried calyces of *Hibiscus sabdariffa* Linn. on the 2, 4 -dinitrophenylhydrazine (2, 4 -DNPH)-induced cytotoxic effects in rabbits. Twenty male adult rabbits used for the study were divided into four groups. Group 1, the control took only water while animals in groups 2 and 4 received 100 mg/kg body weight of the anthocyanin extract of *H. sabdariffa* once daily for 28 days. After the 22nd day of treatment, the rabbits in groups 3 and 4 received 28 mg/kg body weight of DNPH for the remaining 5 days of treatment, after which the animals were sacrificed. Relative to control, DNPH caused significant (p < 0.05) increase in the formation of malondialdehyde (MDA) in serum, liver and brain and decreased the levels of reduced glutathione (GSH) in liver and brain. Also, DNPH caused a significant (p < 0.05) elevation in the activity of glucose-6-phosphate dehydrogenase (G6PD) in the serum and liver. However, pretreatment with *H. sabdariffa* anthocyanin extract significantly (p < 0.05) reduced MDA formation, increased the levels of GSH and maintained at normalcy the activity of G6PD in the tissues, thereby effectively ameliorated the toxic effects of DNPH. These findings indicate that anthocyanin extract from dried calyces of *H. sabdariffa* protects the rabbit against 2, 4 -DNPH lipoperoxidative and cytotoxic effects.

Key words: Anthocyanin extract, 2, 4-dinitrophenylhydrazine, glucose-6-phosphate dehydrogenase, *Hibiscus sabdariffa*, rabbit, reduced glutathione, malondialdehyde.

INTRODUCTION

The human body has a complex system of natural enzymatic and non-enzymatic antioxidant defenses which counteract the harmful effects of free radicals and other oxidants. Protection against free radicals can be enhanced by ample intakes of dietary antioxidants, of which the best studied are vitamins C and E as well as carotenoids (Vertuani et al., 2004). There is a considerable amount of epidemiological evidence revealing an association between diets rich in fruits and vegetables and a decreased risk of cardiovascular disease and certain forms of cancer (Ames, 1983; Block, 1992; Hertog and Feskens, 1993; Wang et al., 2000). It is generally assumed that the active dietary constituents

contributing to these protective effects are antioxidant nutrients such as α - tocopherol and β - carotene. However, recent investigations have revealed that polyphenolic components of plants do exhibit antioxidant properties and do contribute to the anticarcinogenic or cardioprotective actions brought about by diet (Wang et al., 2000; Stanner et al., 2004). In particular some beverages such as red wine and tea have been shown to elicit antioxidant properties in both *in vitro* and *in vivo* systems (Kanner et al., 1994). Among the more than 300 species of *Hibiscus* plant is *Hibiscus sabdariffa* L., which has many medicinal uses (Morton, 1987; Gill, 1992). The dried calyces contain the flavonoids - gossypetin, sabdaretin, hibiscetin and anthocyanins (Pietta, 2000). Flavonoids are phenolic substances that act in plants as antioxidants. Antioxidant vitamins such as vitamins C and E along with flavonoids have been shown to be effective in reducing atherosclerosis along with many other

---

*Corresponding author. E-mail: oluologundudu@yahoo.com.

diseases (Jackson et al., 1993; Gaxlane et al., 1994; Amin and Buratovich, 2007). There are indications that the extract from the red calyces of *H. sabdariffa* possess antioxidant principles (Tseng et al., 1997; Wang et al., 2000; Ologundudu and Obi, 2005; Ologundudu et al., 2006a, b; Ologundudu et al., 2009a, b). This research was therefore carried out to evaluate the protective effect of *H. sabdariffa* anthocyanins using the model of 2, 4 - dinitrophenylhydrazine-induced oxidative stress in rabbits.

MATERIALS AND METHODS

Experimental animals and materials

Male rabbits (weight range 800 - 1000g and four months old) used for this study were purchased from a local breeder in Benin City, Nigeria. 2, 4 -Dinitrophenylhydrazine, trichloroacetic acid, sodium chloride and diethyl ether were purchased from BDH Chemical Company (Poole, England), 2,-thiobarbituric acid from Koch-Light Laboratories (England). Hydrochloric acid and absolute ethanol were obtained from WN Laboratories (US) and glucose- 6 - phosphate dehydrogenase kit was obtained from Randox Laboratories, UK. Chow (Growers mash) was obtained from Bendel Feed and Flour Mills, Ewu, Edo State, Nigeria.

Preparation of anthocyanin extract

Anthocyanin extract from *H. sabdariffa calyces* was prepared according to the method described by Hong and Wrolstad, (1990a). 1 kg of *H. sabdariffa* calyces was pulverized and extracted with 10 l of 0.1% trifluoroacetic acid (TFA) solution for 12 h at 40 °C. The extract was filtered through filter paper (Advantech filter paper no. 5C). The filtrate was applied to sepabeads SP-207 resin column (Mitsubishi Chemicals, Japan). The resin was washed with 3 l of water and then eluted with 50% ethanol solution containing 0.1% TFA. The eluate was dried under vacuum at 40 °C. The concentrated eluate was then subjected to high-speed liquid chromatography (HPLC) in order to identify its active principles.

HPLC analysis

The HPLC system consisted of a horizontal flow-through planar centrifuge with a multilayer coil (Pharma-Tech Research Co., Model CCC-1000, MD,USA), a pump (JASCO, 880-PU), a microflow pH sensor (Broadley-James, Model 14, CA, USA), a manual injection valve with a 20 ml loop and a fraction collector (JASCO, SF-212N). The upper phase, consisting of a mixture of *tert*-butylmethylether: 1-butanol: MeCN: water (2:2:1:5 v/v) containing 0.2% of TFA, was used as the stationary phase, while the lower phase was as the mobile phase. A total of 300 mg of crude anthocyanin extract was dissolved in 20 ml of a mixture of the stationary phase: mobile phase (3:1 v/v) and introduced through the injection port. The mobile phase was pumped at 2.5 ml/min, while centrifugation was carried out at 1000 rpm. 4 ml of each fraction was collected. A multi-wavelength detector (Waters, 490E) monitored the absorbance of the effluent at 515 nm.

Treatment of animals

Experimental rabbits were divided into four groups, 5 rabbits each and housed in standard cages. Rabbits were given free access to feed and water throughout the experiment period that lasted for 28 days. Rabbits in group 1 were given a daily dose of 2.5 ml H_2O/kg body weight by gavage for 4 weeks. Similar treatment with anthocyanin extract at dose of 100 mg/kg body weight was given to rabbits in groups 2 and 4. After the 22nd day of the experiment, rabbits in groups 3 and 4 received a dose of 28 mg/kg body weight of 2, 4 -dinitrophenylhydrazine intraperitoneally, for 5 consecutive days.

By the end of the experimental period, the animals were anaesthetized in a diethyl ether saturated chamber. While under anesthesia the abdominal regions were opened exposing the heart and liver. Blood was obtained by cardiac puncture by means of a 5 ml hypodermic syringe and needle and placed in heparinized bottles, centrifuged at 3500 rpm for 10 min (SM 9026B bench centrifuge, Surgifriend Medicals, England). From each rabbit, liver and brain samples were also obtained and 1 g portion of each homogenized in ice-cold saline (1:4, ^{w}/v) and centrifuged at 3500 rpm for 10 min to obtain a clear supernatant.

Biochemical assay protocol

Lipid peroxidation was determined spectrophotometrically by thiobarbituric acid reactive substances (TBARS) method as described in Varshney and Kale (1990). Results were expressed in terms of malondialdehyde (MDA) formed per mg protein. Reduced glutathione concentration in the blood was determined using the method of Jollow et al. (1974). The total activity of glucose-6-phosphate dehydrogenase (EC 1.1.1.49) was determined using assay kit from Randox.

Statistical analysis

The data obtained were subjected to standard statistical analysis of variance (ANOVA) using the procedure of SAS (SAS Inst. Inc.1999). Treatment means were compared using the Duncan procedure of the same software. The significance level was set at P < 0.05.

RESULTS

Figure 1 shows the profile displayed on a multiwave length detector used to monitor the absorbance of the effluent of *H. sabdariffa* extract at 515 nm. The peaks on the graph indicate the different anthocyanins present in the *H. sabdariffa* extract in form of their glucosides. The anthocyanins were identified by extrapolating from the graph shown in Figure 2 which is the HPLC chromatogram of known anthocyanins. The result showed that *H. sabdariffa* calyces contained several anthocyanins but the main ones were delphinidin-3-monoglucoside, cyanidin-3-monoglucoside and petunidin-3-monoglucoside.

As shown in Table 1, the results obtained in this study indicated that intoxication with a single dose of 28 mg/kg DNPH (Group 3) daily for 5 consecutive days resulted in a significant increase (p < 0.05) in plasma, liver and brain malondialdehyde (MDA) concentrations compared to control. However, pretreatment with 100 mg/kg body weight of anthocyanin extract of *H. sabdariffa* for 22 days before DNPH intoxication (Group 4) significantly (p<0.05)

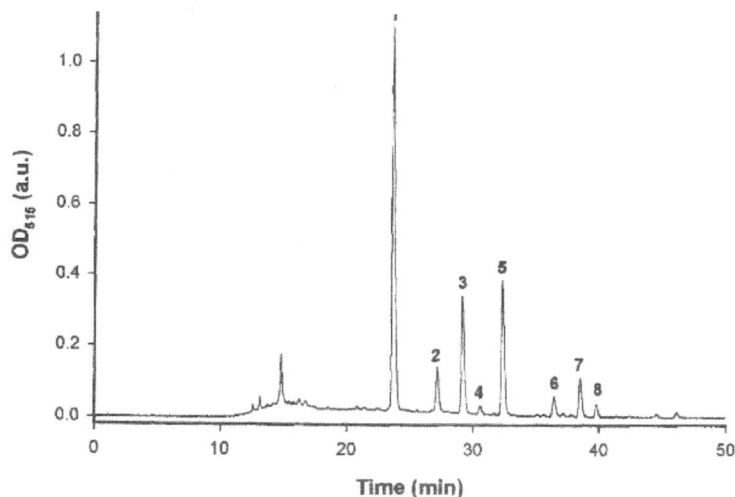

Figure 1. HPLC chromatogram of *Hibiscus sabdariffa* anthocyanins.

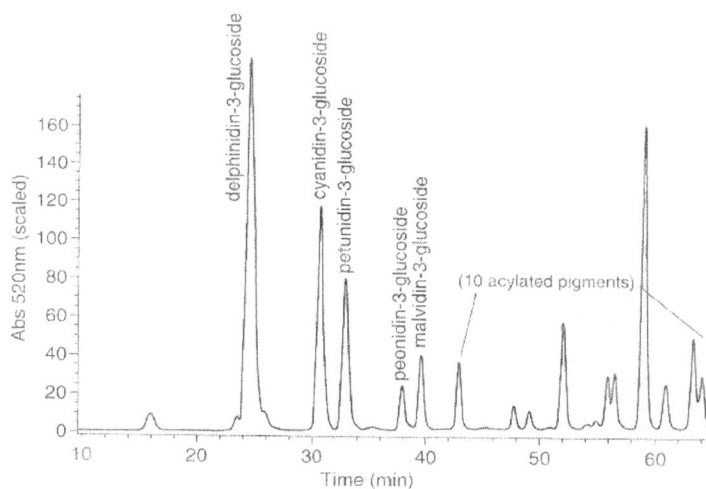

Figure 2. HPLC chromatogram of standard anthocyanins.

decreased by 22.5, 9.2 and 27.2% plasma, liver and brain respectively the MDA concentration compared to control. DNPH administration without pretreatment with anthocyanins extract produced a 51.6, 17.5 and 31.5% significant rise in plasma, liver and brain respectively in MDA concentration compared to control while a non-significant increase was observed in plasma, liver and brain MDA level in group 4 (AN+DNPH) compared to control.

Table 2 shows that DNPH intoxication significantly (p < 0.05) reduced the liver and brain levels of reduced glutathione (GSH) by 34.0 and 28.6% respectively compared to control while treatment with *H. sabdariffa* anthocyanins extract for 28 days without DNPH

intoxication (group 2) increased significantly (p < 0.05) the GSH levels (by 8.6%) in the liver compared to control. Pretreatment with *H. sabdariffa* anthocyanins prior to DNPH intoxication (group 4) caused a significant restoration (almost to control level) of the DNPH depleted GSH concentration in both liver and brain.

Table 3, shows that the activity of glucose-6-phosphate dehydrogenase (G6PD) was significantly (p < 0.05) elevated in the serum following DNPH intoxication as observed when group 3 is compared to control. However, treatment with *H.* anthocyanins alone (Group 2) had no significant (p > 0.05) effect on G6PD activity in the serum and liver compared to control. Pretreatment with the extract prior to DNPH administration (Group 4) maintained

Table 1. Effect of 2, 4 -dinitrophenylhydrazine and *Hibiscus* anthocyanins on the tissue levels of malondialdehyde (μmol per mg protein) of rabbits.

Rabbit group	Treatment	Serum	Liver	Brain
1	2.5 ml H_2O/kg bd. wt. (control)	1.38 ± 0.02	2.71 ± 0.19	3.24 ± 0.13
2	100 mg AN/kg bd. wt.	1.07 ± 0.9^a	2.46 ± 0.20	2.36 ± 0.35^a
3	28 mg DNPH/kg bd. wt.	8.50 ± 0.64^b	8.45 ± 0.78^a	13.43 ± 0.48^b
4	100 mg AN/kg bd. wt. + 28 mg DNPH/kg bd. wt	1.44 ± 0.19	3.70 ± 0.12	3.33 ± 0.38

Results are means of 5 determinations ± SEM. Statistical comparison is strictly within the same tissue. Values carrying notations are statistically ($p < 0.05$) significantly different from control (group 1) while values carrying different superscripts are statistically significantly different from another.

Table 2. Effect of 2, 4-dinitrophenylhydrazine and *Hibiscus* anthocyanins on the tissue levels of reduced glutathione (nmol per g protein) of rabbits.

Rabbit group	Treatment	Liver	Brain
1	2.5 ml H_2O/kg bd. wt. (control)	24.84 ± 1.33	24.48 ± 0.21
2	100 mg AN/kg bd. wt.	26.39 ± 0.25^a	24.71 ± 0.69
3	28 mg DNPH/kg bd. wt.	16.39 ± 0.92^b	17.48 ± 0.19^a
4	100 mg AN/kg bd. wt. + 28 mg DNPH/kg bd. wt	23.53 ± 0.26	24.10 ± 0.38

Results are means of 5 determinations ± SEM. Statistical comparison is strictly within the same tissue. Values carrying notations are statistically ($p < 0.05$) significantly different from control (group 1) while values carrying different superscripts are statistically significantly different from another.

Table 3. Effect of 2, 4 -dinitrophenylhydrazine and *Hibiscus* anthocyanins on the tissue levels of glucose- 6 -phosphate dehydrogenase (μmol per min. per mg protein) of rabbits.

Rabbit group	Treatment	Serum	Liver
1	2.5 ml H_2O/kg bd. wt. (control)	1.27 ± 0.22	33.33 ± 0.37
2	100 mg AN/kg bd. wt.	1.10 ± 0.13	33.83 ± 0.44
3	28 mg DNPH/kg bd. wt.	4.50 ± 0.98^a	34.07 ± 0.06^a
4	100 mg AN/kg bd. wt. + 28 mg DNPH/kg bd. wt.	1.21 ± 0.00	33.40 ± 0.23

Results are means of 5 determinations ± SEM. Statistical comparison is strictly within the same tissue. Values carrying notations are statistically ($p < 0.05$) significantly different from control (group 1) while values carrying different superscripts are statistically significantly different from another.

at normalcy the levels of the enzyme in the liver and serum.

DISCUSSION

An explosion of interest in examining the involvement of free radicals in carcinogenesis has led to the use of dietary antioxidant treatments in quenching free radical-mediated attacks, hence promoting general human health. The antioxidants may act as free radical scavengers, reducing agents, chelating agents for transition metals, quenchers of singlet oxygen molecules and activators of antioxidative defense enzyme systemsto suppress free radical damage in biological systems (Satue-Gracia et al., 1997; Aviram and Fuhrman, 2002).

In this study, malondialdehyde (MDA), reduced glutathione (GSH) and glucose-6 -phosphate dehydrogenase (G6PD) activity were used to indicate the degree of tissue damage following 2, 4 -dinitrophenylhydrazine (DNPH) intoxication and the levels of protection against such damage offered by pretreatment with *H. sabdariffa* anthocyanins prior to DNPH intoxication.

MDA concentration in serum, liver and brain tissue homogenates in our investigations significantly increased after 5 consecutive days of DNPH administration. Increase in tissue levels of MDA are reliable indices of oxidative stress and lipoperoxidative tissue damage (Clemens et al., 1984; Maduka et al., 2003; Ologundudu and Obi, 2005; Ologundudu et al., 2009a, b). Therefore, the profile of MDA in the tissues of DNPH-treated rabbits is a clear indication that DNPH provokes oxidative stress in rabbits. *In vivo*, this toxicant is believed to undergo auto oxidation and becomes a strong oxidant with the ability to initiate lipid peroxidation in membrane phospholipids (Jain and Hochstein, 1979), once the antioxidant defense system has been overwhelmed. The induction of lipid peroxidation is thought to ultimately cause cytotoxic response (Sipes et al., 1977). The lipid oxidation causes disruption of the bilayer and cell integrity accompanied by leakage of cellular content from the damaged organ into the blood stream. However, this phenomenon was effectively blocked when the animals were pretreated with *H. anthocyanin* extract before DNPH intoxication.

In accord with our earlier reports (Ologundudu et al., 2009a,b), the results presented in this study (Table 2) show that DNPH administration caused a statistically (p < 0.05) significant decrease in reduced glutathione (GSH) concentration in the liver and brain. It is well established that GSH, the most important biomolecule protecting against chemically induced cytotoxicity, can participate in the elimination of reactive intermediates by conjugation or by direct free radical quenching. So the decrease in the levels of GSH in these tissues results in the accumulation of free radicals leading to increased rate of lipid peroxidation in these tissues. This study however, showed that pretreatment with anthocyanin extract of *H. sabdariffa* prior to DNPH intoxication significantly increased the concentration of reduced glutathione in the liver and brain, thus reducing the accumulation of free radicals and decreased rate of lipid peroxidation.

Table 3 shows that DNPH administration significantly (p < 0.05) increased both serum and liver activity of G6PD relative to control (Group 1) and anthocyanin extract-treated DNPH-free rabbits (Group 2). This change in G6PD activity is a toxic response to DNPH intoxication. Pretreatment with *H. anthocyanins* however effectively ameliorated this change in both serum and liver (group 4). G6PD is a widely distributed enzyme in tissues and its role as the anchor of reductive metabolism is well established. It does not only represent the major regulatory enzyme of the pentose phosphate pathway but also its metabolic products: NADPH and phosphogluconolactone are involved in reductive anabolism and nucleotide synthesis respectively. The deficiency of this enzyme has been implicated in favism and hemolytic anemia. Its determination in this research was necessitated by the fact that it represents the chief enzyme that provides the reduced glutathione with hydride ion (H^-) in form of NADPH for sustenance of free radical detoxification (Bergmeyer et al., 1974).

The mechanism by which *H. sabdariffa* anthocyanins prevent DNPH-induced changes is not clear at this stage. However, it is likely that the extract protected the tissues from damage by blocking DNPH-induced free radical formation. The protection may also be due to the impaired free radical propagation and or complementation of the antioxidant defense system. Further investigation is required to be able to establish the precise mechanism operating here.

As indicated earlier in this report, the mechanism of DNPH-mediated tissue damage suggests an underlying process of oxidation. Therefore the hypothesis on which this investigation was based is that if the anthocyanin extract of dried calyces of *H. sabdariffa* possesses antioxidant properties, therefore, it would prevent lipid peroxidation and other metabolic side effects of DNPH caused by its oxidant action. Present results demonstrated reasonably well that treatment of rabbits with *H. anthocyanins* prior to DNPH intoxication significantly inhibited its cytotoxic and other metabolic side effects in tissues.

Conclusion

Antioxidant activity of *H. anthocyanins* seems to play a critical role against the 2, 4 -dinitrophenylhydrazine-induced tissue damage in rabbits.

REFERENCES

Ames BN (1983). Dietary carcinogens and anticarcinogen: Oxygen radicals and degenerative diseases. Science 221: 1256-1264.

Amin A, Buratovich M (2007). The Anti-cancer Charm of Flavonoids: A cup-of-tea will do! Recent Pat. Anti-Cancer Drug Discov. 2(2): 109-117.

Aviram M, Fuhrman B (2002). Wine flavonoids protect against LDL oxidation and atherosclerosis. Ann. NY. Acad. Sci. 957: 146-161.

Bergmeyer HU, Bernt E, Schmidt F, Stork H (1974). D-Glucose: determination with hexokinase and glucose-6-phosphate dehydrogenase. In: Methods of enzymatic analysis 2nd English ed. Bergmeyer, H. U. (Ed) Harcourt Brace New York, NY. 1974: 1196-1201.

Block G (1992). The data support a role for antioxidants in reducing cancer risk. Nutr. Rev. 50: 207-213.

Clemens MR, Reinmer H, Waller HD (1984). Phenylhydrazine-induced lipid peroxidation of red blood cells *in vitro* and *in vivo*: Monitoring by the production of volatile hydrocarbons. Biochem. Pharmacol. 53(11): 1715-1718.

Gaxlane JM, Mansion JE, Hennekens CH (1994). Natural antioxidants and cardiovascular disease. Observational epidemiologic studies and randomized trials. In: B. Frel (Ed), Natural antioxidants in human health and disease. Academic press, San Diego pp. 387-409.

Gill LS (1992). Ethnomedicinal uses of plants in Nigeria. University of Benin Press, Benin City p. 132.

Hertog MGL, Feskens EJM (1993). Dietary antioxidants flavonoids and risk of coronary heart disease: The Zutphen Elderly Study. Lancet 342: 1007-1011.

Hong V, Wrolstad RE (1990a). Use of HPLC separation/photodiode detection for characterization of anthocyanins. J. Agric. Food Chem. 38: 708-715.

Jackson RL, Ku G, Thomas CE (1993). Antioxidant, a biological defense mechanism for the prevention of atherosclerosis. Med. Res. Rev. 13: 161-182.

Jain S, Hochstein P (1979). Generation of superoxide radicals by hydrazine: Its role in phenylhydrazine-induced hemolytic anemia. Biochem. Biophys. Acta 586: 128-136.

Jollow DJ, Mitchel JR, Zampaghonic A, Gillette JR (1974). Bromobenzene-induced live necrosis; protective role of glutathione and evidence for 3, 4-bromobenzeneoxide as the hepatotoxic metabolite. Pharmacol. 11: 151-169.

Kanner J, Frankel E, Granit R, German B, Kinsella J (1994). Natural antioxidants in grapes and wines. J. Agric. Food Chem. 42: 64-69.

Maduka HCC, Okoye ZSC, Eje A (2003). The influence of *Sacoglottis gabonensis* stem bark extract and its isolate bergenin, Nigerian alcoholic additive, on the metabolic and hematological side effects of 2, 4-dinitrophenylhydrazine-induced tissue damage. Vascular Pharmacol. 39: 317-324.

Morton JF (1987). Rosella. In: Fruits of Warm Climates. Darling, C.F. (Ed). Media Inc. Green-shore, pp. 281-286.

Ologundudu A, Lawal AO, Adesina OG, Obi FO (2006a). Effect of ethanolic extract of Hibiscus sabdariffa L on 2, 4-dinitrophenylhydrazine-induced changes in blood parameters in rabbits. Global J. Pure Appl. Sci. 12(3): 335-338.

Ologundudu A, Lawal AO, Adesina OG, Obi FO (2006b). Effect of ethanolic extract of Hibiscus sabdariffa L on 2, 4-dinitrophenylhydrazine-induced low glucose level and high malondialdehyde levels in rabbit brain and liver. Global J. Pure Appl. Sci. 12(4): 525-529.

Ologundudu A, Obi FO (2005). Prevention of 2, 4-dinitrophenylhydrazine-induced tissue damage in rabbits by orally administered decoction of dried flower of *Hibiscus sabdariffa L* J. Med. Sci. 5(3): 208-211.

Ologundudu A, Ologundudu AO, Ololade IA, Obi, FO (2009a). Effect of *Hibiscus sabdariffa* anthocyanins on 2, 4-dinitrophenylhydrazine-induced hematotoxicity in rabbits. Afr. J. Biochem. Res. 3(4): 140-144.

Ologundudu A, Ologundudu AO, Ololade IA, Obi FO (2009b). The effect of *Hibiscus* anthocyanins on 2, 4-dinitrophenylhydrazine-induced hepatotoxicity in rabbits. Int. J. Phys. Sci. 4(4): 233-237.

Pietta PG (2000). Flavonoids as antioxidants. J. Nat. Prod. 63(7): 1035-1042.

SAS Institute Inc. (1999). SAS/STAT User's Guide. Version 8 for Windows. SAS Institute Inc., SAS Campus Drive, Cary, North Carolina, USA.

Satue-Gracia MT, Heinonen M, Frankel E (1997). Anthocyanins as antioxidants on human low-density lipoprotein and lecithin-liposome systems. J. Agric. Food Chem. 45: 3362-3367.

Sipes IG, Krishna G, Gillette JR (1977). Bioactivation of carbon tetrachloride, chloroform and bromotrichloromethane: role of cytochrome P-450. Life Sci. 20: 1541-1548.

Stanner SA, Hughes J, Kelly CN, Butriss J (2004). A review of the epidemiological evidence for the antioxidant hypothesis. Pub. Health Nutr. 7(3): 407-422.

Tseng TH, Kao ES, Chu FP, Lin-Wa HW, Wang CJ (1997). Protective effect of dried flower extracts of *Hibiscus sabdariffa L* against oxidative stress in rat primary hepatocytes. Food Chem. Toxicol. 35(12): 1159-1164.

Varshney R, Kale RK (1990). Effect of calmodulin antagonist on radiation induced lipid peroxidation in microsomes. Int. J. Rad. Biol. 58: 733-743.

Vertuani S, Augusti A, Maufredina S (2004). The antioxidants and pro-oxidants network: An overview. Cur. Pharm. Des. 10(14): 1677-1694.

Wang CJ, Wang JM, Lin WL, Chu CY, Chou FP, Tseng TH (2000). Protective effect of *Hibiscus* anthocyanins against tert-butyl hydroperoxide-induced hepatic toxicity in rats. Food Chem. Toxicol. 38: 411-416.

High urinary iodine content (UIC) among primary school children in Ibadan, Nigeria, a public health concern

Onyeaghala A. A.[1*], Anetor J. I.[1], Nurudeen A.[1] and Oyewole O. E.[2]

[1]Department of Chemical pathology, University College Hospital, Ibadan. Nigeria.
[2]Department of Health Promotion and Education, Faculty of Public Health, University of Ibadan.

Urinary iodine excretion is a good marker for the dietary intake of iodine, and is the index for evaluating the degree of iodine deficiency, correction and toxicity. A study, investigating the random urinary iodine level in school children in Ibadan, a South-Western cosmopolitan city of Nigeria, has not been evaluated, thus the emanation of this study. Random urinary iodine was measured in 300 primary school children in Ibadan after obtaining their consent. The urinary iodine level was measured using the standard method of ammonium persulphate reaction. Classifying the urinary iodine level obtained based on World Health Organization (WHO), United Nation's International Children Emergency Fund (UNICEF) and International Council for the Control of Iodine Deficiency Disorders (ICCIDD) recommendation, it was found that 15 (5%) had moderate iodine deficiency, 15 (5%) had mild iodine deficiency, 69 (23%) fell into the sufficient group and 201 (67%) fell into the excess group, with urinary iodine level greater than 300 µg/L. This study infers that if this trend continues unmonitored, the entire population could be prone to developing iodine induced hyperthyroidism (IIH) with the associated toxicity.

Key words: Iodine, hyperthyroidism, iodine deficiency disorders, Iodine induced hyperthyroidism.

INTRODUCTION

Iodine Deficiency Disorder (IDD) is a major global cause of morbidity, mortality and impaired develop-ment[1]. Universal salt iodisation has been extremely effective in reducing the burden of IDD and represents a major global public health success (Dunn et al., 1998). In Africa, great progress has been made towards the elimination of iodine deficiency, saving millions of children from its adverse affects, largely due to the increased household availability of iodized salt.

The daily recommended intake of iodine: 150 µg for adults, 200 µg during pregnancy, 50 µg for the first year in life, 90 µg for ages 1 to 6, and 120 µg for ages 7 to 12 (Dunn et al., 1998). The trace element iodine is an essential nutrient for human growth and development.

The thyroid gland depends on iodine for production of thyroid hormone and this is one of the physiological functions of iodine. The association between iodine deficiency and endemic goiter has been known for centuries. These disorders are collectively described as iodine deficiency disorders (IDD) and it includes: Endemic goiter, hypo-thyroidism, cretinism and congenital abnormalities (Dunn et al., 1998). While endemic goiter is the most visible consequence of iodine deficiency, the most significant and profound effects are on the developing brain. The potential impact of iodine deficiency on the intellectual development of large segments of the populations in underdeveloped countries is of particular concern, especially when all of the adverse effects of iodine deficiency can be prevented by long-term, sustainable iodine prophylaxis.

*Corresponding author. E-mail: vip4162003@yahoo.co.uk, oaugustine@in.com.

In most countries of the world, universal salt iodization has been employed as a means of eliminating disorders secondary to iodine deficiency. WHO, UNICEF and ICCIDD has brought iodine sufficiency within reach of about 1.5 billion people of the world who were deficient decades ago; and now rely on the urinary iodine concentration as the primary indicator of effectiveness (WHO, ICCIDD, 1999).

In Africa and indeed Nigeria, great progress has been made towards the elimination of iodine deficiency, saving millions of children from its adverse effects, largely due to the increased household availability of iodized salt (International Council for the Control of Iodine Deficiency Disorders, 2003; World Health Organization, 2007; WHO, UNICEF, ICCIDD, 2001; Lantum, 2009). However, the relationship between iodine intake and the risk of thyroid disease is U-shaped, with both low and high iodine intake being associated with thyroid disease (Laurberg et al., 2001). The effect of the consumption of additional iodine is also dependent on the initial status of the population. In populations that are mildly deficient or replete, increases in dietary iodine may induce hypothyroidism, while in populations that were previously severely deficient, increased dietary iodine is associated with hyper-thyroidism (Markou et al., 2001; Stanbury et al., 1998).

In Nigeria, the National Agency for Food and Drug Administration and Control (NAFDAC) has greatly promoted salt iodization using public campaigns (Lantum, 2009). This, coupled with the influx of Western diets into the country has exposed lots of families to consume high levels of iodized salts and iodine containing food. Currently, the programme is now in place to monitor the level of high iodine intake in the population. This study was therefore, carried out to investigate the levels of exposure to iodine in the study population.

MATERIALS AND METHODS

Selection of patients and sample collection

This study was a public health study. The subjects were 300; apparently healthy primary school children, with no history or biochemical marker suggestive of renal failure, whose mean age were 9.45 ± 1.26 years, and comprised both males and females. However, they had been resident in Ibadan for at least five years. Primary school children are appropriate population group for the assessment of iodine status because of their physiological vulnerability.

Furthermore, measurement of urinary iodine levels in schoolchildren is important for public health considerations, as this group effectively reflects the current status of IDD in the general population, as well as the extent to which IDD control measures have had an impact on the population.

All the pupils were in their early primary school years. Prior to the study, consent was received from the school authorities, parents and pupils. Anthropometric measurements of the study population were taken. 10 ml random urine sample was collected into clean and sterile universal bottle from all the pupils who were chosen at random. Since the samples were not analysed immediately, they were stored and frozen at -20°C until they were ready for analysis.

Table 1. Measured parameters in the study population.

Variable	Mean ± SD
Age (yrs)	9.45 ± 1.26
Height (m)	1.31 ± 0.11
Weight (kg)	23.71 ± 5.28
Waist circumference (m)	0.48 ± 0.17
Body mass index (kg/m^2)	14.03 ± 1.87
Mean urinary iodine (µg/l)	317.13 ± 118.29

Table 2. Classification of iodine nutrition of the studied population based on the epidemiological criteria for assessing Iodine nutrition using joint criteria of WHO, UNICEF and ICCIDD (2001).

Range (µg/L)	Distribution (%)
Severe (< 20)	0
Moderate (20 - 49)	15 (5)
Mild (50 - 99)	15 (5)
Sufficient (100 - 199)	69 (23)
Excess (> 300)	201 (67)

Analytical method

The standard method (the ammonium persulphate technique) was used for estimating the level of iodine in the urine (Dunn et al., 1993). Urine is digested with ammonium persulphate. Iodine present in the urine acts like a catalyst in the reduction of ceric ammonium sulphate (yellow) to cerous ammonium sulphate (colourless). The degree of disappearance of the yellow colour is a measure of iodine content in the urine. A standard curve plotted during the analysis was used to extrapolate the concentration of iodine in the urine samples.

RESULTS

Table 1 shows the measured parameters in the studied population. The urinary iodine obtained from the subjects was classified into various degrees of iodine deficiency using the criteria stipulated by the World Health Organization. The results are shown in Table 2. From the table, it was evident that none of the pupils had any degree of iodine deficiency. However, 5% (15) fell into moderate and mild levels of iodine deficiency respectively. Furthermore, 23% (69) had optimal level and 67% (201) were excessively sufficient with iodine.

DISCUSSION

Urinary iodine excretion is a good marker of the dietary intake of iodine, and is the index for evaluating the degree of iodine deficiency, correction and toxicity. Many countries have adopted massive salt iodisation as a means of correcting IDD in countries where they were prevalent. The results from this study show that none of the pupils had any degree of severe iodine deficiency. This is a welcome development in terms of public health and could attest that lots of families have massively embraced salt iodisation. However, that there were still mild to moderate iodine deficiency as shown from our study could imply that not all families have adopted the massive iodization process. This finding is consistent with the findings of Mu et al. (2001), who reported mild to moderate iodine deficiency across the populations they studied including school children. Iodized salt is widely available commercially, but it may not be impossible that only a hand full of all households' purchase iodized salt for domestic use. This however, needs to be evaluated through further study.

The contribution of iodized table salt in the production of iodine nutrition is probably insignificant. Most of the salt in our diet comes from salt added during preparation and processing of food. The assessment, monitoring and evaluation of the iodine content of salts imported and those produced within the country are currently poor. It is not unlikely that household salt may not contain the recommended level of iodine. The World Health Organization has recently recommended that iodization of salts should be maintained at concentration of 20 to 40 ppms (World Health Organization (WHO)/United Nations Children's Fund/International Council for Control of Iodine Deficiency Disorders, 1996).This is in contrast to the present legislation on salt iodization in the country which recommended salt iodization at 50 ppm/kg salt.

The excretion of high urinary iodine found in the studied population is of great public health concern. Our studies showed that 201 (67%) of the pupils had excess urinary iodine. Our finding was consistent with the report of Delange et al. (1999), who reported high concentrations of urinary iodine in some African countries few years after the introduction of massive iodization programme. They concluded that the risk of Iodine Induced Hyperthyroidism (IIH) after correction of iodine deficiency is closely related to a recent excessive increment of iodine supply. Ibadan, a South –Western state of Nigeria is not included in the goiter belt region of the country; and has been assumed to be sufficient in iodine even prior to salt iodization campaign. It is therefore, not impossible that the massive iodization programme embarked by the government has exposed residents to higher concentrations of iodine.

Furthermore, the daily recommended allowance of iodine for children is about 150 µg. Large quantities of iodide are present in drugs, antiseptics, bread, food preservatives and some fast food products. Considering the flair that school children have for bread and fast foods, it is likely that these food products could contribute to the increased level of iodine observed. Further reasons that could be responsible for the high urinary iodine level observed could be due to poor monitoring of the production, quantity and quality of iodine use for food production and preservation. In most instances, the monitoring of iodised salt focuses mainly on whether iodine is present or not but the quantity is not always considered.

The World Health Organization has reported that population that exposed to excess concentrations of urinary iodine could be prone to developing Iodine induced Hyperthyroidism (IIH) and autoimmune thyroid disorders (Dunn et al., 1998). The thyroid gland has intrinsic mechanisms that maintain normal thyroid function even in the presence of iodine excess; however, this mechanism can be depleted at increased iodine levels (Roti and Uberti, 2001). It has widely been reported that administration of iodine in any chemical form could induce the development of IIH. In iodine-sufficient areas, iodine-induced hyperthy-roidism has been reported in euthyroid patients with or without previous thyroid diseases (Stanbury et al., 1998).

IIH is most commonly encountered in older persons with long standing nodular goiter and in regions of chronic iodine deficiency, but instances in the young have been abound (Stanbury et al., 1998). It generally occurs after an incremental rise in mean iodine intake in the course of programmes for the prevention of iodine deficiency. There is evidence, particularly in animals, that high levels of iodine can be a contributing factor in the development of autoimmune thyroid disease. The production of excess iodine–rich thyroglobulins, which may possess some immunogenic properties, has been incriminated to contribute to the development of IIH. Some, but not all, studies on humans have shown that the presence of serum anti-microsomal antibodies is more prevalent in areas having adequate iodine than in areas of mild iodine deficiency. Mutational events in thyroid cells have also been incriminated to be responsible for the development of IIH. These events lead to autonomy of function. When the mass of cells with such an event becomes sufficient and iodine supply is increased, the subject may become thyrotoxic. These changes may occur in localized foci within the gland or in the process of nodule formation.

In conclusion, this study showed that high urinary iodine level is present in the studied population and it has public health implication. Despite remarkable progress in the control of the iodine deficiency disorders (IDD), they remain a significant global public health problem. Assessing the control of IDD and preventing the development of IIH, and monitoring the progress of salt iodisation programmes are cornerstones of a control strategy. Thyroglobulin has become a promising new biochemical marker for the diagnosis of iodine induced hyperthyroidism.

Adequate monitoring using this marker is highly recommended if the development of IIH is to be prevented.

REFERENCES

Delange F, de Benoist B, Alnwick D (1999) Risks of iodine-induced hyperthyroidism after correction of iodine deficiency by iodized salt. Thyroid Jun., 9(6): 545-556.

Dunn JT, Semigran MJ, Delange F (1993). Methods for measuring Iodine in urine. International Council for the Control of Iodine Deficiency Disorders.

Dunn JT, Semigran MJ, Delange F (1998). The prevention and management of iodine-induced hyperthyroidism and its cardiac features. Thyroid, Jan., 8(1): 101-6

International Council for the Control of Iodine Deficiency Disorders. IDD Newsletter (2003). Iodine Nutrition in Africa. IDD Newsletter 19: 1-6.

Lantum DN (2009). Did universal salt iodization help reduce the infant mortality rate in Nigeria? IDD Newslett., May, 32 (2): 6-7.

Laurberg P, Bulow Pedersen I, Knudsen N, Ovesen L, Andersen S (2001). Environmental iodine intake affects the type of nonmalignant thyroid disease. Thyroid, 11: 457-469.

Markou K, Georgopoulos N, Kyriazopoulou V, Vagenakis AG (2001). Iodine-induced hypothyroidism. Thyroid, 11: 501-510.

Mu Li, Gary Ma, Karmala Guttikonda , Steven C Boyages ,Creswell J Eastman (2001). Re-emergence of iodine deficiency in Australia. Asia Pacific J. Clin. Nutr., 10(3): 200-203.

Roti E, Uberti ED (2001). Iodine excess and hyperthyroidism. Thyroid, 11(5): 493-450.

Stanbury JB, Ermans AE, Bourdoux P, Todd C, Oken E, Tonglet R, Vidor G, Braverman LE, Medeiros-Neto G (1998). Iodine-induced hyperthyroidism: occurrence and epidemiology. Thyroid, 8: 83-100.

WHO, UNICEF, ICCIDD (2001). Assessment of the Iodine Deficiency Disorders and monitoring their elimination. Geneva: World Health Organization; 2001. WHO Document WHO/NHD/01.1.

WHO, ICCIDD (1999). Progress towards the elimination of iodine deficiency disorders (IDD) Geneva: WHO, publisher,pp1-13

World Health Organization (2007). Assessment of iodine deficiency disorders and monitoring their elimination a guide for programme managers. – 3rd ed.

World Health Organization (WHO)/United Nations Children's Fund/International Council for Control of Iodine Deficiency Disorders (1996). Recommended Iodine Levels in Salt and Guidelines for Monitoring their Adequacy and Effectiveness. WHO/NUT/96.13. Geneva: WHO.

3

Effect of cashew wine on histology and enzyme activities of rat liver

S. Awe[1]* and E. Tunde Olayinka[2]

[1]Department of Biological Sciences, Ajayi Crowther University, P. M. B. 1066, Oyo, Oyo State. Nigeria.
[2]Department of Chemical Sciences, Ajayi Crowther University, Oyo, Nigeria.

Saccharomyces cerevisae SIL 59703 was used to ferment cashew apple juice supplemented with 30% Sucrose in the laboratory incubated at room temperature (28 ± 2°C) for six days (aerobic) and six weeks (anaerobic) for wine production. Toxicological effect of the wine produced was based on histology and enzyme activities of rats' liver. Forty-two rats were divided into three groups: group I (negative control) received no wine, group II subdivided into three sub groups A, B and C received 6.25 ml/kg body weight of 5, 7.5 and 10% alcohol content of the Red wine (positive control) respectively, while, group III also subdivided into three sub groups A, B and C received 6.25 ml/kg body weight of 5, 7.5 and 10% alcohol content of the cashew wine, respectively for eighteen days. Histological studies of the rat liver fed with both the red wine and cashew wine revealed marked alteration in cellular structure at 7.5 and 10% alcohol content which was not noticeable in the negative group. Enzyme activities indicated that both red wine and cashew wine at 7.5 and 10% alcohol content induced marked liver failure characterized by a significant increase ($p < 0.05$) in serum aspartate ransaminase (AST), alanine transaminase (ALT), Lactate dehydrogenase (LDH) and Gamma glutamyl transpeptidase (GGT) activities. In conclusion, 5% alcohol content of the cashew wine showed no apparent disruptions of the normal liver structure by histological and enzyme activities assessment.

Key words: Enzyme activities, fermentation, histology, serum, wine.

INTRODUCTION

Wine is an alcoholic beverage produced by normal alcoholic fermentation of the juice of ripe grapes by the natural yeasts found on the skin of the fruit or by using selected yeast such as *Saccharomyces cerevisiae* (Okafor, 1978). Other fruits such as apples, berries, blackcurrants, orange, water melon are sometimes also fermented for wine production (Berry, 2000). France, Italy and Germany produced over half the total world output of wine, although other countries in parts of the world are also involved in wine production (Lau and Jaworski, 2003).

Slinkard (2007) suggests that red wine protects against heart attack by interfering with production of a body chemical: endothelin-1 (ET-1) which clogs up arteries. However, the benefits are available only when wine is taken in moderate intake as over consumption of alcohol including wine can cause some diseases such as cirrhosis of the liver and alcoholism (Jordan, 2002).

Cashew (*Anacardium occidentale* L.) is one of the important nut crops, ranking third in the international trade. Cashew apple is not a true fruit but a swollen peduncle to which the cashew nut is attached. It is a soft but fibrous juicy fruit. It possesses exotic flavor characteristics. Based on external color of the fruit, cashew apple can be broadly classified into red and yellow varieties.

Nigeria is blessed with a vast array of seasonal fruits, which are rich in sugars. The fruits are produced yearly in quintiles that are in excess of their consumption. Large quantities of those fruits are disposed off yearly due to non- availability of or poor storage facilities. These results in loss of the vital nutrients such as vitamins that are associated with it and potential revenue sources. If the fruits are put to other use such as wine production, the nutrients that are so lost can be harnessed and made available all year round. This study is aimed at assessing

*Corresponding author. E-mail: asflor5@yahoo.com.

the histological and toxicological effect of wine from cashew fruits compared with imported red wine.

MATERIALS AND METHODS

Fermenting organism

Pure culture of *S. cerevisiae* SIL 59703 was purchased locally and used to ferment the must.

Preparation of cashew wine

Ripe, fresh and healthy cashew apple was weighted (9.8kg), surface sterilized with sodium metabisulphate solution to remove microbial contaminants. The apple was then chopped to bits, crushed with sterilized blender to produce 16 L of juice. The juice was then diluted to 24 L with warm water (45°C) to give the 'must' needed for wine production. Standardized campden tablet, 30% sucrose and yeast nutrient were added and allowed to stay for 24 h after which yeast was added (Berry, 2000).

Aerobic fermentation

Standardized amount of yeast was added to 24 L of must in a fermenting jar by sprinkling it over the surface of the juice. The inoculated must was covered with muslin cloth and incubated at room temperature (28 ± 2°C). The fermenting must was aerated daily by stirring twice to encourage yeast multiplication (Berry, 2000). Aerobic fermentation was terminated after 6 days and the must was sieved to remove the shaft and debris of the crushed fruits.

Anaerobic fermentation

The filtrate obtained after sieving the must was transferred into anaerobic fermentation jar and incubated at room temperature. An air trap was fixed to the fermenting jar. Campden tablet was added to the filtrate to supply sulfur dioxide gas. Fermentation was terminated after six weeks. The wine was then stored to allow the yeast to flocculate. The resulting wine was racked monthly for three months to clear the wine and then aged. After aging for 6 months, the wine was filtered using pressurized filtering kit, decanted into sterile bottles and corked.

Experimental procedure

Animals weighing an average of 160g were bred and housed in the Animal House of the Department of Chemical science, Ajayi Crowther University, Oyo, Nigeria. They were kept in wire meshed cages and fed with commercial rat chow (Bendel feeds Nigeria ltd) and supply water ad libitum._Fourty-two (42) albino rats (wistar strain) were divided into three groups:

Group I consist of only 6 rats served as the negative control to which only distilled water (no wine) was administered orally.
Group II consist of 18 rats divided into 3 subgroups A, B,C received 6.25 ml/kg body weight of 5, 7.5 and 10% alcohol content of the Red wine (positive control) respectively.
Group III consist of 18 rats divided into 3 sub groups A, B, C received 6.25 ml/kg body weight of 5, 7.5 and 10% alcohol content of the cashew wine respectively for eighteen days. After 18 days, the rats were sacrificed approximately 24 h after the last treatment.

Collection of blood samples for serum preparation

The rats were sacrificed by cervical dislocation. Blood samples were collected via ocular punctures into plain bottles. Serum was prepared by aspiration of the clear yellowish liquid after clotting and centrifuged for 10 min at 3000 x g in an MSC (Essex, UK) bench centrifuge. Estimation of enzyme activities was done by using clear supernatant. The liver was immediately removed, blotted and fixed into 10% formaldehyde solution.

Histological procedure

The method of Baker and Silverton (1972) was employed for the histology of the liver. The liver was section transversely in its biggest diameter with a steel blade and fixed in 10% formaldehyde solution for 24 h. After fixation, the liver section were dehydrated, cleared, infiltrated, embedded, sectioned and stained, using haematoxylin. After which the mounting was done using Canada balsam as the mountant after which the slides were examined using Leitz microscope and their photomicrographs taken and examined.

Determination of serum aspartate ransaminase (AST) and alanine transaminase (ALT) activities

Serum AST and ALT activities were determined using Randox diagnostic kits. Determination of AST and ALT activities were based on the principle described by Reltman and Frankel (1957).

Determination of serum lactate dehydrogenase (LDH) activities

Serum LDH activities were determined using Randox diagnostic kits following the principle described by Wroblewski and Ladue (1955).

Determination of serum (gamma glutamyl transpeptidase) GGT activities

Serum GGT activities were determined using Randox diagnostic kits following the principle described by Szasz (1969).

Statistical analysis

The data were analyzed using one way ANOVA followed by Duncan multivariable post-hoc test for comparison between control and treated rats in all groups. P values less than 0.05 were considered statistically significant.

RESULTS

Figure 1 showed the effect of different wines of varying alcoholic contents on serum AST activities, while the effect of different wines of varying alcoholic contents on serum ALT activities was presented in Figure 2. Figures 3 and 4 showed the effect of different wines of varying alcoholic contents on serum LDH activities and GGT activities, respectively. All the results showed that there were a significant increased ($p < 0.05$) in the level of all the four enzymes activities in those rats treated with wines of 7.5 and 10% alcoholic content when compared with the control. However, at 5% alcoholic content the

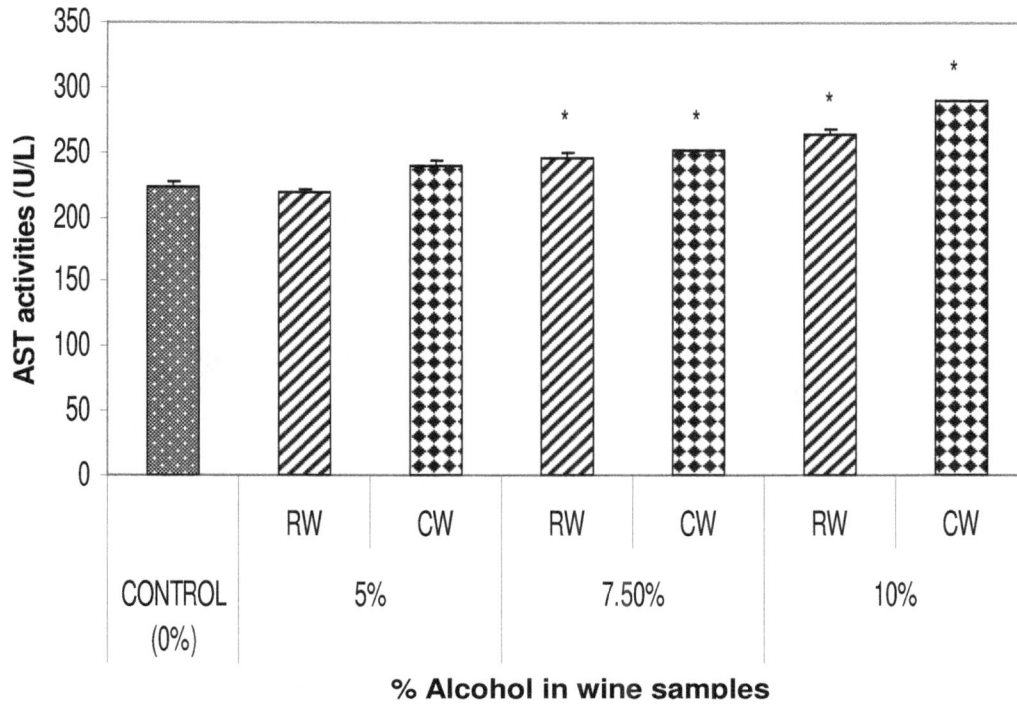

Figure 1. Effect of different wines of varying alcoholic contents on serum AST activities.
Key: RW = Red Wine; CW = Cashew WineThe values are the Means ± SD (range) for six rats in each group. * Significantly different from the control, p<0.05 (Duncan's multiple comparison test).

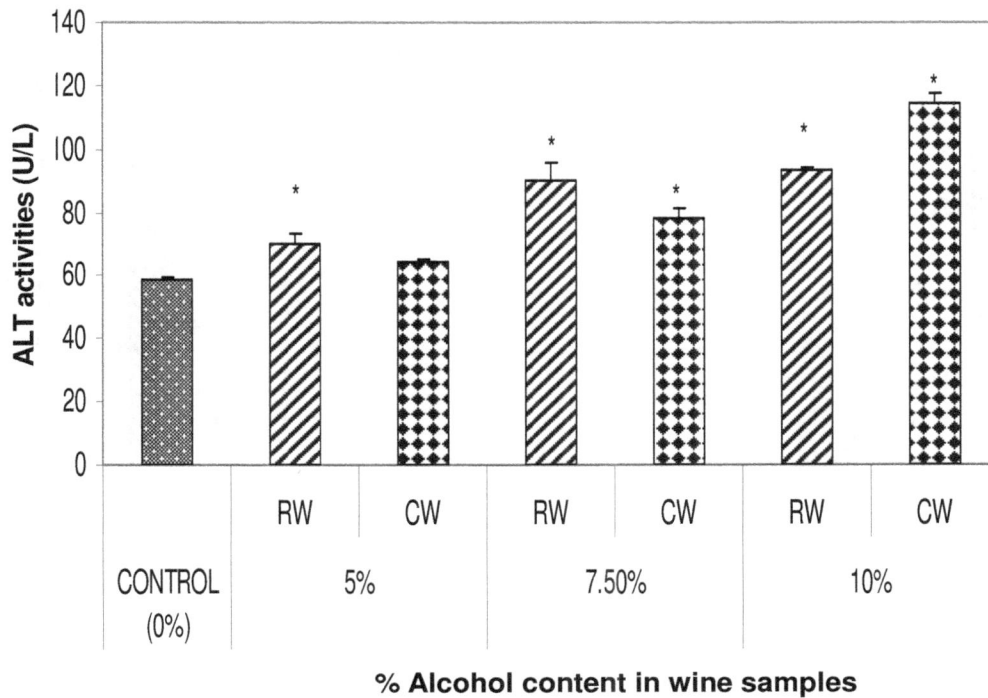

Figure 2. Effect of different wines of varying alcoholic contents on serum ALT activities. Key: RW = Red Wine; CW = Cashew Wine. The values are the Means ± SD (range) for six rats in each group. * Significantly different from the control, p < 0.05 (Duncan's multiple comparison test).

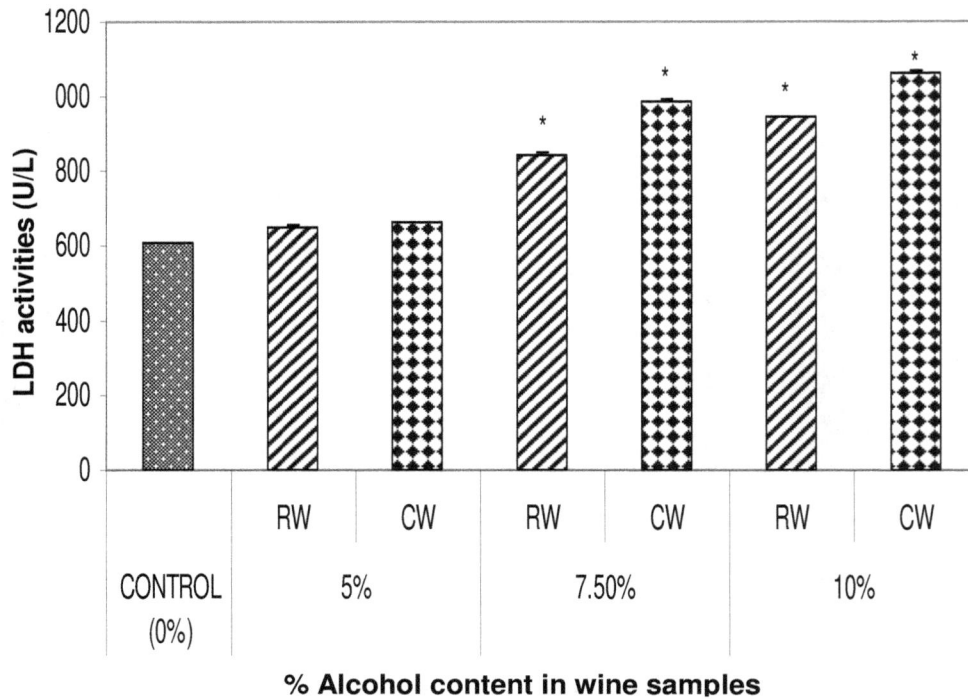

Figure 3. Effect of different wines of varying alcoholic contents on serum LDH activities.
Key: RW = Red Wine; CW = Cashew Wine. The values are the Means ± SD (range) for six rats in each group. * Significantly different from the control, p<0.05 (Duncan's multiple comparison test).

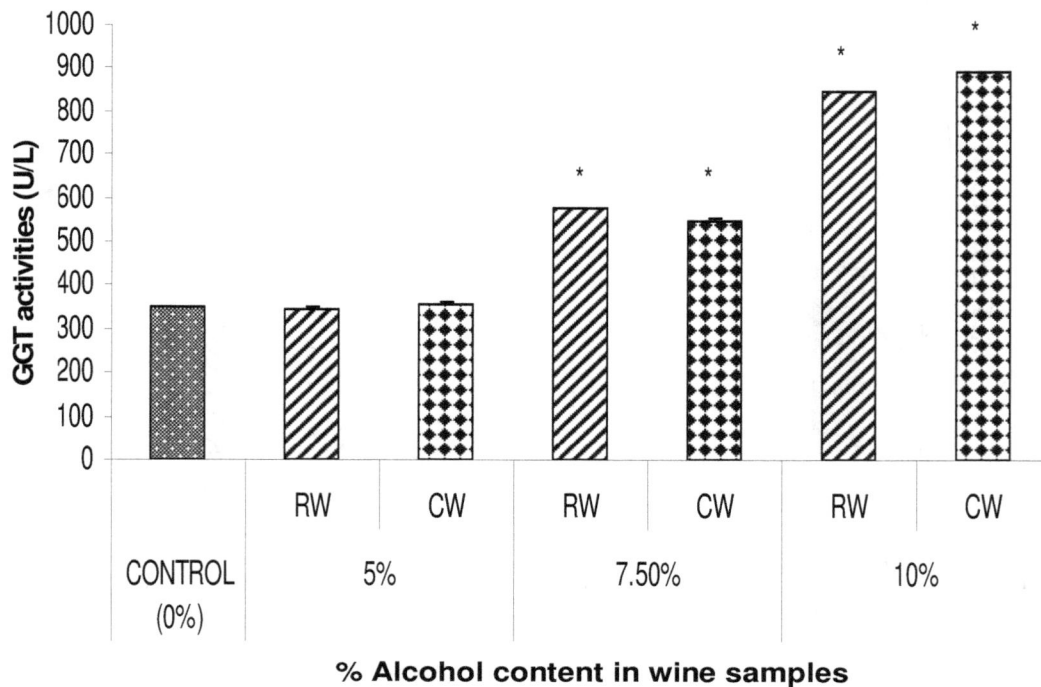

Figure 4. Effect of Different wines of varying alcoholic contents on serum GGT activities.
Key: RW = Red Wine; CW = Cashew Wine.The values are the Means ± SD (range) for six rats in each group. * Significantly different from the control, p < 0.05 (Duncan's multiple comparison test).

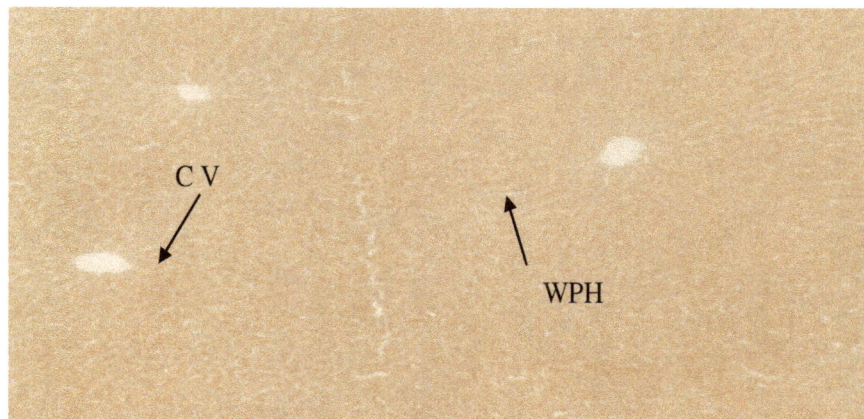

Plate 1. Photomicrograph (X120) showing effect of administration of distilled water (CONTROL) on architecture of liver of rats. (XH stain). Key: CV- central vein, - WPH- well preserved hepatocytes.

wine show no significant effect on the enzymes activities when compared with the control.

Plate 1 showed the photomicrographs of the normal liver architecture (negative control), while Plates 2 and 3 showed the photomicrographs of the liver cell treated with red wine and cashew wine as positive control at different alcohol content levels. The liver histology was significantly affected by administration of the wine at 7.5 and 10% alcohol level as characterized by mild and severe hyperplasia of central vein, distorted hepatocytes and degenerated area of fatty acid, while at 5.0% the liver structure was well preserved.

DISCUSSION

Histopathological change in tissues is a late manifestation of a chemical, physical, mechanical or inflammatory assault on the tissue and usually complements enzyme studies. Administration of red wine and cashew wine at 5% alcohol content showed minimal effect on the liver histology (Plate 2 and 3). There were no apparent disruptions of the normal liver structure. However, administration at 7.5 and 10% alcohol content showed mild and serve hyperplasia. These suggest that administration of the wines at 7.5 and 10% alcohol contents could cause major histopathological changes. Jacquelyn and Maher (1997) reported that liver damage caused by alcohol and by-product of the metabolism such as acetaldehyde and highly reactive molecules such as free radicals. Similar effects were observed with the red wine and cashew wine produced that indicated that the wine comparable in terms of histopathological effects on the liver, hence the quality of the wines can be ascertained. The measurement of the various enzyme activities in the tissues and body fluids play a significant role in disease investigation, diagnosis and detection of tissue cellular damage (Akanji and Ngaha, 1989;

Malomo, 2000). Tissues enzyme assay revealed that damages even before structural damages are detected by conventional histological techniques (Akanji, 1986). The marker enzymes assayed are specifically located in some cell; however, they can leak into the serum or other parts as a result of injury to the cell where they are located (Ngaha, 1982; Adesokan and Akanji, 2003; Jensen and Freese, 2009).

Low levels of AST is normally found in the blood, however, when the liver or heart is damaged additional AST is released into the bloodstream. It rises within 6 to 10 h and remains high for 4 days. ALT is produced within the cells of the liver and is the most sensitive marker for liver cell damage (Tietz, 1991; Jensen and Freese, 2009). Any form of hepatic cell damage can result in an elevation in the ALT; as the cells are damaged, the ALT leaks into the bloodstream leading to a rise in the serum levels. Therefore, the high AST and ALT levels in the serum of rats to which the wines (red wine and fruits wine produced) were administered are indications of leakage into the bloodstream due to liver damage (Sampson, 1980; Jacobs, 1996). AST and ALT are predominantly cytosolic enzymes but some are found in the mitochondria that are involved in transamination reactions in amino acid metabolism.

Adesokan and Akanji (2003) reported alterations in the concentration of these enzymes, following administration of drug and chemical agents, including alcohol. LDH is an enzyme found in the cells of many body tissues, including the heart and liver (Janis, 2006). Owing to its widespread distribution in the tissues, elevation of the total LDH in the serum is generally of little value in diagnosis. High values commonly occur after myocardiac infarction, in megaloblastic anemia, progressive muscular dystrophies and in neoplastic diseases especially widely dissemi-nated forms (Curtis and Roth, 1974; Ziegenhorn et al., 1978). Serum LDH is usually within the normal range in chronic renal disease associated with uremia, high

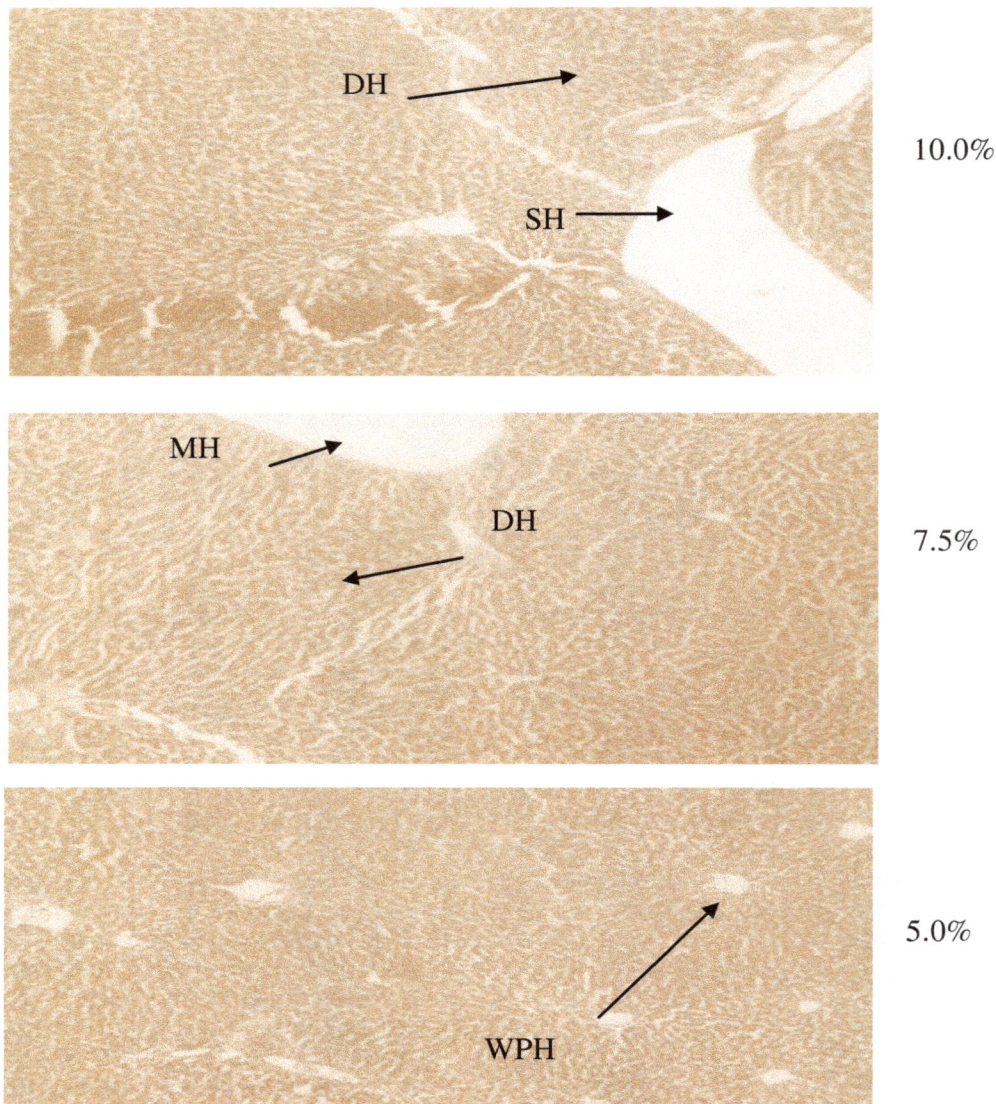

Plate 2. Photomicrograph (X120) showing effect of administration of red wine at different alcohol levels (5, 7.5 and 10%) on architecture of liver of rats. (XH stain), WPH- well preserved hepatocyte, SH-sever hyperplasia, MH-mild hyperplasia central vein, dh- distorted hepatocyte.

activities usually occur in infective hepatitis, mononucleosis and toxic jaundice in which hepatocellular damage occur.

The results of GGT activities in the rats serum showed that the activities were elevated in those that were administered 7.5 and 10 % wines, except those that were administered wine of 5% alcohol content. GGT is a membrane bound glycoprotein found in cells with a high secretion or absorption activity (e.g. liver, kidney, pancreas and intestine) produced by the bile ducts It is believed to be involved in the amino acids and peptides transportation into cells as well as glutathione metabolism. Jensen and Freese (2009) reported that certain GGT levels reflect rare forms of liver disease and

alcohol can cause increases in the GGT. The significant elevation of the enzyme activities by the wine in the serum needs further assessment through liver test functions. The result of the enzyme activities clearly demonstrated that the effect of the wines was dose dependent. Generally the wine consumption affected the liver structure or cells which confirmed by Akanji (1986) studies that tissue enzyme assay detect liver damage before structural damages are detected by conventional histological technique.

It is concluded that the toxicological effect exhibited by 5% alcohol content of the cashew wine with no apparent disruptions of the normal liver structure by histological and enzyme activities assessment.

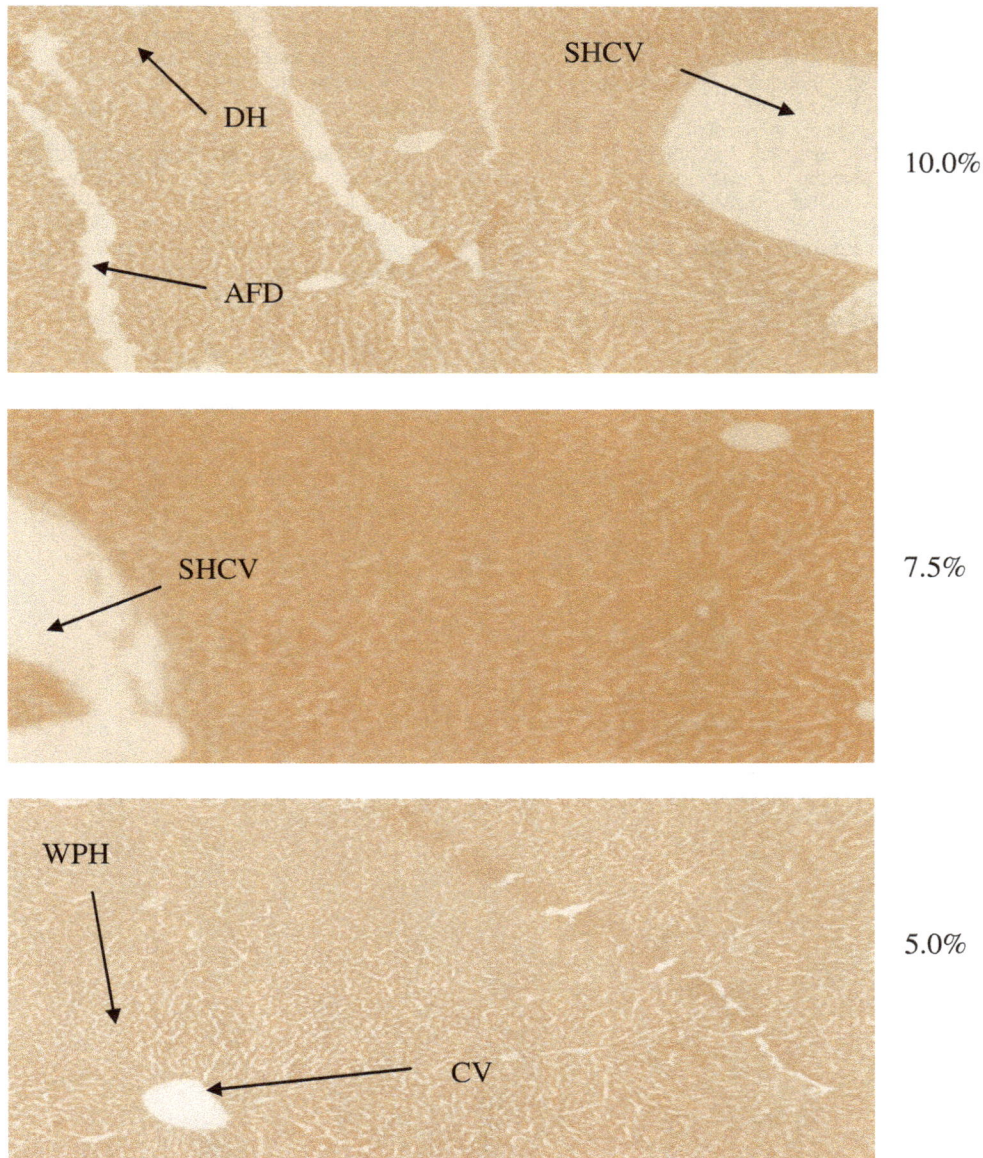

Plate 3. Photomicrograph (X120) showing effect of administration of cashew wine at different alcohol levels (5, 7.5 and 10%) on architecture of liver of rats. (XH stain).WPH- Well preserved Hepatocyte, SHCV- sever hyperplasia central vein, MH-mild hyperplasia central vein, DH-distorted hepatocyte, AFD- area of fatty acid degeneration, CV- central vein.

REFERENCES

Adesokan AA, Akanji MA (2003). Effect of administration of aqueaous extract of *Enantia chlorantha* on the activities of some enzymes in the small intestine of rats. Nigeria J. Biochem. Mole. Biol., 18(2): 103-105.

Akanji MA (1986). A Comparative Biochemical Study of Interaction of Some Trypanocides with Rat Tissue Cellular System Ph.D Thesis. University of Ife, Ile-Ife.

Akanji MA, Ngaha EO (1989).Effect of repeated administration of beremil on urinary excretion with corresponding tissue pattern in rats. Pharmacal. Toxication 64: 272-275.

Baker FJ, Silverton RE, Luckcock ED (1972). Haematological Indices in: Introduction to Medical Laboratory Technology. Fourth Ed. Butterworth and Co. Ltd., p. 558-562.

Berry CJJ (2000). First steps in wine making. Published by GW Kent, Inc. 3667 Morgan Road, Ann Arbor M I 48108, p. 235.

Curtis HC, Roth M (1971). Clinical implication of lactate dehydrogenase, clinical biochemistry principles and methods vol 2 ch II. Water deGruyter, Berlin, p. 614-617.

Jacquelyn J, Maher MD (1997). Exploring Alcohol's Effect on Liver Function. Alcohol Health Res. Word, 21: 1.

Janis OF (2006). Gale Encyclopedia of Medicine. Published by the Gale Groups. pp. 216-221.

Jensen JE, Freese D (2009). Liver Function Tests. Colorado Center for Digestive disorders 205s.Suite A Longmont Co. 80501. Pp 151-163

Jacobs MB (1996). The Chemistry Analysis of food and food product. 3rd Edition CBS publishers and Distributors, New Delhi India, p. 250.

Jordan D (2002). An Offering of Wine, Doctoral Thesis, The department of SemiticStudiesUniversityofSydney.(http:llses.library.usyd.edu.au/bitstream/2123/482/1/ at-NU2003 1211.15583702 whole.pdf.).

Lau CK, Jaworski JF (2003). Industrial Sustainability through Biotechnology ASM News, 69: 110-111.

Malomo SO (2000). Toxicological implication of Ceftriaxome administration in rats. Nig. J. Biochan. Mol. Bial., 15(1): 33-35.

Ngaha EO (1982). Further studies on the in VWO effect of Cephaloridine on the stability of rat kidney lysosomes. Biochem. Physical, 8: 1843-1847.

Okafor N (1978). Industrial Microbiology. University of Ife Press Ltd., p. 413.

Reltman S, Frankel S (1957). A colorimetric method for the determination of serum ALT and AST. Am. J. Clin. Pathol., 28: 56-63.

Sampson EJ, Wheeler VS, Buntis CA, Mc Knenily SS, Fast DM, Dayse DD (1980). An Inter-Laboratory Evaluation of the IFCC Method for Aspatate Aminotransferase with use of purified Enzyme Materials Clin. Chem., 26: 1156-1164.

Slinkard Stacy (2007). Wine: Red for health – The Guardian (life), I 2(14): 10-11.

Szasz G (1969).Reaction rate method for gama- glutamyl transeptidase activities in serum. J. Clin. Chem., 22: 205-210.

Tietz NW (1991). Fundamental of Clinical Chemistry. 3rd Edition. W.B. Sanders Co. Philadelphia, pp. 391-395.

Ziegenhorn J, Brandhuber M, Bartl K (1978). Enzyme in health and Diseases. Soc. Clin. Enzymol. Karger, Basel, p. 131.

4

The correlation between the malignant proliferation on cholangiocarcinoma cell lines and in bile from *Helicobacter pylori* infected patients with biliary tract stones

Xudong Xu[1], Quan Sun[2], Zhisu Liu[2], Lin Zhang[1], Zhiyong Luo[1], Yun Xia[1] and Yaqun Wu[1]*

[1]Department of General Surgery, Tongji Hospital, Tongji Medical College, Huazhong University of Science and Technology, Wuhan 430030, P.R. China
[2]Department of General Surgery, Zhongnan Hospital, Wuhan University, Wuhan 430071, P.R. China.

Helicobacter pylori has been detected in human tissue and is a candidate for etiologic investigations on the causes of hepatic and biliary tract diseases, but reliable serologic tests need to be developed in order to pursue such investigations. The aim of this study was to assess the correlation between the infection of *H. pylori* in bile from patients with biliary tract stones and the proliferation of human cholangiocarcinoma and its mechanism. Choledocholithiasis bile with *positive H. pylori (PCB)*, Choledocholithiasis bile with *negative H. pylori* (NCB) and *normal bile* (NB) were part of the study. Cholangiocarcinoma cell lines QBC939 and TFK-1 were analyzed. The proliferative effects were measured by methabenzthiazuron (MTT) assay. Cell cycle and apoptosis were analyzed by flow cytometry. Compared with NB and NCB, PCB significantly promoted the proliferation of cholangiocarcinoma cell lines QBC939 and TFK-1. The proliferative index in PCB group was obviously higher than that in NCB group after being treated with 1% PCB for 48 h (p < 0.05). As far as the apoptosis rate was concerned, there were no obvious differences between PCB group and NCB group (p < 0.05), same as between PCB group and NB group (p < 0.05). The percentage of S phase increased remarkably in PCB group compared with NCB group, while the percentage of G_0/G_1 phase decreased remarkably in PCB group compared with NCB group. It was suggested that PCB can greatly promote the malignant proliferation of human cholangiocarcinoma cell lines QBC939 and TFK-1, and the mechanism was affected by the changes of cell cycle. So we can predict that there was perhaps a close relation between *H. pylori* infection and cholangiocarcinoma.

Key words: Malignant proliferation, *Helicobacter pylori*, cholelithiasis, bile, cholangiocarcinoma.

INTRODUCTION

Cholangiocarcinoma (CCA) is a devastating malignancy that appears late, which is notoriously difficult to diagnose, and is associated with a high mortality rate (Briggs et al., 2009). The incidence of intrahepatic cholangiocarcinoma is increasing worldwide. The cause of this rise is unclear, although it could be related to an interplay between pre-disposing genetic factors and environmental triggers. CCA was first reported by Durand-Fardel in 1840 (Olnes and Erlich, 2004). The tumor arises from the ductular epithelium of the biliary tree, either within the liver (intrahepatic CCA) or more commonly from the extrahepatic bile ducts (extrahepatic CCA). The disease is usually fatal because of its late clinical presentation and the lack of effective non-surgical therapeutic modalities (Hong et al., 2009). Most patients have unresectable disease at presentation

---

*Corresponding author, E-mail: wyqmd@126.com.

and die within 12 months from the effects of cancer cachexia and a subsequent rapid decline in performance status. Overall survival rate including resected patients is poor, with less than 5% of patients surviving to 5 years, which has not changed significantly over the past 30 years (Shaib and El-Serag, 2004). Although CCA is a relatively rare tumor, interest in this disease is rising as incidence and mortality rates increase markedly (Khan et al., 2002). In recent years, the etiology research on CCA has been the focus of attention (Anderson et al., 2004; Chen et al., 2005; Enjoji et al., 2005). According to the epidemiology data, cholelithiasis is a dangerous factor to CCA, and the infection of *Helicobacter pylori* is closely related to the occurrence of cholelithes and chronic inflammation of biliary epithelium (Boomkens et al., 2004; Chang et al., 2005; Monstein et al., 2002; Tsai et al., 2004).

H. pylori is a noninvasive, nonspore-forming, and spiral-shaped Gram-negative rod bacteria measuring approximately 3.5 × 0.5 µm. This bacterium induces infiltration of the biliary mucosa by neutrophils, macrophages, and T and B lymphocytes. However, this immune and inflammatory response to *H. pylori* infection suggests there is an increase in both epithelial all proliferation, as well as cell death by apoptosis. It is transmitted from human to human, usually in early childhood, and it is always transferred to biliary tract by oddi sphincter. It colonizes the biliary epithelium for a lifetime in the absence of specific antimicrobial therapy. Recent studies reveal that *H. pylori* injects bacterial proteins which greatly damage biliary epithelium and cause chronic inflammation to the cytosol of the host cell and regulates the intracellular signal transduction in the host cell (Boomkens et al., 2005). This mechanism provides a noble means of resolving how *H. pylori* survives in human biliary tract. Some molecular studies have recently shown that *H. pylori* DNA and intracellular expression of *H. pylori* virulence genes are still detectable in biopsies of patients with precancerous lesions (Clyne and Drumm, 1996). *H. pylori* infection's ability to foster a chronic inflammatory response is the best explanation of the bacterium's carcinogenic potential. Preliminary experiments testified that the proliferation of CC cell lines QBC939 and TFK-1 had something to do with the infection of *H. pylori*, but other CC cell lines did not have type characteristics of CCA like QBC939 and TFK-1. Therefore we conducted the study to emphasize the discussion on the proliferative effects of choledocholithiasis bile (CB) infected by *H. pylori* on the culture *in vitro* of human cholangiocarcinoma cell lines QBC939 and TFK-1.

MATERIALS AND METHODS

Collection and treatment of samples

The experimental objects are divided in 3 groups, *H. pylori* positive choledocholithiasis bile (PCB) group, *H. pylori* negative choledocholithiasis bile (NCB) group, and normal bile (NB) group, each taking 120 samples. The 120 *H. pylori* PCB samples are selected from surgical patients in Tongji Hospital, including 67 males, 53 females, aged between 24 and 62. Among the sample patients of this group, 78 samples have choledocholith with calculus of intrahepatic duct and 42 samples have simple choledocholith. The 120 *H. pylori* NCB samples are from surgical patients in Tongji Hospital, including 81 males, 39 females, aged between 25 and 58. Among the sample patients of this group, 42 samples have choledocholith with calculus of intrahepatic duct, 57 samples have simple choledocholith, and 21 samples have choledocholith with cholelithiasis. The 120 NB samples are from patients without liver or gallbladder diseases during the corresponding period, including 71 males and 49 females, aged between 25 and 56. All patients are informed of relative information and they all express consent and show their cooperation in our study. Bile is sampled during the examination or during the operation because of hepatobiliary diseases. Especially in reference to normal bile, samples are taken from normal people during the examination or in other diseases during the operation. The choledocholithiasis bile (CB) samples have all passed ELISA test and PCR test. The type of ELISA is double antibody sandwich. The PCR primers used in this study are 5'-GCCAATGGTAAATTAGTT-3'(ureA1)   and 5'-CTCCTTAATTGTTTTTAC-3' (ureA2). UreA1 and ureA2 primers are amplified 411 bp fragment of the ureA gene. The cycles consist of 2 min at 94℃ for the first cycle, 1 min at 94℃, 1 min at 42℃ and 1 min at 72℃ for the next 40 cycles and 10 min at 72℃ for the last cycle. The PCR products are analyzed by means of electrophoresis of a 20 µl aliquot through a 1% agarose gel containing 0.5 µg/ml ethidium bromide. If results from both tests are positive, the sample will be defined as *H. pylori* positive; otherwise, it will be defined as *H. pylori* negative(Gieseler et al., 2005; Rahn et al., 2004). And results from both tests are negative in normal bile. The bile samples are centrifuged rapidly at a speed of 4000 rpm for 5 min after extraction. The supernatant is taken and reserved in the refrigerator at −80□ (Iwata et al., 2002; Zampa et al., 2004).

Cell culture and reagent

These studies are performed with human CCA cell lines QBC939 and TFK-1. Human CCA cell lines QBC939 and TFK-1 were established and donated by Professor Wang Shuguang from Southwest Hospital of Third Military Medical University. They were constructed with cloning technique such as recombinant plasmid and so on. The CCA cell lines QBC939 and TFK-1 have type characteristics of CCA which is highly invasive. RPM11640 culture medium and 10% fetal calf serum are purchased from the American Gibco Industries, Inc. Meanwhile, penicillin and streptomycin are added into culture medium. The volume ratios are both 1 versus 1000.

Cytotoxicity test

We test the cytotoxicity of the gradient dilution from the bile samples (Kanbar et al., 2004; Ricci et al., 2002; Yao et al., 2005). When the final concentration of the samples is 1% (10 µl bile sample/ml culture medium), there will be no obvious toxicity to CCA cell lines QBC939 and TFK-1. And 1% samples are used for cytotoxicity test.

Methabenzthiazuron (MTT) assay

We used 2.5 g/L trypsin to digest CCA cell lines QBC939 and TFK-1 in logarithmic growth phase, making single-cell suspension. CCA cell lines QBC939 and TFK-1 are inoculated in 96-well plates and cell counting gives   a number of 1.0×10^5 cells/ml. 1%PCB, 1%NCB,

Table 1. Variation of optical density (OD 490 nm) values at different time points during the co-culture of choledocholithiasis bile and QBC939.

Time	OD 490 nm		
	NB	NCB	PCB
1 d	0.15±0.02	0.19±0.03	0.20±0.03
2 d	0.20±0.03	0.33±0.04 *#	0.48±0.05 **#Δ
3 d	0.24±0.03	0.52±0.06 *##	0.82±0.09 **##ΔΔ

Note: compared with 1 d, *p < 0.05, **p < 0.01; compared with NB group, #p < 0.05, ##p < 0.01 ; compared with NCB group, Δp < 0.05, ΔΔp < 0.01.

Table 2. Variation of optical density (OD 490 nm) values at different time points during the co-culture of choledocholithiasis bile and TFK-1.

Time	OD 490 nm		
	NB	NCB	PCB
1 d	0.14±0.02	0.18±0.03	0.21±0.03
2 d	0.22±0.04	0.35±0.06 *#	0.47±0.08 **#Δ
3 d	0.25±0.04	0.51±0.06 *##	0.83±0.11 **##ΔΔ

Note: compared with 1 d, * p<0.05, ** p < 0.01; compared with NB group, # p < 0.05, ## p < 0.01 ; compared with NCB group, Δp < 0.05, ΔΔp < 0.01.

1%NB are added respectively after cells are adhered. After incubating for 1 d (24 h), 2 d (48 h), 3 d (72 h) respectively, 20 µl of MTT stock solution (4 mg/ml) is added to each well, containing about 0.25 ml medium, and the mixture is incubated for 3 h. The supernatant solution is removed with a pipette. 150 µl of DMSO is added rapidly to each well (within 3 s) at 37□and oscillated for 5 min in order to fully dissolve the crystallisate. The MTT solution is carefully decanted off and formazan is extracted from the cells in each well. The optical density, OD value (490 nm) is detected on the enzymes immunoassay analyzer (Pouns et al., 2009). All MTT assays are tripled.

Cell cycle and apoptosis analyzed by flow cytometry

The flow cytometer used is a Coulter EPICS-XL (Coulter, Miami, FL, USA). Instrument calibration is performed daily according to the recommendation of the manufacturer. An average of 1×10^4 cells from each sample are counted in the flow cytometer. After incubated with 1%PCB, 1%NCB, and 1%NB for 48 h, QBC939 and TFK-1 cell solutions are respectively collected to centrifuge at 800 rpm for 5 min. Then, they are fixed by 80% ethanol pre-cooled at -20°C. The cells are incubated in the dark at room temperature and then diluted to 1 ml in harvest buffer and stored in ice until FACS analysis. To differentiate viable from dead cells, 1 mg/ml propidium iodide is added to samples approximately 3 min before analysis. This protocol works for freshly prepared human cells and cell lines. Cell cycle and apoptosis analysis are measured by flow cytometry (Tian et al., 2009). The same condition is used in order to ensure the credibility of experimental results. When cell cycle is arrested in G_0/G_1 phase, cell apoptosis is strengthened and we can see blue

particulate fluorescence signal with the help of fluorescence microscopy.

Statistical analysis

Mean±SD was used for measurement of data; all the data are analyzed with t-test in Statistical Package for Social Science (SPSS11.5, license code: 30001359390), and independent t-test is selected to analyze the data. The confidence interval is calculated at the 95% level. All P values are 2-tailed and P values <0.05 are considered to be statistically significant.

RESULTS

Effects of choledocholithiasis bile (CB) infected by helicobacter pylori on the proliferation of QBC939 and TFK-1 Cells

After co-culturing bile and CCA cell lines for 1 d, the variation of optical density among PCB group, NCB group and NB group shows no significant difference (p > 0.05), the difference between PCB group and NCB group is not significant either (p > 0.05). After co-culturing bile and CCA cell lines for 2 d, the variation of optical density among PCB group, NCB group and NB group shows significant difference (p < 0.05), and the difference between PCB group and NCB group also shows significance (p < 0.05). After co-culturing bile and CCA cell lines for 3 d, the variation of optical density among PCB group, NCB group and NB group shows significant difference (p < 0.01), and the difference between PCB group and NCB group also shows significance (p < 0.01). Therefore both PCB group and NCB group can promote the proliferation of QBC939 and TFK-1 cells. This promotion is time dependent. Compared with NCB group, PCB group has a more apparent promotional function on the proliferation of human CCA cell lines QBC939 and TFK-1 (p < 0.05) (Tables 1 and 2).

Cell cycle and apoptosis analyzed by flow cytometry

Proliferative indexes (PI) of PCB group, NCB group and NB group about CCA cell line QBC939 are 72±6, 50±4 and 30±3, respectively (Figure 1). And the proliferative indexes (PI) of PCB group, NCB group and NB group about CCA cell line TFK-1 are 74±7, 49±4 and 28±3 respectively (Figure 2). Compared with NB group, the difference within PCB group and NCB group has prominent significance (p < 0.05), and the difference between the two groups also has marked significance (p < 0.05), PI = (S+G_2/M). The proportion of CCA cell line QBC939 in PCB group and NCB group in S phase are 58±5% and 38±4% respectively, which shows a marked increase on the basis of NB group 20±3%, and the difference between PCB group and NCB group shows significance (p < 0.01) (Figure 3). The proportion of CCA

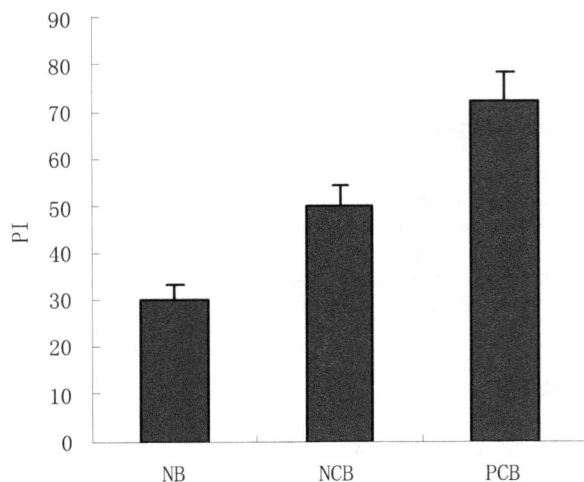

Figure 1. Comparison of proliferative indexes (PI) about QBC939 cell.

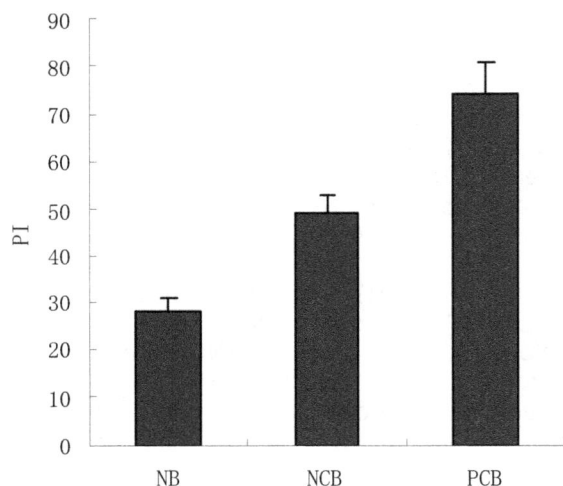

Figure 2. Comparison of proliferative indexes(PI) about TFK-1 cell.

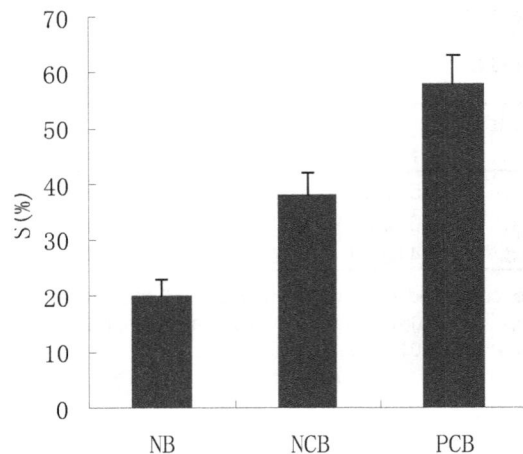

Figure 3. Comparison of cell proportions in S phase about QBC939 cell.

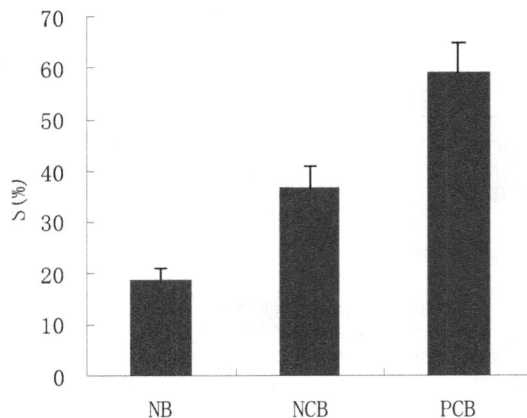

Figure 4. Comparison of cell proportions in S phase about TFK-1 cell.

cell line TFK-1 in PCB group and NCB group in S phase are 59±6% and 37±4% respectively, which shows a marked increase on the basis of NB group 19±2%, and the difference between PCB group and NCB group shows significance (p < 0.01) (Figure 4). The proportion of CCA cell line QBC939 in PCB group and NCB group in G_0/G_1 phase are 28±3% and 48±5% respectively, which shows a remarkable decrease on the basis of NB group 68±8%, and the difference between PCB group and NCB group shows significance (p < 0.01) (Figure 5). The proportion of CCA cell line TFK-1 in PCB group and NCB group in G_0/G_1 phase are 29±3% and 51±6% respectively, which shows a remarkable decrease on the basis of NB group 70±8%, and the difference between PCB group and NCB group shows significance (p < 0.01) (Figure 6). The

apoptosis rates (AR) of QBC939 cell in PCB group, NCB group and NB group are 22±3%, 28±6% and 24±4% respectively (Figure 7), while the apoptosis rates (AR) of TFK-1 cell in PCB group, NCB group and NB group are 23±3%, 27±6% and 25±4% respectively (Figure 8), the comparison among the three groups shows no significant difference (p > 0.05).

DISCUSSION

The vast majority of experts and scholars acknowledge that *H. pylori* could cause cell cancerization through inducing the activation of oncogene and inactivation of anti-oncogene (Nakajima et al., 2008; Nakajima et al., 2009). Since Lin et al. (1995) employed PCR method in finding *H. pylori* in the bile duct of patients with carcinoma of the head of the pancreas and CCA for the first time in

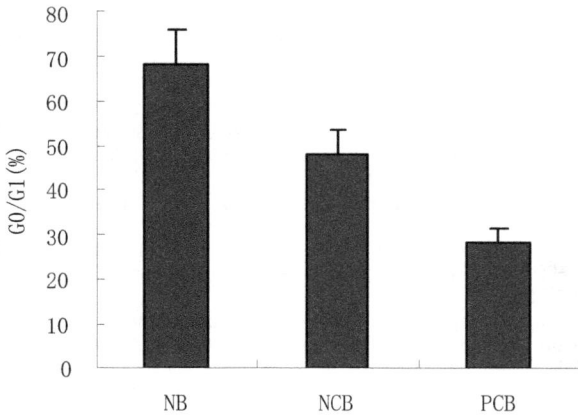

Figure 5. Comparison of cell proportions in G_0/G_1 phase about QBC939

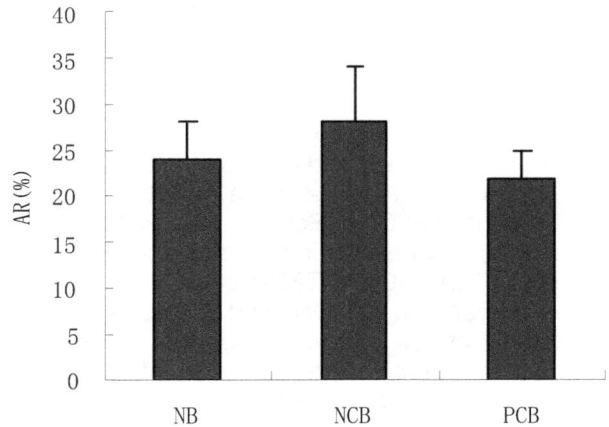

Figure 7. Comparison of apoptosis rates about QBC939.

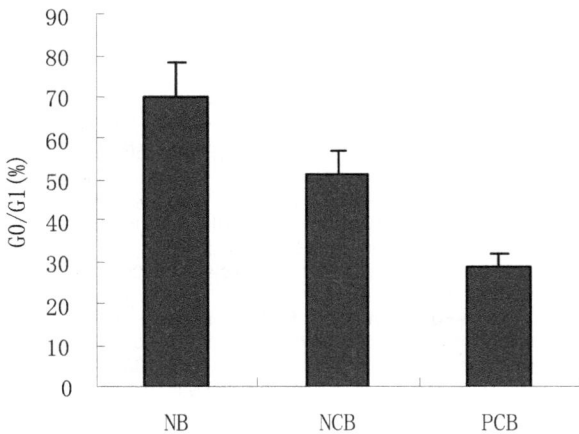

Figure 6. Comparison of cell proportions cell in G_0/G_1 phase about TFK-1 cell.

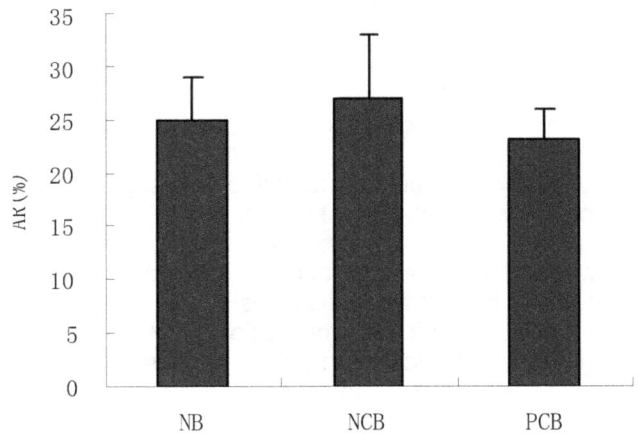

Figure 8. Comparison of apoptosis rates cell about TFK-1 cell.

1995; the relation between *H. pylori* and malignant disease of biliary tract has drawn people's attention. In 2000, Paziak-Domanska et al. (2000) co-cultured cagA[+] *H. pylori* mutant strain at certain concentration and lymphocyte in peripheral blood of healthy people for 96 h, and employed flow cytometry in measuring T lymphocyte and B lymphocyte in the culture media. They found that cagA[+] *H. pylori* strain could significantly inhibit the proliferation of T lymphocyte, namely, cagA[+] *H. pylori* has the ability to inhibit the immune function of human cells, which led to the further deduction that *H. pylori* could indirectly promote the occurrence and development of CCA through hypoimmunity and reducing the immunity monitoring function. In the meantime, a lot of scholars have explored the relation of other bacteria or cytokine present in bile and CCA. Until now, there is not an identical conclusion. No evidence is shown that CCA has something to do with other bacteria or cytokine. In 2002, Matsukura et al. (2002) measured the bile of patients with

benignant and malignant biliary tract diseases, and found *H. bilis* closely related to CCA and other malignant biliary tract diseases in some populations, while it has nothing to do with benign biliary tract diseases. Further studies suggest that there is a relationship between bacteria load and the effect on CCA cells proliferation and apoptosis. It is also confirmed by preliminary experiment. In a way, the changes of QBC939 and TFK-1 cells proliferation and apoptosis are on the premise of bacteria load. That is to say, there are perhaps some intimidating correlations between *H. pylori* infection of biliary tract and CCA. However, there are no clear conclusions for the mechanisms of malignancy of biliary tract cells induced by *H. pylori* infection. It needs further intensive research.

The results from this experiment indicate that both PCB group and NCB group could promote the proliferation of CCA cell lines, but the proliferative effects of PCB group are more obvious than that of NCB group. That is to say, patients with *H. pylori* positive cholelithiasis are more liable to develop CCA. This experimental result can well explain the phenomena such as the high prevalence rate of CCA among patients with cholelithiasis and the uneven probability of CCA among patients with cholelithiasis. Therefore, *H. pylori* infection of biliary tract perhaps plays a role in the occurrence and development of CCA. It is yet to be studied and there is not a consistent conclusion about the causes of CCA (Boberg et al., 2006).

How does *H. pylori* infection promote the proliferation of human CCA QBC939 and TFK-1 cells? Flow cytometry is adopted in analyzing cell cycle and apoptosis in this study. The analysis of cell cycle indicates that, compared with NB group, the proliferative indexes of QBC939 and TFK-1cells increased obviously after being treated with PCB group and NCB group for 48 h, the percentage of S phase increased markedly, while the percentage of G_0/G_1 phase decreased remarkably, and the difference between PCB group and NCB group is significant. Data of cell apoptosis shows that PCB group and NCB group have no obvious impact on the apoptosis rate of QBC939 and TFK-1 cells. This group study verifies that bile from patients with choledocholithiasis infected by *H. pylori* promotes the proliferation of CCA cell lines QBC939 and TFK-1 through changing the cell cycle rather than the cell apoptosis.

This study conducts a tentative investigation of the relation between choledocholithiasis and CCA and the internal relation between *H. pylori* infection and occurrence and development of CCA, which illustrates the mechanism that choledocholithiasis bile infected by *H. pylori* could promote the proliferation of CCA cell through changing the cell cycle. Further research is still needed in regards to how *H. pylori* infection induces the changing of the cell cycle.

ACKNOWLEDGEMENTS

The work was supported by the National Natural Science Foundation of China (No. 30672426 and No. 30570908), and Project of Scientific Commission of Hubei province (No. 2003AA301C27 and No. 2004AA301C28).

REFERENCES

Anderson CD, Rice MH, Pinson CW, Chapman WC, Chari RS, Delbeke D (2004). Fluorodeoxyglucose PET imaging in the evaluation of gallbladder carcinoma and cholangiocarcinoma. J. Gastrointest. Surg. 8: 90-97.

Boberg KM, Jebsen P, Clausen OP, Foss A, Aabakken L, Schrumpf E (2006). Diagnostic benefit of biliary brush cytology in cholangiocarcinoma in primary sclerosing cholangitis. J. Hepatol. 45: 568-574.

Boomkens SY, de Rave S, Pot RG, Egberink HF, Penning LC, Rothuizen J, Zondervan PE, Kusters JG (2005). The role of Helicobacter spp. in the pathogenesis of primary biliary cirrhosis and primary sclerosing cholangitis. FEMS Immunol. Med. Microbiol. 44: 221-225.

Boomkens SY, Kusters JG, Hoffmann G, Pot RG, Spee B, Penning LC, Egberink HF, van den Ingh TS, Rothuizen J (2004). Detection of Helicobacter pylori in bile of cats. FEMS Immunol. Med. Microbiol.

42:307-311.

Briggs CD, Neal CP, Mann CD, Steward WP, Manson MM, Berry DP (2009). Prognostic molecular markers in cholangiocarcinoma: a systematic review. Eur. J. Cancer. 45: 33-47.

Chang FY, Chen CY, Lu CL, Luo JC, Lu RH, Lee SD (2005). Response of blood endothelin-1 and nitric oxide activity in duodenal ulcer patients undergoing Helicobacter pylori eradication. World J. Gastroenterol. 11: 1048-1051.

Chen CY, Lin XZ, Wu HC, Shiesh SC (2005). The value of biliary amylase and Hepatocarcinoma-Intestine-Pancreas/Pancreatitis-associated Protein I (HIP/PAP-I) in diagnosing biliary malignancies. Clin. Biochem. 38: 520-525.

Clyne M, Drumm B (1996). Cell envelope characteristics of Helicobacter pylori: their role in adherence to mucosal surfaces and virulence. FEMS Immunol. Med. Microbiol. 16: 141-155.

Enjoji M, Nakamuta M, Yamaguchi K, Ohta S, Kotoh K, Fukushima M, Kuniyoshi M, Yamada T, Tanaka M, Nawata H (2005). Clinical significance of serum levels of vascular endothelial growth factor and its receptor in biliary disease and carcinoma. World J. Gastroenterol. 11: 1167-1171.

Gieseler S, Konig B, Konig W, Backert S (2005). Strain-specific expression profiles of virulence genes in Helicobacter pylori during infection of gastric epithelial cells and granulocytes. Microbes. Infect. 7: 437-447.

Hong SM, Pawlik TM, Cho H, Aggarwal B, Goggins M, Hruban RH, et al. (2009). Depth of tumor invasion better predicts prognosis than the current American Joint Committee on Cancer T classification for distal bile duct carcinoma. Surgery, 146: 250-257.

Iwata K, Shijo H, Kamimura S, Uehara Y, Kitamura Y, Iida T, Okada Y, Akiyoshi N, Watanabe H, Sakisaka S (2002). Effects of esophageal varices obliteration by endoscopic variceal sclerotherapy on asialoscintigraphy and liver function test. Hepatol. Res. 22: 45-51.

Kanbar G, Engels W, Nicholson GJ, Hertle R, Winkelmann G (2004). Tyramine functions as a toxin in honey bee larvae during Varroa-transmitted infection by Melissococcus pluton. FEMS Microbiol. Lett. 234: 149-154.

Khan SA, Taylor-Robinson SD, Toledano MB, Beck A, Elliott P, Thomas HC (2002). Changing international trends in mortality rates for liver, biliary and pancreatic tumours. J. Hepatol. 37: 806-813.

Lin TT, Yeh CT, Wu CS, Liaw YF (1995). Detection and partial sequence analysis of Helicobacter pylori DNA in the bile samples. Dig. Dis. Sci. 40: 2214-2219.

Matsukura N, Yokomuro S, Yamada S, Tajiri T, Sundo T, Hadama T, Kamiya S, Naito Z, Fox JG (2002). Association between Helicobacter bilis in bile and biliary tract malignancies: H. bilis in bile from Japanese and Thai patients with benign and malignant diseases in the biliary tract. Jpn. J. Cancer Res. 93: 842-847.

Monstein HJ, Jonsson Y, Zdolsek J, Svanvik J (2002). Identification of Helicobacter pylori DNA in human cholesterol gallstones. Scand. J. Gastroenterol. 37: 112-119.

Nakajima T, Enomoto S, Ushijima T (2008). DNA methylation: a marker for carcinogen exposure and cancer risk. Environ. Health Prev. Med. 13: 8-15.

Nakajima T, Yamashita S, Maekita T, Niwa T, Nakazawa K, Ushijima T (2009). The presence of a methylation fingerprint of Helicobacter pylori infection in human gastric mucosae. Int. J. Cancer. 124: 905-910.

Olnes MJ, Erlich R (2004). A review and update on cholangiocarcinoma. Oncology, 66: 167-179.

Paziak-Domanska B, Chmiela M, Jarosinska A, Rudnicka W (2000). Potential role of CagA in the inhibition of T cell reactivity in Helicobacter pylori infections. Cell Immunol. 202:136-139.

Pouns O, Mangas A, Covenas R, Geffard M (2009). Circulating antibodies directed against "polycyclic aromatic hydrocarbon-like" structures in the sera of cancer patients. Cancer Epidemiol. 33: 3-8.

Rahn W, Redline RW, Blanchard TG (2004). Molecular analysis of Helicobacter pylori-associated gastric inflammation in naive versus previously immunized mice. Vaccine 23: 807-818.

Ricci V, Sommi P, Fiocca R, Necchi V, Romano M, Solcia E (2002). Extracellular pH modulates Helicobacter pylori-induced vacuolation and VacA toxin internalization in human gastric epithelial cells.

Biochem. Biophys. Res. Commun. 292: 167-174.

Shaib Y, El-Serag HB (2004). The epidemiology of cholangiocarcinoma. Semin. Liver. Dis. 24: 115-125.

Tian HL, Zhao D, Ren LM, Su XH, Kang YH (2009). Effects of (-)doxazosin on histomorphologic and cell apoptotic changes of the hyperplastic prostate in castrated rats. Am. J. Med Sci. 338: 196-200.

Tsai CC, Huang LF, Lin CC, Tsen HY (2004). Antagonistic activity against Helicobacter pylori infection in vitro by a strain of Enterococcus faecium TM39. Int. J. Food Microbiol. 96: 11-12.

Yao Q, Chen J, Cao H, Orth JD, McCaffery JM, Stan RV, McNiven MA (2005). Caveolin-1 interacts directly with dynamin-2. J. Mol. Biol. 348: 491-501.

Zampa A, Silvi S, Fabiani R, Morozzi G, Orpianesi C, Cresci A (2004). Effects of different digestible carbohydrates on bile acid metabolism and SCFA production by human gut micro-flora grown in an *in vitro* semi-continuous culture. Anaerobe 10: 19-26.

5

Frequently-used agrochemicals lead to functional and morphological spermatozoa alterations in rats

Hurtado de Catalfo Graciela, Astiz Mariana, Alaniz María J. T. de and Marra Carlos Alberto*

INIBIOLP (Instituto de Investigaciones Bioquímicas de La Plata), CCT La Plata, CONICET-UNLP, Cátedra de Bioquímica y Biología Molecular, Facultad de Ciencias Médicas, Universidad Nacional de La Plata, 60 y 120 (1900) La Plata, Argentina.

It is known that agrochemicals alter male reproductive functions. Previous studies from our lab have demonstrated the relationship between male fertility dysfunction and oxidative-nitrative stress. In this work, morphological and functional spermatozoa parameters were studied in a rat model sub-chronically (5 weeks) intoxicated with low doses (i.p. $1/250$ LD_{50}) of a mixture of dimethoate (D), glyphosate (G) and zineb (Z). The cytological assays showed alterations in spermatozoa morphology and in plasma membrane integrity. Modifications in the fatty acid composition were also shown. RIA analyses demonstrated androgenic hormone imbalance in plasma and testes. The acrosome reaction was also altered. Free thiols (positively correlated with DNA denaturation) and fructose levels were elevated in seminal vesicles from treated rats. Taking into account the low doses of pesticides that provoke these alterations, it was assumed that the environmental pollution may play a key role as a causative factor for fertility abnormalities.

Key words: Dimethoate, zineb, glyphosate, rat, spermatozoa, fertility/sperm abnormalities.

INTRODUCTION

Health disorders associated with environmental pollution are a cause of international concern. Among them, the incidence of male fertility disturbance has been increasing since the 1980's probably due to multi-factorial events involving both environmental and genetic factors (Giwercman et al., 1993; Petrelli and Mantovani, 2002;

Pflieger-Bruss et al., 2004). Several studies concerning the effect of agrochemicals on the reproductive system suggest various possible mechanisms of toxicity. The most documented ones are the hormonal disruption (Sarkar et al., 2000; Bhatnagar, 2001; Basrur, 2006; Joshi et al., 2007) and the alteration of the antioxidant defense system (Sikka, 2001; Sheweita et al., 2005; Aitken and Baker, 2006; Kesavachandran et al., 2009).

So far, pesticides are widely used for agricultural purposes. Thus, residues of many of them (and possibly their metabolites) could remain as pollutants in water, air and food. In this work, we studied three of the most commonly used pesticides worldwide: zineb (Z), glyphosate (G) and dimethoate (D) in combination. Dimethoate is an organophosphorus insecticide of systemic action extensively used in pest treatment in onions, tomatoes, and citric fruits among others (Sharma et al., 2005b). Glyphosate is a systemic herbicide used to

*Corresponding author. E-mail: contactocarlos@hotmail.com, camarra@med.unlp.edu.ar.

Abbreviations: c, Control group; **D,** dimethoate; **FSH,** folicle-stimulating hormone; **G,** glyphosate; **LH,** luteinizing hormone; **OPs,** organophosphorus pesticides; **OS,** oxidative stress; **PEG-400,** polyethyleneglycol-400; **RNS,** reactive nitrogen species; **ROS,** reactive oxygenated species; **ZGD,** treated group; **Te,** testosterone; **Z,** zineb.

control undergrowth before seeding in corn, soybean, vine, etc. (Daruich et al., 2001) and zineb is a contact fungicide used to control pests in carrots, onions, citric fruits and potatoes (Heikkila et al., 1976).

It was previously reported that dimethoate impairs spermatozoa motility, decreases serum testosterone levels and testicular weight, and increases the percentages of dead and abnormal spermatozoa in rats, rabbits (Salem et al., 1988; Walsh et al., 2000b; Afifi et al., 1991) and mice (Farag et al., 2007). Moreover, a previous work from our lab demonstrates that, dimethoate displays a complex mechanism of action involving disturbances in the hormone production (at both systemic and Leydig cell levels). We found alterations in the antioxidant defense system, decreased phospholipids araquidonate content, inhibition of StAR protein expression with simultaneous stimulation of COX-2 (overproduction of PGF2α) and also the inhibition of steroidogenic enzymes 17βHSD and 3βHSD (Astiz et al., 2009b).

Recent studies in rats suggest that, the exposure to glyphosate during the pre- and post-natal periods induces adverse effects on male reproductive performance (Dallegrave et al., 2007). The exhaustive bibliographic revision made by Basrur (2006) showed that, glyphosate can act as a sexual differentiation disruptor and as an estrogen-like compound in domestic animals and humans. Moreover, epidemiological evidence indicates that women which couples were in contact with glyphosate had difficulty to conceive and also showed a higher rate of miscarriage (Arbuckle et al., 2001; Caglar and Kolankaya, 2008). In addition, zineb produces a decrease in mouse fertility performance due to alterations in male and female pronuclei formation (Rossi et al., 2006).

Traditionally, male infertility diagnosis depends on microscopic and biochemical assays to assess spermatozoa concentration, morphology, and motility. Over the past decades, several in vitro tests have been developed to evaluate aspects involved in sperm functional competence which included movement ability, cervical mucus penetration, capacitation, acrosome reaction, sperm-oocyte fusion, redox status, and integrity of nuclear and mitochondrial DNA. These methodologies should be useful to predict either in vitro or in vivo sperm fertilizing ability with relatively high accuracy (Aitken, 2006; Lewis, 2007). However, most (if not all) investigations mentioned earlier concerning reproductive toxicity, were performed in single intoxication animal models using high doses (near LD$_{50}$) of pesticides rarely observed in daily exposure. For this reason, the present work was undertaken to evaluate the chronic effect of low doses (1/250 LD$_{50}$) of the aforementioned agrochemicals administered, as a mixture. The doses selected for our experiments were chosen according to previous reports on their toxic effects (John et al., 2001; Sharma et al.,

2005a; Beuret et al., 2005; Nielsen et al., 2006; Patel et al., 2006).

Our aim was to explore the impact of such association in testicular performance by a more realistic experimental model, because many pollutants reach our tissues as mixtures present in environment to which animals and humans are inevitably exposed during prolonged periods (Cory-Slechta, 2005). Our specific targets were (i) to evaluate alterations in the morphological and functional sperm characteristics, and (ii) to study biochemical parameters related with fertility performance such as capacitation and acrosome reaction, hormonal levels, and thiol and fructose contents. The results may contribute to the risk assessment of involuntarily and chronically pesticide-exposed populations.

MATERIALS AND METHODS

Chemicals

Most chemicals used were of reagent grade and obtained from Sigma Chem. Co. (CA, USA, or Buenos Aires, Argentina) or Merck Laboratories (Darmstadt, Germany). Organic solvents were from Carlo Erba (Milano, Italy). Other chemicals were purchased from local commercial sources and they were of analytical grade. The pesticides, dimethoate (O,O-dimethyl-S-methyl-carbamoyl-methyl phosphorodithioate), zineb (zinc ethylene-bis-dithiocarbamate) and glyphosate (N-phosphonomethyl-glycine) were obtained as a gift from Instituto Nacional de Tecnología Agropecuaria (INTA, National Institute for Agrochemical Technology, Castelar, Argentina).

Animal care and treatment

Male Wistar rats weighing 190 ± 20 g were breaded by the Laboratory Animals service from the Veterinarian School at La Plata National University, they had a certified pathogen-free status. Before starting with the experiment, they were allowed to acclimatize for a week. The rats were maintained under controlled temperature (25 ± 3°C), and with a normal photoperiod of 12 h darkness and 12 h light. They were fed with standard Purina chow from Ganave S.A. (Santa Fe, Argentina) and given water ad libitum in agreement with the American Institute of Nutrition (Reeves et al., 1993). Clinical examinations and body weight evaluations were performed every week during the experiment. The animals were in very good conditions along the experimental period. No sign of toxic effect was observed. None of the treatments influenced water consumption (approx. 15 ml/day), final body weights, body weight gain rate, food efficiency ratio, and testicular weight (Table 1). Pesticide exposure did not affect the animal behavior. Also, no visible signs of toxicity and/or cholinergic effects were observed during the entire experimental period. There was no mortality.

The animals were randomly divided into two groups of five animals each, assigned as control rats (C) injected intraperitoneally (i.p.) with polyethyleneglycol-400 (PEG-400), and treated (ZGD) rats injected i.p. with a combination of 15 mg dimethoate (D)/kg of body weight (bw), 15 mg zineb (Z) /kg bw and 10 mg glyphosate (G)/ kg bw dissolved in PEG-400. All animals were injected three times a week for five weeks. Dosing schedule was selected in view of previous experimental protocols used by us and other researchers of the area (Bagchi et al., 1995; John et al., 2001;

Table 1. Main feeding parameters associated to experimental treatments.

Parameters	Treatments	
	C	ZGD
Initial body weight (g)	181.0 ± 3.7	174.0 ± 7.3
Final body weight (g)	307.5 ± 13.7	288.8 ± 20.9
Body weight gain (g)	126.5 ± 3.4	114.8 ± 5.5
Rate of body weight gain (g/day)	3.6 ± 0.1	3.3 ± 0.2
Food efficiency ratio[a]	9.5 ± 0.2	8.7 ± 0.3
Absolute testicular weight (g)	2.8 ± 0.1	2.9 ± 0.1
Relative testicular weight (mg/g)[b]	9.1 ± 0.3	10.0 ± 0.4

C, Control rats; ZGD, treated rats. Values represent the mean ± SD ($n = 8$). [a] Food efficiency ratio = [body weight gain (g) / food intake (g)].10^2. [b] Relative testicular weight = testis weight (mg)/ body mass (g).

Nielsen et al., 2006; Sivapiriya et al., 2006; Astiz et al., 2009a, b, c and d). We think that such low doses would be reached in certain circumstances, especially in rural zones exposed to frequently agrochemical spraying from airplanes, or even by ingestion of fruits and vegetables not previously analyzed. Most of the previous reports in the field of pesticides and oxidative stress were performed in single intoxication animal models. For this reason, the present investigation was undertaken to evaluate the effect of a sub-chronic exposure to low doses of agrochemicals administered in combination. This may contribute to the understanding of the effect of a simultaneous exposition, and it may have interest for the evaluation of the damage degree to which humans are potentially exposed due to environmental pollution in a more similar way as it happens in real life exposure. Of course, it is impossible to reproduce exactly what happens to humans; however, with this model we can compare at least the effects of a frequent mixture. We did not pretend to establish a toxicosis with clinical manifestation.

The aim of the study was to explore the effect of sub-clinical and sub-symptomathologycal doses of the agrochemicals tested. So, clinical examinations were performed in order to exclude any sign of toxicosis. A veterinarian observes for the presence of miosis, mouth smacking, salivation, or lacrimation. The rats were also placed in an open field for observation of tremors or gait abnormalities according to Moser et al. (2006). There were no significant differences between controls and treated animals.

Animal maintenance and handling were in accordance with the NIH guide for the care and use of laboratory animals published in 1985. All procedures were approved by the local laboratory animal bioethical committee of Facultad de Ciencias Médicas (UNLP, Argentina).

Sample collection

At the end of the treatment all rats were killed by rapid decapitation. Blood was collected using heparin as anticoagulant (10 UI/ml) and plasmas were immediately prepared by centrifugation (4000 g, 10 min) and stored at -80°C until analyzed. Both testicular epididymis were quickly excised and the caudal zones were transferred into flasks containing 8 ml of Biggers, Witten and Whittingham medium (BWW) (Biggers et al., 1995).

Tissues were minced and incubated (35°C, 20 min) under gentle

and constant shaking. Under these conditions epididymal content was spontaneously released into the medium minimizing cell damage (Klinefelter et al., 1991). After spontaneous decanting (10 min) the remaining tissue was removed, while cells were transferred into plastic tubes and centrifuged (700 g; 15 min). Pelleted spermatozoa from both experimental groups were washed twice with BWW and resuspended in 10 ml fresh medium. Both seminal vesicles of each rat were excised, washed and homogenized (glass/teflon homogenizer) with 3 ml ice-cold phosphate buffer 100 mM, pH 7.40 with 6 mM of EDTA. Homogenates were stored at -80°C until analyzed.

Spermatozoa morphology evaluation

Morphology was evaluated microscopically as described by Filler (1993). We studied epididymal spermatozoa shape and morphology by examination of head, mid-piece and tail sections. The staining procedure was similar to the one described by Larson and Miller (1999). In brief, aliquots (1 ml) from each spermatozoa suspension were incubated (25°C, 60 min) with 100 µl of eosin-Y (1% w/v in NaCl 0.9%). Smears were prepared, air dried, covered with Biopack (Synthetic Canadian Balsam 130305 from Biopack S.A., Buenos Aires, Argentina) and a cover slip. Five microscopic fields of each slide (approx. 500 cells each) were analyzed using light microscope (400 X). Spermatozoa were classified as follows: N, normal (without alterations); MZ, with altered head and/or middle section; F, flagellum or tail alterations; or M, mixed (combined) alterations. Results were expressed as percentage of the total spermatozoa count following the normal morphology criteria defined by World Health Organization (1999).

Spermatozoa vitality and plasma membrane integrity

Spermatozoa vitality was estimated using the eosin-nigrosin (EN) supravital stain technique as described by Eliasson and Treichl (1971). 100 µl of each spermatozoa suspension was gently mixed with 100 ml of eosin-nigrosine solution (eosin-Y 5% (w/v) and nigrosine 10% (w/v) in NaCl 0.9%). This dye exclusion technique differentiates vital spermatozoa from dead. Intact spermatozoa (considered alive) were not stained, while spermatozoa with altered cell plasma membrane integrity (considered dead) took up the dye. Smears were prepared on acetone-pretreated slides repeatedly washed with distilled water. They were dried at 37°C and then observed under light microscope (400 X). At least two fields/slide (200 cells/field) of each sample were analyzed and classified as follows: alive (non-stained); dying (partially stained) or dead (completely stained).

Spermatozoa capacitation and acrosome reaction *in vitro*

Spermatozoa capacitation and the acrosome reaction are both essential processes for oocyte fertilization. Acrosome reaction could be induced *in vitro* by adding the Ca-ionosphore A-23187 (Ionomicine) according to the protocol described by Beitbart and Naor (1999). 1 ml of each spermatozoa suspension in BWW containing Bovine Serum Albumin (BSA) 4 g/L and sodium bicarbonate 25 mM was treated with 1 µl of Ionomicine (1 µM final concentration). An equivalent aliquot of each suspension (in BWW free of BSA and sodium bicarbonate) was incubated without Ionomicine as control assay. After a 30 min-incubation (at 35°C), sub-aliquots (0.5 ml) were taken and mixed with 25 μl of formaldehyde 40%. The remaining suspensions were incubated for further 30 min and then mixed with 25 μl of formaldehyde 40%. All samples were centrifuged (1000 g, 10 min) and fixed cell pellets were resuspended in PBS (0.5 ml). Smears were prepared and stained with Coomasie Brilliant Blue G-250 (0.22% in methanol 50% with 10% of concentrate acetic acid) for 2 min, washed with distilled water, mounted with Biopack® and observed under light microscope (400x). The spermatozoa that underwent acrosome reaction had the head zone unstained, while those without acrosome reaction were completely stained in blue. Results at both both incubation times (30 and 60 min) were expressed as percentage.

Analytical determinations

Fatty acid composition of caudal epididymal spermatozoa

Since spermatozoa lipid composition plays an important role for a successful fertilization, we analyzed the fatty acid (FA) composition of total lipids obtained from caudal epididymal spermatozoa of control and treated rats as described by Aksoy et al. (2006). An aliquot of the spermatozoa suspension (1 ml) was centrifuged (1500 g; 5 min). 5 ml of Folch reagent was added to the pelleted cells (Folch et al., 1957) and shook vigorously for 1 min.

After partitioning, the extracted lipids were saponified and esterified under N_2 atmosphere to obtain the corresponding fatty acyl methyl esters (FAMEs) using boron trifluoride (14%) according to Morrison and Smith (1964) with minor modifications, as described in detail in a previous paper (Hurtado and Gómez, 2002). FAMEs mass-composition was analyzed by c-GLC (capillary-gas liquid chromatography) using a HP6890 GL chromatograph equipped with an Omegawax 250 fused silica column (30 m x 0.25 mm with 0.25 μm phase from Supelco, Bellefonte, PA). Data were processed electronically and fatty acids were identified on the chromatograms by means of their relative retention times compared with pure fatty acid mixtures processed in parallel (Sigma Chem. Co., Buenos Aires, Argentina). Results were expressed as mol % of total FA.

Free thiol determination

DNA integrity in mature spermatozoa is correlated with the levels of free thiol groups (Zini et al., 2001). Each spermatozoa suspension (8 ml) was centrifuged (600 g; 5 min) and the pellet resuspended in 2 ml of cold TNE buffer (Tris-HCl 0.01 M, pH 7.40 with NaCl 0.15 M and EDTA 1 mM). The suspensions were centrifuged again and resuspended in 500 μl of TNE buffer with 10% glycerol. 200 μl of each final spermatozoa suspension was incubated with 10 μl of SDS 10% (10 min; 25°C), and then treated with 800 μl of DTNB (5,5'-dithiobis(2-nitrobenzoic acid) 0.6 mM in potassium phosphate buffer 0.1 M, pH 7.0 with 6% ethanol). After mixing, the samples were centrifuged at 10000 g for 10 min. The supernatant optical density was measured at 405 nm vs blank tubes processed in identical manner but omitting the sample. Thiol concentrations were obtained from a calibration curve using GSH (10 mM in distilled water) as standard solution.

Fructose level

Fructose is produced in seminal vesicles during spermatozoa maturation. Due to its inverse relationship with spermatozoa motility this sugar is considered as a biomarker of seminal vesicle function

and motility (Gonzales, 2001). Its concentration was determined in seminal vesicle homogenates (100 μl) after deproteinization with 100 μl of sulfosalicylic acid (10%). Supernatants from centrifugation (14000 rpm; 10 min) were employed for the assay, performed according to Somani et al. (1987). Briefly, 10 μl of each supernatant was incubated (55°C, 90 min) with 1 ml of ATS reagent (10% (w/v) of anthrone and 10% (w/v) tryptophan in sulphuric acid 75% (v/v) plus 440 μl distilled water. The optical densities were measured at 520 nm against a reaction blank. Fructose concentrations were obtained from a calibration curve using a fructose solution (222 μM in distilled water) as standard.

Other analytical determinations

All plasma hormone measurements (luteinizing (LH), follicle-stimulating (FSH), total and free testosterone (Te), and estradiol) were performed by radioimmunoassay (RIA) using commercial kits (KP7CT, KP6CT, KS24CT, and KS33CTN, respectively) from Radim (Radim SpA, Pomezia, Italy). To measure Te in testicular homogenates, crude preparations were previously centrifuged (2000 g; 15 min) and the supernatants were used as samples for the RIA assay using the kit KS24CT from Radim. Protein content was determined according to Bradford (1976).

Statistical analysis

Results were analyzed by one way analysis of variance (ANOVA) followed by Tukey multiple comparison procedure or Student-t test where appropriate. Data were expressed as the mean ± SD of at least five independent determinations. They were considered different with respect to the control data at two levels of significance $^*p < 0.05$ and $^{**}p < 0.01$.

RESULTS

It is well-known that the lipid composition from spermatozoa membranes plays an important role during the fertilization process (Aksoy et al., 2006; Tavilani and Doosti, 2007; Aitken, 1995; Coniglio, 1994). Thus, we studied the fatty acid composition of total lipids from control (C) and treated (ZGD) rats. Figure 1 shows the differences between both groups. Treated rats exhibited a relative decrease of palmitic (16:0) (p<0.05) and palmitoleic (16:1) acids content while polyunsaturated fatty acids (PUFAs) 22:4 n-3, 22:5 n-6 and 22:6 n-3 were increased compared to C. Thus, the treated group showed a significant increase in the insaturation index (153.2 ± 18.8 vs 121.4 ± 9.8). PUFAs from both essential series (n-6 and n-3) were elevated in ZGD rats, showing a significant increase in the proportion of n-3 fatty acyl chains compared to control (p<0.05). In spite of that, the ratio n-6/n-3 was not significantly different between both groups (5.5 ± 1.2 and 4.0 ± 2.5, respectively) (Table 2).

We also performed morphological studies. The shape and morphology of mature spermatozoa were studied analyzing head, middle zone and tail features. Results were expressed as percentage of total spermatozoa count

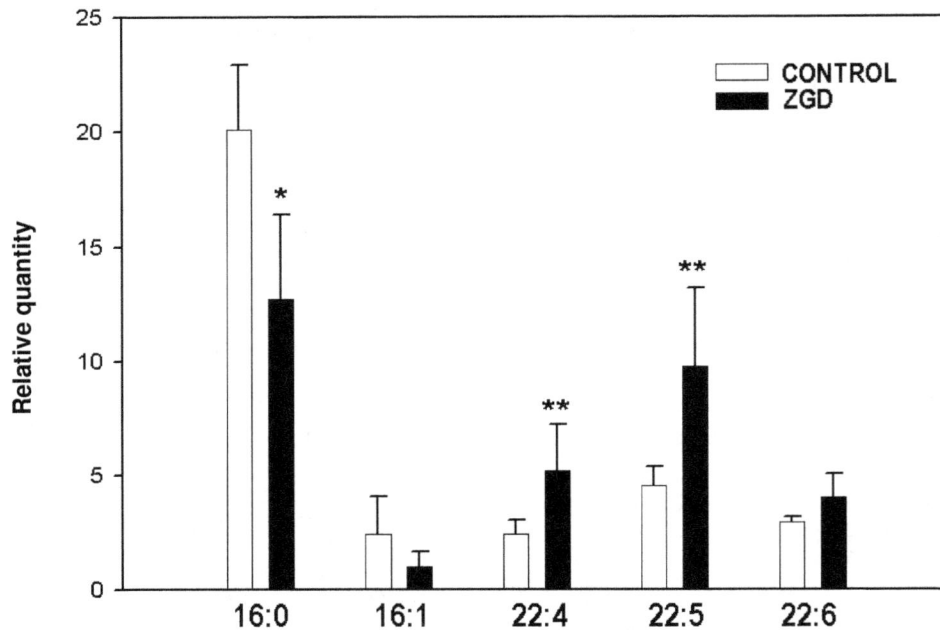

Figure 1. Fatty acid composition of total lipids from epididymal sperm obtained from control (white bars) or (ZGD)-treated (black bars) rats. Animals were treated as described under materials and methods. The figure shows only those fatty acids that showed significant differences compared to controls. Data were expressed as mol percentage. Each bar represents the mean of five independent determinations ± SD. Significant differences from control data were indicated with asterisks (*; $p < 0.05$ and **; $p < 0.01$).

Table 2. Fatty-acids composition indexes of epididymal spermatozoa from control and treated rats.

Analytical indexes	Treatment	
	Control	ZGD
Σ Triethilenic fatty acids	2.9 ± 0.7	5.9 ± 2.1 *
Σ Tetraethilenic fatty acids	3.1 ± 0.3	4.9 ± 0.8 **
Σ Pentaethilenic fatty acids	0.9 ± 0.2	10.5 ± 3.5 **
Insaturation index (II)	121.4 ± 9.8	153.2 ± 18.8 *
Σ n-6	33.0 ± 5.7	38.0 ± 1.5
Σ n-3	6.0 ± 0.3	10.0 ± 3.1 *
Σ n-6/Σn-3	5.5 ± 1.2	4.0 ± 0.9

Results were expressed in percentages as the mean ± SD of five independent assays. Values significantly different compared to control group were indicated as *$p < 0.05$ or **$p < 0.01$.

(Figure 2). The exposure to the agrochemical mixture (ZGD) significantly reduced the proportion of spermatozoa with normal morphology (N) with a concomitant increase of cells with morphological aberrations (in MZ, F and in both zones) ($p < 0.01$). The agrochemical-induced alterations were similarly distributed among the spermatozoa zones (Figure 2A). Figure 2B is a representative microscopic preparation obtained from a treated rat smear where some evident morphological alterations were observed (arrows). Taking into account that, the evaluation of the spermatozoa membrane integrity is a reliable biomarker of spermatozoa viability currently assessed by staining techniques, we have studied the response of spermatozoa to the eosin-nigrosine (E-N) supra-vital assay. Results obtained are shown in Table 3. Rats treated with the agrochemical

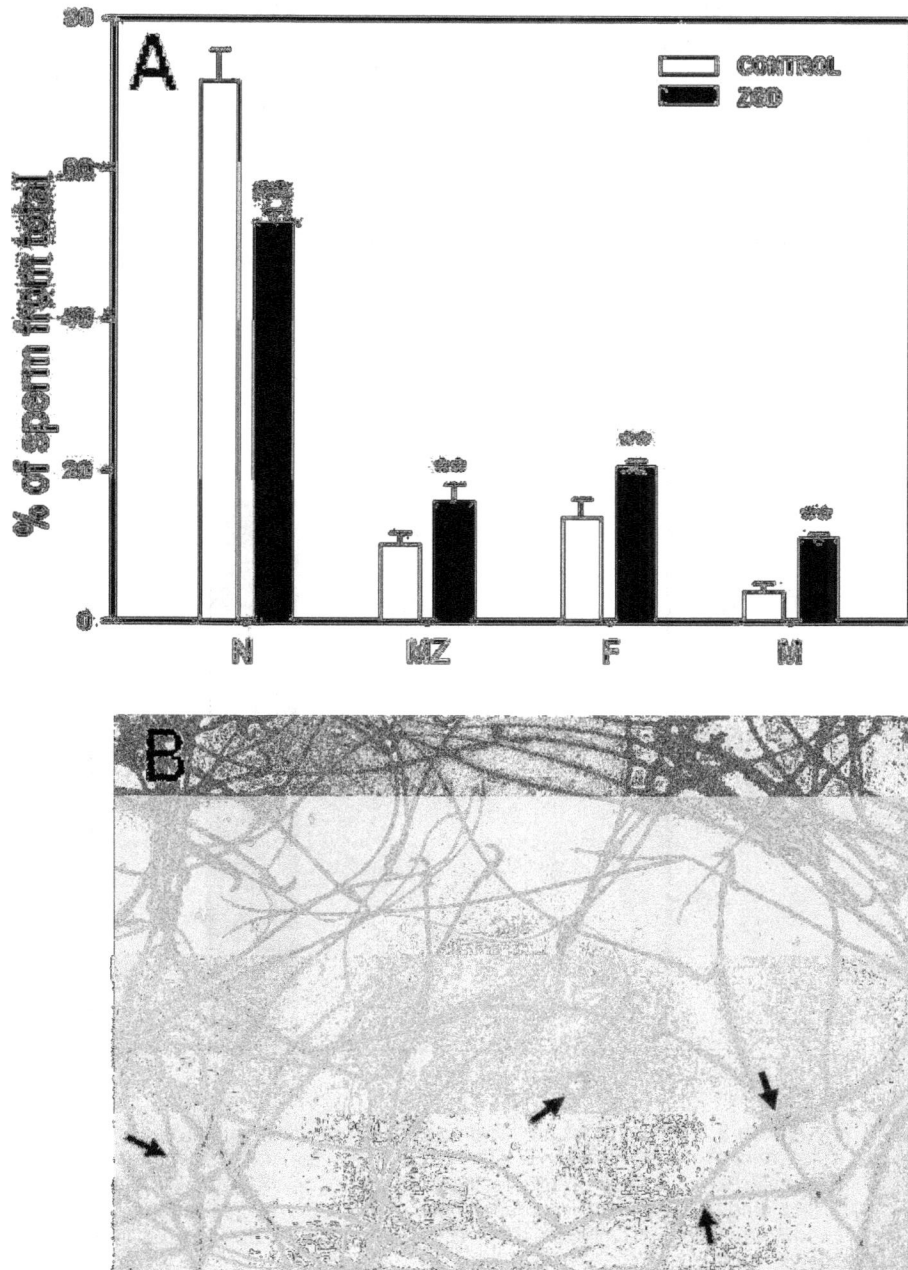

Figure 2. Mature sperm morphology (A) from control (white bars) or (ZGD)-treated (black bars) rats. Animals were treated as described under materials and methods. Briefly, spermatozoa suspension were incubated (25 °C, 60 min) with 100 µl of eosin-Y (1% w/v in NaCl 0.9%). Smears were prepared, air dried, covered with Biopack and a cover slip. Five microscopic fields of each slide (approx. 500 cells each) were analyzed using light microscope (400 X). Spermatozoa were classified as follows: N, normal (without alterations); MZ, with altered head and/or middle section; F, flagellum or tail alterations; or M, mixed (combined) alterations. Results were expressed as percentage of the total spermatozoa count following the normal morphology criteria defined by World Health Organization (1999). Each bar represents the mean of five independent determinations ± SD. Significant differences compared to control data were indicated with asterisks (**; p<0.01). In panel B a representative microscopic field of a smear from a treated rat, where some evident morphological alterations were indicated (arrows).

Table 3. Functional integrity of sperm cell membrane in control and treated rats.

Condition	Treatment	
	Control	ZGD
Alive	97.85 ± 0.65	95.51 ± 0.33 **
Dying	1.23 ± 0.48	2.31 ± 0.78 **
Died	0.91 ± 0.59	2.17 ± 0.73 **

Results were expressed as the mean ± SD of five independent assays in percentages. Values significantly different compared to control group were indicated with asterisks (**; p< 0.01).

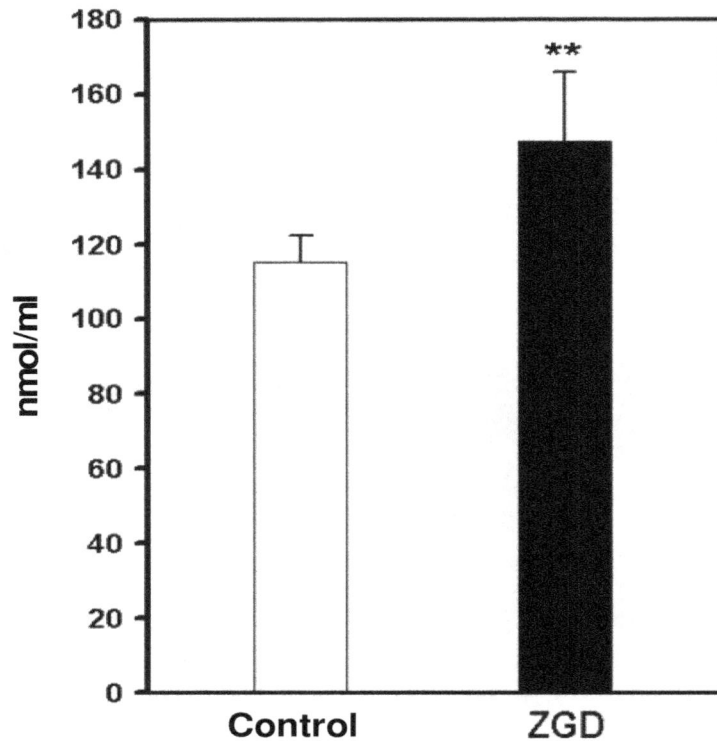

Figure 3. Content of free thiols in mature sperm from caudal epididyms of control (white bar) or (ZGD)-treated (black bar) rats. Animals were treated as described under materials and methods. Data are expressed as nmol/ml. Each bar represents the mean ± SD of five independent determinations. Significant differences from control data were indicated with asterisks (**; p< 0.01).

mixture showed a significant increase of both dying (p<0.01) and dead (p<0.01) spermatozoa, compared to control rats and aconcomitant decrease in the percentage of the living ones (p<0.01).

It is known that the content of spermatozoa free thiols correlates positively with DNA denaturation (Folch et al., 1957). We found that the concentration of this biomarker (expressed as nmol/ml) in epididymal spermatozoa from treated rats was significantly higher compared with that of the control group (p<0.01; Figure 3). Fructose production is usually employed as a marker of seminal vesicle function due to its inverse relationship with spermatozoa motility. Fructose levels in seminal vesicle homogenates were significantly higher in ZGD compared to C rats (p<0.01; Figure 4). Figure 5 shows the percentage of positive acrosome reactions either in control or treated

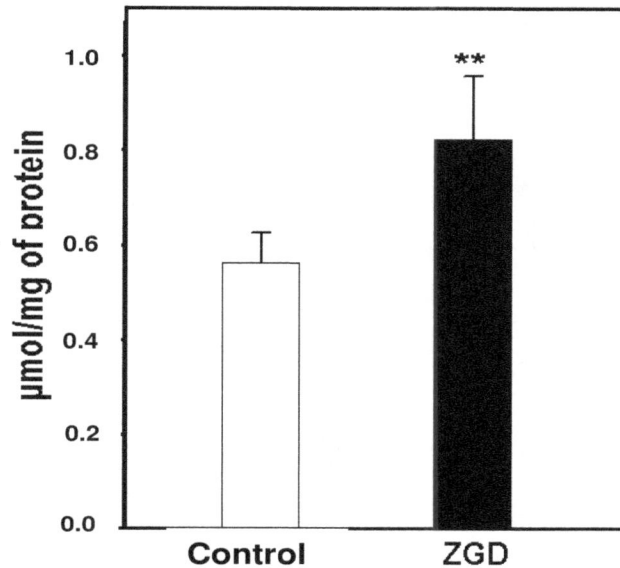

Figure 4. Fructose content in seminal vesicle homogenates from control (white bar) or (ZGD)-treated (black bar) rats. Animals were treated as described under materials and methods. Data were expressed as µmol/mg of protein. Each bar represents the mean ± SD of five independent determinations. Significant differences from control data were indicated with asterisks (**; p< 0.01).

group. The ZGD group showed a significant decrease of spermatozoa with positive acrosome reaction at both times assayed (30 and 60 min) compared to their respective controls. Control spermatozoa reacted in a time-dependent fashion under conditions of pro-activation (Ca^{2+}-ionophore addition). In addition, suppression of Ca^{2+} strongly decreased the percentage of activated spermatozoa at both times. Agrochemical treatment significantly depressed (p<0.01) spermatozoa activation (43 and 56% decrease compared to the corresponding control at 30 and 60 min, respectively). Similarly, reacted spermatozoa were strongly reduced compared to control assay under non pro-activating conditions (p<0.05). In this case, the reaction time-dependence was completely abolished in both experimental groups (Figure 5).

In order to explore the effects of the agrochemical treatment on the androgenic function, hormonal parameters were determined in both plasma and testis homogenates from control or intoxicated rats using RIA methodologies (Table 4). Free and bound testosterone as well as estradiol were diminished in plasma from treated rats (p<0.01), whereas the ratio free/bound Te was indistinguishable between groups. LH and FSH levels in treated rats were significantly increased (p<0.01). We also observed a great reduction (approximately 50%) of Te production in testis homogenates obtained from

treated rats (p<0.01).

DISCUSSION

Several studies suggest that, the decay of human semen quality may be related to the occupational or involuntary exposure to pesticides (Gray et al., 2001; Sanderson, 2006; Recio-Vega et al., 2008; Yucra et al., 2008; Perry, 2008). However, to establish a perfect association with a specific pollutant is almost impossible. More likely, the sum of many environmental contaminants seems to be responsible for the increased incidence of male reproductive dysfunction (Saradha and Mathur, 2006). In addition, the interpretation of experimental results obtained from single intoxication models, often leads to uncertain conclusions mainly associated with the effects displayed by the same pollutant under different contexts (Murono et al., 2001). In this work we used a more realistic animal model sub-chronically exposed to low doses of three of the most frequently used agrochemicals worldwide -administered in combination (ZGD) - to explore the effect of the mixture on semen alterations. This kind of experimental strategy is currently adopted in order to approach the situation in real life, and also to discover and validate robust biomarkers for health risk

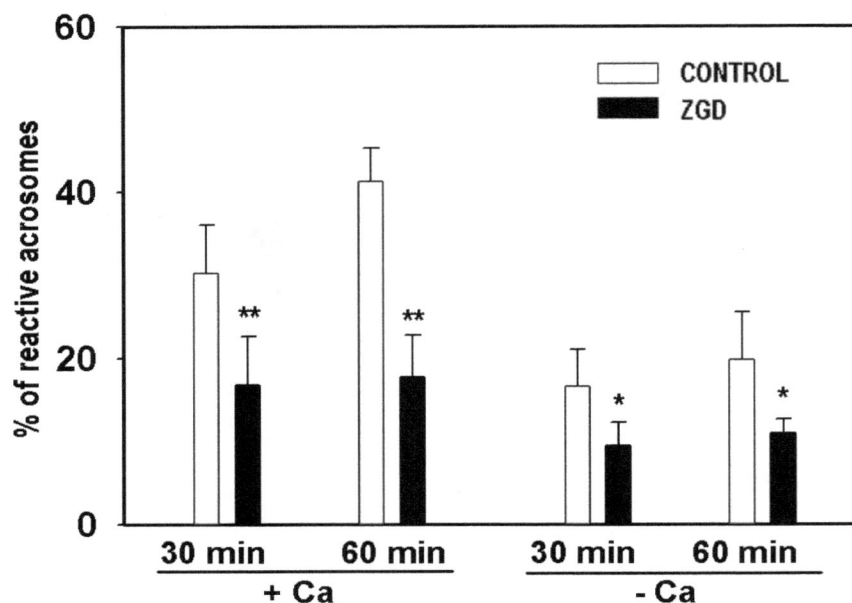

Figure 5. *In vitro* sperm acrosome reaction of control (white bars) and (ZGD)-treated (black bars) rats. Animals were treated as described under materials and methods. Suspensions were incubated in the presence (+ Ca) or in the absence (– Ca) of Ca++ ions for 30 or 60 min at 35°C under gentle shaking. Data were expressed as percentage of reactive acrosomes. Each bar represents the mean ± SD of five independent determinations. Significant differences compared to the corresponding control data were indicated with asterisks (*; $p < 0.05$) and **; $p < 0.01$).

assessment (Knudsen and Hansen, 2007).

We have already reported a profound alteration in the redox status of various tissues isolated from rats treated with this agrochemical mixture (Astiz et al., 2009a). Both types of oxidative stress biomarkers - enzymatic and non-enzymatic- were severely altered in plasma and testis indicating a significant unbalance between free radical production and antioxidant defenses. Plasma hormonal levels were also altered. Free and bound testosterone, as well as estradiol levels were depressed in plasma from treated rats while LH and FSH content were increased compared to controls. These results could be interpreted as an adaptive feed-back response of the gonadal-pituitary axis induced by the decay of testosterone level. In line with this fact, we have demonstrated in a recent paper that dimethoate (D) by itself inhibits testosterone biosynthesis in interstitial (Leydig) cells by a mechanism that involves COX-2 and StAR expression, even at the low doses used in our experimental model (Astiz et al., 2009a). The oral administration of technical dimethoate also produces adverse effects on male reproductive performance in mice (Farag et al., 2007).

In addition, two independent reports demonstrated that glyphosate (the active ingredient of Roundup formulation)

and octylphenol (a surfactant additive frequently used in many industrial applications) are both inductors of steroidogenic dysfunction (Walsh et al., 2000a; Murono et al., 2000). These results emphasize the importance of performing studies with mixtures of agrochemicals, instead of single-drug experimental models. Traditionally, spermatozoa count is considered as a biomarker of semen quality; however, there are numerous factors that directly or indirectly affect reproduction and reduce the level of desirable statistical power of these results (Seed et al., 1996). In contrast, more recent research determined a direct (positive) relationship between plasma membrane integrity and spermatozoa motility and viability (Dougherty et al., 1975; Vetter et al., 1998; Pesch and Bergmann, 2006). Other authors strongly support the correspondence between oxidative stress biomarkers and spermatozoa characteristics (Abarikwu et al., 2009, El-Taieb et al., 2009), or spermatozoa morphology and genomic integrity of spermatozoa (Zini et al., 2009). All this experimental evidence is in agreement with findings concerning the relationship between oxidative stress, semen characteristics, and clinical diagnosis in men undergoing infertility (Pasqualotto et al., 2000).

Our results clearly indicated that, pesticides provoked

Table 4. Hormonal parameters in plasma and testicular homogenates from control and treated rats.

Determinations	Treatment	
	Control	ZGD
Plasma		
Free testosterone (µM)	4.7 ± 0.2	3.2 ± 0.1**
Bound testosterone (µM)	22.2 ± 1.5	15.3 ± 0.8**
Free/bound testosterone[1]	212 ± 15	209 ± 9
Estradiol (µg/ml)	15.1 ± 0.6	9.3 ± 0.2**
LH (mU/ml)	7.7 ± 0.2	13.8 ± 0.3**
FSH (mU/ml)	8.1 ± 0.4	17.1 ± 0.5**
Testicular homogenates		
Testosterone (µM)	27.9 ± 1.3	11.4 ± 1.5**

Hormone levels were analyzed using RIA kits commercially available from Radim as indicated in Materials and Methods section. Results were expressed as the mean ± standard deviation (SD) of five independent determinations assayed in triplicate. [1][Free/bound testosterone]$\cdot 10^3$. Results significantly different respect to control values were indicated with asterisks (**; $p < 0.01$).

severe alterations in spermatozoa plasma membrane integrity that could be attributed to a free radical attack. It is well known that epididymal spermatozoa membranes are particularly susceptible to oxidative stress due to their high content of polyunsaturated fatty acids and their lack of Sertoli cell barrier protection (Aitken, 1995). Interestingly, we found that in spite of the pro-oxidant environment induced by pesticide administration the insaturation index of spermatozoa was conserved and even increased compared to controls. This fact is important considering the sub-chronical characteristics of our experimental protocol that may provoke an adaptative response in order to compensate the loss of PUFA from spermatozoa membranes. It is likely that in acute models, the induced oxidative stress may cause peroxidation of PUFAs in the whole testis including the spermatozoa. However, due to the importance of this kind of fatty acyl chains in events associated to fertility performance, chronic exposures may activate a sort of adaptative mechanism (induced biosynthesis, decreased catabolism, selective sequestration, or their combination). Unfortunately, we could not find analytical data reported by other laboratories using a similar experimental condition. Thus, this explanation remains to be further explored.

In line with this, previous work from our laboratory demonstrates that the content of 22:6 n-3 fatty acids in testis from rats under oxidative stress is approximately constant and independent of the pro-oxidant condition (Hurtado de Catalfo et al., 2008; Hurtado de Catalfo et al., 2009). On the other hand, the content of

plasmalogens decreases significantly in testis homogenates (Pesch and Bergmann, 2006) but it remains constant in the spermatozoa fraction (Snyder et al, 2001; Nagan and Zoeller, 2001). So, another possible mechanism that justifies the increment in C_{22} PUFAs could be associated to a selective elevation of plasmalogen content in spermatozoa, since this lipid sub fraction is enriched in PUFAs (especially those from the n-3 series) (Coniglio, 1994; Snyder et al., 2001) and it acts as potent anti-oxidant moieties (Snyder et al, 2001; Nagan and Zoeller, 2001). In agreement with this hypothesis we observed a particular increment of the n-3 fatty acyl chains proportion that is intimately associated to the conservation of both the spermatozoa vitality and functionality (Aksoy et al., 2006; Snyder et al., 2001; Furland et a., 2007).

Over the past decades new laboratory tests have been developed to determine properties of spermatozoa function including capacitation, basal and induced acrosome reaction, sperm-zona pellucida interactions and nuclear DNA damage. It was clearly demonstrated that damages to spermatozoa DNA may result in male infertility (Zini et al., 2001; Agarwal and Said, 2003). This could be in part attributed to the reduced ability of mature spermatozoa to repair their own DNA (Van Loon et al., 1991). Our results concerning the levels of spermatozoa protamine free thiol (-SH) groups showed that the combined agrochemical treatment induced a significant increment of these free thiols that correlated with spermatozoa DNA denaturation. This finding is in agreement with those reported by other authors (Zini et al., 2001). Moreover, this biomarker has a positive correlation with the infertility incidence in men (Zini et al, 2001). In addition, it has been recently shown that spermatozoa nuclear and mitochondrial DNA integrity could be a sensitive biomarker of spermatozoa health (Erenpreiss et al., 2006; Lewis et al., 2008). This direct relationship found between free thiols and DNA damage should be considered another putative biomarker with clinical utility for the screening of pesticide-exposed populations.

It is well known that the function of seminal vesicles is under androgen control. Also, it has been demonstrated a direct association between serum testosterone levels, seminal fructose levels, and spermatozoa motility/fertility (Gonzales, 2001). In line with this, we found higher fructose levels in intoxicated rats which could be associated with a poor consumption due to a lower spermatozoa motility induced by agrochemicals. In fact, the epididymal spermatozoa observed under light microscopy, suggested a decreased motility in the treated groups compared to the control one. Epididymal maturation is also an essential process in the transformation of testicular spermatozoa to mature gametes capable of fertilization by the gain of functional competence (Cooper, 1995). Mature spermatozoa undergo the acrosome reaction when interacting with the zona pellucida of the egg, event that enables it to penetrate the oocyte. Before this binding, spermatozoa cells undergo several biochemical transformations in the female reproductive tract collectively called capacitation. This process involves lipid transfer across plasma membrane (particularly efflux of cholesterol), phospholipids remodeling in plasma membrane (increasing plasma membrane permeability), and changes in protein phosphorylation status, as well as modifications in the intracellular levels of Ca^{2+} and other ions (Abou-haila and Tulsiani, 2009). We have reproduced the acrosomal reaction by an *in vitro* assay, that has been recognized as a predictor of the fertilizing ability of spermatozoa either *in vitro* or *in vivo* (Dematteis et al., 2008). Ca/Ionosphore-induced acrosome reaction was significantly altered in spermatozoa suspensions obtained from treated rats. This assay is easy to be performed as a screening procedure in exposed populations and, as stated before, it should be taken into account as a biomarker of pollutant effects on male reproductive performance.

Results reported in the present study showed that, the mixture of the most frequently used agrochemicals administered at very low doses produced significant detrimental effects on both the spermatozoa characteristics and functional parameters. Therefore, more studies are necessary to attribute the increasing fertility problems as a consequence of the involuntary exposure to low doses of pesticide mixtures over long periods of time. Moreover, we suggested that some of the biomarkers described in this work should be validated and after that, implemented (at least in occupationally exposed populations) in order to establish their cut-off points for their further use in prevention and/or clinical practice.

ACKNOWLEDGEMENTS

This study was supported by a grant from Consejo Nacional de Investigaciones Científicas y Técnicas (CONICET), Argentina. We would like to thank Eva Illara de Bozzolo and Norma Cristalli for their excellent technical assistance, and Norma Tedesco for language revision.

REFERENCES

Abarikwu SO, Adesiyan AC, Oyeloja TO, Oyeyemi MO, Farombi EO (2009). Changes in sperm characteristics and induction of oxidative stress in the testis and epididymis of experimental rats by a herbicide, Atrazine. Arch. Environ. Contam. Toxicol. In press; DOI: 10.107/s00244-009-9371-2.

Abou-haila A, Tulsiani DR (2009). Signal transduction pathways that regulate sperm capacitation and the acrosome reaction. Arch. Biochem. Biophys., 485: 72-81.

Afifi NA, Ramadan A, El-Aziz MI, Saki EE (1991). Influence of dimethoate on testicular and epididymal organs, testosterone plasma level and their tissue residues in rats. Dtsch. Tierarztl. Wochenschr, 98: 419-423.

Agarwal A, Said TM (2003). Role of sperm chromatin abnormalities and DNA damage in male infertility. Hum. Reprod. Update, 9: 331-345.

Aitken RJ, Baker MA (2006). Oxidative stress, sperm survival and fertility control. Mol. Cell. Endocrinol., 250: 66-69.

Aitken RJ (1995). Free radicals, lipid peroxidation and sperm function. Reprod. Fertil. Dev., 7:659- 668.

Aitken RJ (2006). Sperm function tests and fertility. Int. J. Androl., 29: 69-75.

Aksoy Y, Aksoy H, Altinkaynak K, Aydin HR, Ozkan A (2006). Sperm fatty acid composition in sub-fertile men. Prostaglandins Leukot. Essent. Fatty Acids, 75: 75-79.

Arbuckle TE, Lin Z, Mery LS (2001). An exploratory analysis of the effect of pesticide exposure on the risk of spontaneous abortion in an Ontario farm population. Environ. Health Perspect, 109: 851-857.

Astiz M, de Alaniz MJT, Marra CA (2009a). The impact of simultaneous intoxication with agrochemicals on the antioxidant defence system in rat. Pest. Biochem. Physiol., 94: 93-99.

Astiz M, Hurtado de Catalfo GE, de Alaniz MJT, Marra CA (2009b). Involvement of lipids in dimethoate-induced inhibition of testosterone biosynthesis in rat interstitial cells. Lipids, 44: 703-718.

Astiz M, Alaniz MJT de, Marra CA (2009c). Antioxidant defense system in rats simultaneously intoxicated with agrochemicals. Environ. Toxicol. Pharmacol., 28: 465-473.

Astiz M, Alaniz MJT de, Marra CA (2009d). Effect of pesticides on cell survival in rat liver and brain. Ecotoxicol. Environ. Saf., 72(7): 2025-203.

Bagchi D, Bagchi M, Hassoun EA, Stohs SJ (1995). In vitro and in vivo generation of reactive oxygen species, DNA damage and lactate dehydrogenase leakage by selected pesticides. Toxicol., 104: 129-140.

Basrur PK (2006). Disrupted sex differentiation and feminization of man and domestic animals. Environ. Res., 100: 18-38.

Beuret CJ, Zirulnik F, Giménez MS (2005). Effect of the herbicide glyphosate on liver lipoperoxidation in pregnant rats and their fetuses. Reprod. Toxicol., 19: 501-504.

Bhatnagar VK (2001). Pesticide pollution: Trends and perspectives. ICMR Bull., 31: 87-88.

Biggers JD, Witten WK, Whittingham DG (1995). The culture of mouse embryos in vitro. In: Methods in mammalian embryology, ed. Daniel, JC., pp. 86-116.

Bradford MM (1976). A Rapid and sensitive method for the quantification of microgram quantities of protein utilizing the principle of protein-dye binding, Anal. Biochem., 72: 248-254.

Beitbart H, Naor Z(1999). Protein kinases in mammalian sperm capacitation and the acrosome reaction. Rev. Reprod., 4: 151-159.

Caglar S, Kolankaya D (2008). The effect of sub-acute and sub-chronic exposure of rats to the glyphosate-based herbicide Roundup. Environ. Toxicol. Pharmacol., 25: 57-62.

Coniglio JG (1994). Testicular lipids. Prog. Lipid Res., 33: 387-401.

Cooper TG (1995). Role of the epididymis in mediating changes in the male gamete during maturation. Adv. Exp. Med. Biol., 377: 87-101.

Cory-Slechta DA (2005). Studying toxicants as single chemicals: does this strategy adequately identify neurotoxic risk?. Neurotoxicol., 26: 491-510.

Dallegrave E, Mantese FD, Oliveira RT, Andrade AJ, Dalsenter PR, Langeloh A (2007). Pre- and postnatal toxicity of the commercial glyphosate formulation in Wistar rats. Arch. Toxicol., 81: 665-673.

Daruich J, Zirulnik F, Gimenez MS (2001). Effect of the herbicide glyphosate on enzymatic activity in pregnant rats and their fetuses. Environ. Res., 85: 226-231.

Dematteis M, Miranda SD, Novella ML, Maldonado C, Ponce RH,

Maldera JA, Cuasnicu PS, Coronel CE (2008). Rat caltrin protein modulates the acrosomal exocytosis during sperm capacitation. Biol. Reprod., 79: 493-500.

Dougherty KA, Emilson LB, Cockett AT, Urry RL (1975). A comparison of subjective measurements of human sperm motility and viability with two live-dead staining techniques. Fertil. Steril., 26: 700-703.

Eliasson R, Treichl L (1971). Supravital staining of human spermatozoa. Fertil. Steril., 22: 134-137.

El-Taieb MA, Herwig R, Nada EA, Greilberger J, Marberger M (2009). Oxidative stress and epididymal sperm transport, motility and morphological defects. Eur. J. Obstet. Gynecol. Reprod. Biol., 144: S199-S203.

Erenpreiss J, Spano M, Erenpreisa J, Bungum M, Giwercman A (2006). Sperm chromatin structure and male fertility: Biological and clinical aspects Asian Androl., 8: 11-29.

Farag AT, El-Aswad AF, Shaaban NA (2007). Assessment of reproductive toxicity of orally administered technical dimethoate in male mice. Reprod. Toxicol., 23: 232-238.

Filler R (1993). Methods for evaluation of rat epididymal sperm morphology. In Methods in Toxicology, eds. Chapin, RE, and Heindel JJ, pp. 334-343.

Folch J, Lees M, Sloane Stanley GH (1957). A simple method for the isolation and purification of total lipids from animal tissues. J. Biol. Chem., 226: 497-509.

Furland NE, Oresti GM, Antollini SS, Venturino A, Maldonado EN Aveldaño MI (2007). Very long-chain polyunsaturated fatty acids are the major acyl groups of sphingomyelins and ceramides in the head of mammalian spermatozoa. J. Biol. Chem., 282: 18151-18161.

Giwercman A, Carlsen E, Keiding N, Skakkebæk NE (1993). Evidence for increasing incidence of abnormalities of the human testis: A review. Environ. Health Perspect. Suppl., 101: 65-71.

Gonzales GF (2001). Function of seminal vesicles and their role on male fertility. Asian J. Androl., 3: 251-258.

Gray LE, Ostby J, Furr J, Wolf CJ, Lambright C, Parks L, Veeramachaneni DN, Wilson V, Price M, Hotchkiss A, Orlando E, Guillette L (2001). Effects of environmental antiandrogens on reproductive development in experimental animals. Hum. Reprod. Update, 7: 248-264.

Heikkila RE, Cabbat FS, Cohen G (1976). In vivo inhibition of superoxide dismutase in mice by diethyldithiocarbamate. J. Biol. Chem., 251: 2182-2185.

Hurtado de Catalfo GE, de Gomez Dumm IN (2002). Polyunsaturated fatty acid biosynthesis from [1-14C]20:3 n-6 acid in rat cultured Sertoli cells, Linoleic acid effect. Int. J. Biochem. Cell. Biol., 34: 525-532.

Hurtado de Catalfo GE, de Alaniz MJT, Marra CA (2008). Dietary lipids modify redox homeostasis and steroidogenic status in rat testis. Nutrition, 24: 717-726.

Hurtado de Catalfo GE, de Alaniz MJT, Marra CA (2009). Influence of commercial dietary oils on lipid composition and testosterone production in interstitial cells isolated from rat testis. Lipids, 44: 345-357.

John S, Kale M, Rathore N, Bhatnagar D (2001). Protective effect of vitamin E in dimethoate and malathion induced oxidative stress in rat erythrocytes. J. Nutr. Biochem., 12: 500-504.

Joshi SC, Mathur R, Gulati N (2007). Testicular toxicity of chlorpyrifos (an organophosphate pesticide) in albino rat. Toxicol. Ind. Health, 23: 439-444.

Kesavachandran CN, Fareed M, Pathak MK, Bihari V, Mathur N, Srivastava AK (2009). Adverse health effects of pesticides in agrarian populations of developing countries. Rev. Environ. Contam. Toxicol., 200: 33-52.

Klinefelter GR, Gray LE, Suarez JD (1991). The method of sperm collection significantly influences sperm motion parameters following ethane dimethanesulphonate administration in the rat. Reprod. Toxicol., 5: 39-44.

Knudsen LE, Hansen AM (2007). Biomarkers of intermediate endpoints in environmental and occupational health. Int. J. Hyg. Environ.

Health, 210: 461-470.

Larson JL, Miller DJ (1999). Simple histochemical stain for acrosomes on sperm from several species. Mol. Reprod. Dev., 52: 445-449.

Lewis SE, Agbaje I, Alvarez J (2008). Sperm DNA tests as useful adjuncts to semen analysis. Syst. Biol. Reprod. Med., 54: 111-125.

Lewis SE (2007). Is sperm evaluation useful in predicting human fertility? Reproduction, 134: 31-40.

Morrison WR, Smith LM (1964). Preparation of fatty acid methyl esters and dimethylacetals from lipids with boron fluoride-methanol. J. Lipid Res., 5: 600-608.

Moser VC, Simmons JE, Gennings C (2006). Neurotoxicological interactions of a five-pesticide mixture in preweanling rats. Toxicol. Sci., 92: 235-245.

Murono EP, Der RC, de León JH (2001). Differential effects of octylphenol, 17β-estradiol, endosulfan, or bisphenol A on the steroidogenic competence of cultured adult rat Leydig cells. Reprod. Toxicol., 15: 551-560.

Murono EP, Derk RC, de León JH (2000). Octylphenol inhibits testosterone biosynthesis by cultured precursor and immature Leydig cells from rat testes. Reprod. Toxicol., 14: 275-288.

Nagan N, Zoeller RA (2001). Plasmalogens: biosynthesis and functions. Prog. Lipid Res., 40: 199-229.

Nielsen BS, Larsen EH, Ladefoged O, Rye Lam H (2006). Neurotoxic effect of maneb in rats as studied by neurochemical and immunohistochemical parameters. Environ. Toxicol. Pharmacol., 21: 268-275.

Pasqualotto FF, Sharma RK, Nelson DR, Thomas AJ, Agarwal A (2000). Relationship between oxidative stress, semen characteristics, and clinical diagnosis in men undergoing infertility investigation. Fertil. Steril., 73: 459-464.

Patel S, Singh V, Kumar A, Gupta YK, Singh MP (2006). Status of antioxidant defence system and expression of toxicant responsive genes in striatum of maneb- and paraquat- induced Parkinson´s disease phenotype in mouse: Mechanism of neurodegeneration. Brain Res., 1081: 9-18.

Perry MJ (2008). Effects of environmental and occupational pesticide exposure on human sperm: a systematic review. Hum. Reprod. Update, 14: 233-242.

Pesch S, Bergmann M (2006). Structure of mammalian spermatozoa in respect to viability, fertility and cryopreservation. Micron, 37: 597-612.

Petrelli G, Mantovani A (2002). Environmental risk factors and male fertility and reproduction. Contraception, 65: 297-300.

Pflieger-Bruss S, Schuppe HC, Schill WB (2004). The male reproductive system and its susceptibility to endocrine disrupting chemicals. Andrologia, 36: 337-345.

Recio-Vega R, Ocampo-Gómez G, Borja-Aburto VH, Moran-Martínez J, Cebrian-García ME (2008). Organophosphorus pesticide exposure decreases sperm quality: Association between sperm parameters and urinary pesticide levels. J. Appl. Toxicol., 28: 674-680.

Reeves PG, Nielsen FH, Fahey GC (1993). AIN-93 purified diets for laboratory rodents: Final report of the American Institute of Nutrition ad hoc writing committee on the reformulation of the AIN-76A rodent diet. J. Nutr., 123: 1939-1951.

Rossi G, Buccione R, Baldassarre M, Macchiarelli G, Palmerini MG, Cecconi S (2006). Mancozeb exposure in vivo impairs mouse oocyte fertilizability. Reprod. Toxicol., 2: 216-219.

Salem MH, Abo-Elezz Z, Abd-Allah GA, Hassan GA, Shaker N (1988). Effect of organophosphorus (dimethoate) and pyrethroid (deltamethrin) pesticides on semen characteristics in rabbits. J. Environ. Sci. Health, B 23: 279-290.

Sanderson JT (2006). The steroid hormone biosynthesis pathway as a target for endocrine-disrupting chemicals. Toxicol. Sci., 94: 3-21.

Saradha B, Mathur PP (2006). Effect of environmental contaminants on male reproduction. Environ. Toxicol. Pharmacol., 21: 34-41.

Sarkar R, Mohanakumar KP, Chowdhury M (2000). Effects of an organophosphate pesticide, quinalphos, on the hypothalamo-pituitary-gonadal axis in adult male rats. J. Reprod. Fertil., 118: 29-38.

Seed J, Chapin RE, Clegg ED, Dostal LA, Foote RH, Hurtt ME, Klinefelter GR, Makris SL, Perreault SD, Schrader S, Seyler D, Sprando R, Treinen KA, Veeramachaneni DN, Wise LD (1996). Methods for assessing sperm motility, morphology, and counts in the rat, rabbit, and dog: a consensus report. Reprod. Toxicol.. 10: 237-244.

Sivapiriya V, Jayanthisakthisekaran S, Venkatraman S. (2006). Effects of dimethoate (O,O-dimethyl S-methyl carbamoyl methyl phosphorodithioate) and ethanol in antioxidant status of liver and kidney of experimental mice. Pestic. Biochem. Physiol., 85: 115-121.

Sharma Y, Bashir S, Irshad M, Gupta SD, Dogra TD (2005a). Effects of acute dimethoate administration on antioxidant status of liver and brain of experimental rats. Toxicol., 206: 49-57.

Sharma Y, Bashir S, Irshad M, Nag TC, Dogra TD (2005b). Dimethoate-induced effects on antioxidant status of liver and brain of rats following sub chronic exposure. Toxicol., 215: 173-181.

Sheweita SA, Tilmisany AM, Al-Sawaf H (2005). Mechanisms of male infertility: role of antioxidants. Curr. Drug Metab., 6: 495-501.

Sikka SC (2001). Relative impact of oxidative stress on male reproductive function. Curr. Med. Chem., 8: 851-862.

Snyder F, Lee TC, Wykle RL (2001). Ether-linked lipids and their bioactive species. In Biochemistry of lipids, lipoproteins and membranes, eds Vance DE, Vance JE, Amsterdam. Elsevier Sci. BV, pp. 233-262.

Somani BL, Khanade J, Sinha R (1987). A modified anthrone-sulfuric acid method for the determination of fructose in the presence of certain proteins. Anal. Biochem., 167: 327-330.

Tavilani H, Doosti M, Nourmohammadi I, Mahjub H, Vaisiraygani A, Salimi S, Hosseeinipanah SM (2007). Lipid composition of spermatozoa in normozoospermic and asthenozoospermic males. Prostaglandins Leukot. Essent. Fatty Acids, 77: 45-50.

Van Loon AA, Den Boer PJ, Van der Schans GP, Mackenbach P, Grootegoed JA, Baan RA, Lohman PH (1991). Immunochemical detection of DNA damage induction and repair at different cellular stages of spermatogenesis of the hamster after *in vitro* or *in vivo* exposure to ionizing radiation. Exp. Cell Res., 193: 303-309.

Vetter CM, Miller JE, Crawford LM, Armstrong MJ, Clair JH, Conner MW, Wise LD, Skopek TR (1998). Comparison of motility and membrane integrity to assess rat sperm viability. Reprod. Toxicol., 12: 105-114.

Walsh LP, McCormick C, Martin C, Stocco DM (2000a). Roundup inhibits steroidogenesis by disrupting steroidogenic acute regulatory (StAR) protein expression. Environ. Health Perspect, 108: 769-776.

Walsh LP, Webster DR, Stocco DM (2000b). Dimethoate inhibits esteroidogenesis by disrupting transcription of the esteroidogenic acute regulatory (StAR) gene. J. Endocrinol., 167: 253-263.

World Health Organization (1999). WHO Laboratory Manual for the examination of human semen and sperm-cervical mucus interactions. 4th ed. Cambridge, United Kingdom: Cambridge University Press.

Yucra S, Gasco M, Rubio J, Gonzales GF (2008). Semen quality in Peruvian pesticide applicators: association between urinary organophosphate metabolites and semen parameters. Environ. Health, 17: 7-59.

Zini A, Kamal K, Phang D, Willis J, Jarvi K (2001). Biologic variability of sperm DNA denaturation in infertile men. Urol., 58: 258-261.

Zini A, Phillips S, Courchesne A, Boman JM, Baazeem A, Bissonnette F, Kadoch I.J, San Gabriel M (2009). Sperm head morphology is related to high deoxyribonucleic acid stability assessed by sperm chromatin structure assay. Fertil. Steril., 91: 2495-2500.

6

Some priority heavy metals in children toy's imported to Nigeria

Sindiku O. K.* and Osibanjo O.

Department of Chemistry, University of Ibadan, Oyo State, Nigeria.

A total number of 51 toys manufactured from different countries were purchased and analysed to determine the level of lead, cadmium, chromium and nickel in the plastic components which was digested with concentrated HNO$_3$ (CPSC-CH-E1002-08 method) and analyzed using atomic absorption spectrophotometer (AAS) to determine heavy metal concentration. The results obtained show that lead, cadmium, chromium and nickel were high and ranged 28.5 to 12600 mg/kg Pb; 0.15 to 9.55 mg/kg Cd; 1.30 to 394.50 mg/kg Cr, and 5.9 to 1911 mg/kg Ni. A comparison of the mean concentration of these metals in the toys sample analyzed showed the following pattern: Pb>Ni>Cr>Cd. Compared with the elemental concentration threshold limits concentration (TTLC) of 90, 75 and 60 mg/kg for lead, cadmium and chromium respectively, Consumer Product Safety Commission, USA, Bureau of Indian Standard and Thailand Industrial Standard for Toys suggest that these toys are hazardous and therefore not safe for children use. This underscores the need for urgent national policy and resolution control on the removal of heavy metals especially lead from children toys.

Key words: Heavy metals, total threshold limit concentration, plastic toys, atomic absorption spectroscopy.

INTRODUCTION

Heavy metal poisoning has become an increasingly major health problem, especially since the industrial revolution. Heavy metals are in the water we drink, the foods we eat, the air we breathe, our daily household cleaners, our cookware and our other daily tools. A heavy metal has a density of at least 5 times that of water and cannot be metabolized by the body, therefore accumulating in the body. Heavy metal toxicity can cause our mental functions, energy, nervous system, kidneys, lungs and other organ functions to decline. Heavy metals include mercury (Hg), cadmium (Cd), arsenic (As), chromium (Cr), thallium (Tl), and lead (Pb) and nickel (Ni) (Duffus, 2002).

Lead is a toxic metal and is ubiquitous in the human environment as a result of industrialization. Children are particular susceptible to its toxic effects. Lead poisoning is widespread and most poisoned children have no symptoms (CDC, 1991). Lead in the environment may enter the body through inhalation, ingestion or percutaneous absorption (WHO, 1995). Lead causes harm and there is no specific contamination route for lead. Humans can intake lead through inhalation or consumption of food. The World Health Organization's (WHO) "safe" limit for lead in blood, originally set in 1995, is 10 µg/dl. Incidences of lead poisoning in children led to the first public study to ascertain the presence of lead in PVC by Arizona Health Department in 1995 (CPSC, 1997). The chewing and swallowing of toys by children is a common path for lead and cadmium exposure (CPSC, 1997). Lead has also been linked to drops in IQ points, behavioral problems, and attention deficit hyperactivity disorder (ADHD). Lead exposure can also cause anaemia, damage to the gastrointestinal tract, and kidneys. Chronic exposure can even lead to DNA damage (Thuppil, 2007).

Cadmium is also a potential environmental hazard. Human exposures to environmental cadmium are primarily the result of the burning of fossil fuels and municipal wastes. Cadmium exposure produces a wide variety of acute and chronic effects in humans, leading to a build-up of cadmium in the kidneys that can cause kidney disease (UNEP, 2006). Lead and cadmium are

*Corresponding author. E-mail: thayor17@yahoo.com.

known poisons, being neurotoxins and nephrotoxins. Neurotoxins are agents that can cause toxic effects on the nervous system while nephrotoxins are agents that can cause toxic effects on the kidney respectively (ATSDR, 2005).

Chromium metal and chromium (III) compounds are not usually considered health hazards; chromium is an essential trace mineral. Trivalent chromium (Cr (III) or Cr^{3+}) is required in trace amounts for sugar metabolism in humans (Glucose Tolerance Factor) and its deficiency may cause a disease called chromium deficiency. In contrast, hexavalent chromium (Cr(VI) or Cr^{6+}) is very toxic and mutagenic when inhaled, as publicized by the film Erin Brockovich. Cr(VI) has not been established as a carcinogen when in solution, though it may cause allergic contact dermatitis (ACD). The lethal dose of poisonous chromium (VI) compounds is about one half teaspoon of material.

Nickel in small amounts is needed by the human body to produce red blood cells; however, in excessive amounts, can become mildly toxic. Short-term overexposure to nickel is not known to cause any health problems, but long-term exposure can cause decreased body weight, heart and liver damage, and skin irritation. The EPA does not currently regulate nickel levels in drinking water. Nickel can accumulate in aquatic life, but its presence is not magnified along food chains.

Physicians and scientists agree that no level of heavy metals in blood is safe or normal (NRC, 1993). The disturbing fact is that exposure to extremely small amounts can have long-term and measurable effects in children while at the same time causing no distinctive symptoms (CDC, 1991). Another problem of heavy metals exposure is it being cumulative in nature. After they have been absorbed into the blood, some of them are filtered out and excreted, but the rest are distributed in the liver, brain, kidneys and bones. What is more disturbing is what happens when lead gets into the bones. Bone stores lead and stay there for decades. Lead can re-enter the body when bone breaks down as part of a regular metabolic process or due to some specific physiological conditions like osteoporosis, causing re-exposure (ATSDR, 2005).

Lead and other heavy metals are widely used in PVC (poly vinyl chloride or vinyl) toys and other children's products to stabilize polymers in order to avoid degradation from heat, sunlight and wear. Vinyl requires the addition of metal stabilizers because it contains chlorine. Without a stabilizer, the chlorine can degrade the product by forming hydrochloride acid (CPSC, 1997). The consumer Product Safety Commission experimentally demonstrated that light and heat can cause degradation of vinyl toys and liberation of lead dust but unfortunately for children, vinyl toys released lead during normal product use (CPSC, 1997). The chewing and swallowing behavior of children is also a common source of lead exposure (Kelley et.al., 1993).

Given the known potential toxicity, the serious health effect and the ability of heavy metals, to leach out of children's toys through contact, the continued use of lead and other heavy metals in children's toy raises serious concern (CDC, 2004). Lead testing of children's toys and products began in Chicago, and then widened to include 10 major U. S. cities and Montreal, Canada (CSPC, 1997). Also researchers at the state University of New York Syracuse found lead in a vinyl play kit, gloves and basket ball toys (Hunt et al., 1997).

This study determined the current pattern in the use of lead and other heavy metals as stabilizer in PVC toys, using analytical techniques that would yield empirical data. The data collected were used to provide a clear picture of hazardous chemicals in PVC toys and other high contact children toys.

MATERIALS AND METHODS

Toys samples were weighed and recorded (Table 1). Adequate sample preparation was done by separating different parts of the toys sample into two major components: Plastics, which is the malleable parts and the metal, rods, nuts, screws etc.

Toys samples were grounded using a specially fabricated heavy duty harmer mill with uniformly sized holes.

The ground samples were then sieved through < 2 mm nylon sieve to ensure efficient extraction of the stabilizer. Milled samples were mixed thoroughly to achieve homogeneity of samples and appropriate amount of the ground sample was taken for analysis.

The grounded samples were wet digested as described by CPSC-CH-E1002-08. The final processed samples were quantitatively analyzed using Buck Scientific Model 210 VGP Atomic Absorption Spectrometer in the flame mode using appropriate resonance wavelengths. The instrument was first calibrated with standards prepared from stock solution provided by Merck. The final processed samples were quantitatively analysed using AAS. After every ten samples analysed using AAS, the first sample was repeated for quality check.

RESULTS

The result of the concentration of heavy metals in the various toys sample analysed is given in Table 2. The average, range, standard deviation and total threshold limit concentration of Pb, Cd, Cr and Ni concentration are also given.

All statistical analyses were performed using Microsoft excel and SPSS 15.0 and the mean, standard deviation and co-efficient of variation were calculated. A student t-test was also used to test the relationship of the mean of the concentration of the metals and the total threshold limit for each metal at 99% confidence level.

A pie chart showing the comparison of the mean value of Pb, Cd, Cr, and Ni in the toys sample analysed is given in Figure 1.

The result of a test of hypothesis between the concentrations of lead in the toys sample and the total threshold limit showed that the paired sample t-test revealed a statistically reliable difference between the mean of the lead concentration and the threshold limit.

Table 1. Toys sampled in Lagos and used for the study.

Laboratory number	Country of manufacture	Description	Weights (grams)	Colour
A1	China	Caterpillar OP	410.8	Black and ash
A2	China	Toy train	136.2	Brown and black
A3	China	Blue toy gun	38.8	Blue
A4	USA	Bonbon buddies	38.0	Red and black
A5	USA	Toy jewery box	34.1	Purple
A6	China	Vanity toy hand drier	44.0	Purple
A7	China	Barbie doll	104.3	Brown
A8	USA	Toy telephone	73.5	Blue and white
A9	China	Snake toy gun	66.4	Black and gold coated
A10	China	Dolphin rattle	42.5	White
A11	China	Train rail	107.9	Black
A12	China	Darby star train rail	108.2	Black
A13	China	McDonald robot doll	61.0	Green, pink and brown
A14	China	Banana case	60.2	Yellow
A15	China	Alien flying gun	562.5	Blue and silver coated
A16	China	Children toaster	129.8	Light blue
A17	China	Bendi frendi box	301.4	Pink
A18	China	Toy warship	15.8	Yellow
A19	China	NYPD Bike	69.5	White paint
A20	China	Water gun	101.2	Yellow
A21	China	Jelly fish toy	56.2	Blue, white, and yellow
A22	China	Harry potter lamp	82.7	Green and purple
A23	China	Toy heroes badge	334.0	Green
A24	China	High school cabinet	118.8	Lilac
A25	USA	Viewmaster telescope	119.2	Red, blue and white
A26	Vietnam	Dolphin tomy	57.4	Orange
A27	China	Sleepingbeauty castle	37.4	Grey and purple
A28	China	Puzzle box MP	81.9	Blue
A29	China	Toy powerbike	175.8	Blue and silver coated
A30	China	Golden stallion horse	104.3	Golden brown
A31	China	Painted toy house	168.4	Red, white and yellow
A32	China	Toy feeding bottle	44.1	Transparent plastic
A33	China	Fischerprice rattle	37.7	Red and black
A34	China	3 secret box	191.2	Pink and purple
A35	Romania	Stallion horse	164.5	Peach with silvery hair
A36	China	Barbie microphone	35.2	Lilac
A37	China	Moffett lorry car	32.7	Red and black
A38	China	Cargo toy train	148.2	Brown and black
A39	China	Police car	31.2	Blue and black
A40	China	Rabbit rattle	83.4	Red and white
A41	China	Armoured tank car	54.8	Green
A42	China	Space car with fan	58.0	Black and grey
A43	U.K	Baby bed toys	469.0	Yellow
A44	China	Baby rattle	26.6	White
A45	U.k	Baby show off	284.4	Yellow
A47	China	Toy gun	102.1	Black and blue
A48	China	Aeroplane	125.2	Red and blue
A49	China	Toy feeding bottle	45.1	White
A50	China	Toy jewery	21.7	Silver and pink
A51	China	Children tiara jewery	24.3	Silver

Table 2. Concentrations (mg/kg) of Pb, Cd, Cr and Ni in the Nigerian toys samples analysed.

Toy type	Laboratory number	Pb	Cd	Cr	Ni
Caterpillar toy car	A1	155.50	1.60	1.55	23.25
Toy train	A2	290.00	0.85	2.95	30.65
Blue toy gun	A3	178.50	1.05	15.55	50.95
Bonbon buddies box	A4	68.00	2.60	4.60	18.40
Toy jewery box	A5	49.00	1.30	5.50	11.10
Vanity toy handrier	A6	459.50	0.5	6.40	36.60
Barbie doll	A7	327.00	0.55	4.95	113.80
Toy telephone	A8	45.00	0.65	4.70	12.30
Snake toy gun	A9	648.00	3.30	22.60	35.20
Dolphin rattle	A10	93.00	1.55	5.45	10.95
Playmobil train rail	A11	165.00	1.20	2.95	17.30
Darby star train rail	A12	105.50	1.05	4.65	19.30
Mcdonald robot doll	A13	100.50	1.60	18.45	13.55
Banana toy case	A14	50.00	0.40	13.05	13.30
Alien flying gun	A15	137.50	0.15	13.95	72.85
Toy toaster	A16	538.50	3.30	95.10	14.85
Bendifriendi box	A17	125.00	2.25	8.90	35.75
Toy warship	A18	76.00	1.65	7.50	31.00
NYPD toy bike	A19	660.00	2.15	11.75	39.50
Water gun	A20	115.50	1.25	4.90	61.00
Toy jelly fish	A21	126.50	3.30	5.20	21.00
Harrypotter toy lamp	A22	765.50	1.70	4.75	16.95
Heroes badge	A23	36.50	2.20	4.70	9.75
Highschool toy cabinet	A24	260.50	3.60	5.85	71.35
Viewmaster telescope	A25	1505.00	1.60	394.50	14.20
Dolpin toy	A26	47.50	1.15	3.95	10.70
Sleeping beauty castle	A27	68.50	9.55	8.65	9.55
Puzzle box	A28	44.50	0.65	4.25	6.75
Toy power bike	A29	224.50	1.00	3.50	20.10
Painted toy house	A30	111.00	1.00	3.65	19.75
Golden stallion horse	A31	12600.00	1.70	10.05	59.70
Toy feeding bottle 1	A32	284.00	1.40	1.30	68.95
Fischer-price rattle	A33	386.00	1.50	3.20	48.65
3 secret box	A34	825.00	2.35	9.00	58.90
Stallion horse 2	A35	639.00	1.35	10.70	46.15
Barbie microphone	A36	28.50	0.95	4.10	6.80
Moffetti toy lorry car	A37	241.00	9.00	34.75	31.65
Toy cargo train	A38	144.00	1.15	6.55	11.70
Police toy car	A39	173.00	2.20	37.65	1911
Rabbit rattle	A40	3220.00	1.20	2.40	37.25
Armoured toy tank	A41	41.50	0.80	19.75	27.85
Toy space car	A42	870.00	1.95	3.05	6.45
Toy baby bed holder	A43	89.50	1.15	7.25	121.65
Jemina baby rattle	A44	185.50	0.60	25.45	23.95
Bluebird toys	A45	29.00	1.20	10.85	31.90
Toy canoe	A46	282.50	1.15	5.60	20.40
Toy gun	A47	173.00	1.40	3.35	5.90
Toy aeroplane	A48	113.50	1.00	14.25	15.50
Toy feeding bottle 2	A49	251.00	0.60	3.25	44.45
Toy jewery 1	A50	621.00	1.20	5.45	431.10

Table 2. Contd.

Children tiara jewery	A51	680.00	1.10	5.30	43.50
Mean		577.53	1.76	17.99	77.09
Range		28.50 - 12600	0.15 – 9.55	1.30 - 394.50	5.9 - 1911
Standard deviation		1789.50	1.72	55.66	2368.98
Total threshold limit concentration (TTLC)		90	75	60	_

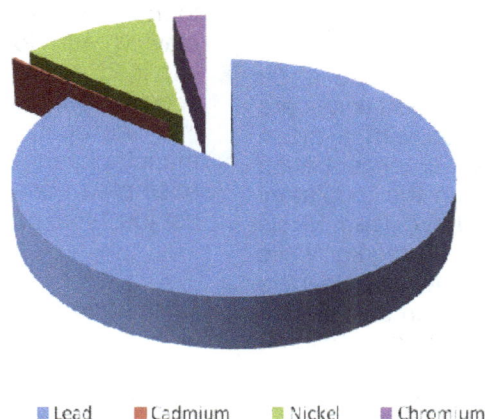

Figure 1. Comparison of the mean values of Pb, Ni, Cd and Cr in the toy samples analysed.

The result of the test of hypothesis between the concentrations of cadmium and chromium in the toy samples and the total threshold limit showed that the paired sample test failed to reveal a statistically reliable difference between the their means and the threshold limit.

DISCUSSION

This study investigated the use and presence of some heavy metals in children toys imported into Nigeria. Pb, Cd, Cr and Ni were found in all the tested toy samples in varying concentrations.

It was observed that majority (92.2%) of the toys sample analyzed in this study were made of plastics of different colours. A full breakdown of the country of manufacture of toys shows that 84.3% were made from China, 8% from USA, 4% from UK, 2% from Vietnam and 2% from Romania. This was attributed to the fact that China is the largest toy seller in the world with about 8,000 toy factories that employ up to 3 million workers (The Children's Toy Business, 2009).

The results of heavy metals analysis in the toy-samples tested were compared with the total threshold limit concentration (TTLC). The total threshold limit concentration for cadmium is 75 mg/kg (Bureau of Indian standard for toys, 1999; Thailand industrial standard for

toys, 1997). The cadmium level in the toy-samples was observed to be far below the threshold limit of 75 mg/kg), with mean concentration of 1.76 mg/kg, and range concentration of 0.15 to 9.55 mg/kg. This can be attributed to the fact that cadmium is one of the six substances banned by the European Union's Restriction on Hazardous Substances directive (UNEP Chemicals, 2006). Cadmium and several cadmium-containing compounds have been reported to be carcinogenic and can induce many types of cancer (Wallace, 2001).

The total threshold limit concentration for chromium is 60 mg/kg (Bureau of Indian standard for toys, 1999; Thailand industrial standard for toys, 1997). Also, the chromium levels in almost all the toy-samples was observed to be far below the threshold limit of 60 mg/kg, with a mean concentration of 18.0 mg/kg and a range concentration between 1.3 and 394.5 mg/kg (Table 2). 96% of the toy samples have chromium levels far below the threshold limit while 4% (two samples; Toy toaster and view master telescope) were above the threshold limit. This is due to the fact that hexavalent chromium (Cr(VI) or Cr^{6+}) is very toxic and mutagenic when inhaled and is also one of the substances whose use is restricted by the European Restriction of Hazardous Substances Directive (UNEP Chemicals, 2006).

The threshold limit concentration for nickel was not stated. It was observed that nickel was present in all the toy samples analysed in this study with a mean concentration of 77.1 mg/kg and a range concentration between 5.9 and 1911 mg/kg. This may be attributed to the fact that small amounts of nickel are needed by the human body to produce red blood cells; however, in excessive amounts, can become mildly toxic. Also, short-term overexposure to nickel is not known to cause any health problems, but long-term exposure can cause decreased body weight, heart and liver damage, and skin irritation. The EPA does not currently regulate nickel levels in drinking water because nickel can accumulate in aquatic life, but its presence is not magnified along food chains (U.S Environmental Protection Agency, 2002).

The total threshold limit concentration for lead is 90 mg/kg (CPSC, 2009). It was observed that lead levels are very high in almost all the toys samples with a mean concentration of 577.5 mg/kg and a range concentration between 28.5 to 12600 mg/kg, exceeding the regulatory limit. 76% of the toy samples tested have lead levels exceeding the current threshold limit of 90 mg/kg; while 24% of the toy samples tested have lead levels below the threshold limit. Such high quantities of lead in toys pose a threat to children's health. High levels of lead in children body may damage their brains, nervous system, and kidney, reduce intelligence quotient, slow down growth, and cause hearing problems. It can also cause behavioral and learning problems and can result in coma, convulsions, and even death (EPA, 2007; CDC, 1991).

According to CPSC (1997), lead is the most commonly used stabilizer in plastic toys, due to the fact that it is readily available and it is the cheapest. A comparism of the mean values of the heavy metals in the toy samples was analysed as shown in Figure 1 shows that lead is the most commonly used metal stabilizer (86%), followed by nickel (11%), chromium (3%) and cadmium (0%). Also, lead is mined in more than forty countries of the world, used and traded globally as a metal in various products. There is extensive global trade of lead raw materials (UNEP, 2006).

A consistent correlation was also found between the country of manufacture and the presence of heavy metals in the toys tested. 81% (35 products) of toys from China had lead levels above 90 ppm, while 19% (8 products) had lead levels below 90 ppm. This may be attributed to the fact that China is the world's major producer and user of lead (UNEP, 2006). Four (4) toys manufactured in the U.S.A were sampled and 25% (1 toy) of those had lead level above 90 ppm, while 75% had lead levels below 90 ppm. Also, 25% (stallion horse from Romania) of toys from other countries had levels above the threshold limit. This suggests that lead poisoning from toys is not just from China.

Previous study by Greenpeace in USA (1997), reported that the lead and cadmium levels exceeded the TTLC in children products and toys purchased at national chain stores like Kmart, Wal-mart, toys R Us in the US with mean concentration of 2893.14 and 70.34 mg/kg for lead and cadmium respectively and range concentration of 104 to 22550 and 0 to 344 mg/kg for lead and cadmium

respectively (Joseph, 1997). This value is at a higher concentration than the average reported in this present study. This discrepancy is likely due to the awareness created against the use of lead and other heavy metals as stabilizers in plastics toys and also the recall incidence of some of China toys suspected of containing lead by CPSC in 2007 (Denver post, 2007; CPSC, 2007). This may also due to the fact that lead and cadmium are known poisons, being neurotoxins and nephrotoxins (ATSDR, 2005).

It must be noted that exposure from lead is in addition to that of cadmium, chromium and nickel. Hence, children playing with toys having both lead and cadmium are exposed to both toxic metals. This is an important concern and any regulatory mechanism must take this into account. These data demonstrated that toys manufacturers who seek to design product that will be exempted from current recall crisis will need to address not just lead, but most importantly, the nickel levels.

Conclusion

Children's toys imported into Nigeria irrespective of the country of manufacture especially toys from China contain toxic heavy metals, such as lead, nickel, chromium and cadmium in varying concentrations and some even showing high lead concentration that may pose hazards to children's health and create a major environmental health hazard in its use and disposal. Especially alarming is the presence of high lead levels in a China made toy feeding bottle designed to put into children's mouth and a stallion horse toy from China.

REFERENCES

Agency for Toxic Substances and Disease Registry Atlanta (ATSDR). (2005). Toxicological Profile for Lead. U.S Department of Health and Human Services. Public Health Service Agency for Toxic Substances and Disease Registry. Agency for Toxicology and Environmental Medicine/Applied Toxicology Branch 600 Clifton Road NE, Mailstop F 32 Atlanta, Georgia 30333, pp. 29-31.

Bureau of Indian Standard (BIS) (1999). Indian Standard Safety Requirement for Toys. specification. ISO 8124-3: 1997 Superceding IS 5411 (Part 1): 1974 and IS 5411 (Part 2): 1972.

Duffus JH (2002). Heavy metals: A meaningless term? (IUPAC Technical Report). Pure Appl. Chem., 74: 793-807. doidoi:10.1351/pac20027405079310.1351/pac200274050793.

Hunt A, Burnett BR, Basford TM, Abraham JL (1997). Lead and other materials in play kit and craft items composed of vinyl and leather. Am. J. Public Health, 87: 1724-1727.

Joseph Di, Gangi J (1996). Lead and cadmium in children's vinyl products.A Greenpeace Study, 1996; http://composite.about.com/gi/dynamic/offsite.htm?site=http://www.gr eenpeaceusa.org (accessed in June 2006).

Kelley M, Watson P, Thorton D, Halpin TJ (1993). Lead intoxication associated with chewing plastic wire coating. MMWR, 42: 465-467.

Thuppil V (2007). Effect of environmental lead on the health status of women and children in developing countries. Presented at the International Conference on children, health and environment, June 2007, Vienna, pp. 1-34.

U.S Consumer Products Safety Commission (CPSC) (1997). Staff Report on Lead and Cadmium in Children's Polyvinylchloride (PVC) Products, from http://www.cpsc.gov/cpscpub/pubs/pbcdtoys.html

Report 21.

U.S Consumer Products Safety Commission (CPSC) (2009). Standard Operating Procedure for Determining Total Lead (Pb) in Non-Metal Children's Products. Directorate for Laboratory Sciences, Div. Chem., pp. 1-7.

U.S Consumer Product Safety Commission (CPSC) (2007). Guidance for Lead (Pb) in consumer products, from http://www.cpsc.gov/BUSINFO/leadguid.html. pp. 1.

UNEP Chemicals, (2006). Interim Review of Scientific Information on Cadmium and Lead. Retrieved October 2010, from http://www.unepchemicals.ch/pb_and_cd/SR/Files/Interim_reviews/UNEP_Cadmium_review_Interim_Oct 2006.pdf., p. 46.

Wallace HA (2001). Principles and Methods of Toxicology. 4th ed. Taylor and Francis Publishing Inc., Philadelphia, pp. 301.

World Health Organization (WHO) (1995). Inorganic lead Environmental Health Criteria Number 165 (Geneva: World Health Organization), pp. 188.

Xiz H, Lin P, Xijin X, Liangkai Z, Bo Q, Zongli Q, Bao Z, Dai H, Zhonxian P (2007). Elevated Blood Lead Levels of children in Guiyu, and Electronic Waste Recycling Town in China. Environ. Health Perspect., 115(7): 1113-1117.

7

Toxicologic characterization of a novel explosive, triaminoguanidinium-1-methyl-5-nitriminotetrazolate (TAG-MNT), in female rats and *in vitro* assays

Larry R. Williams*, Cheng J. Cao, Emily M. Lent, Lee C. B. Crouse, Matthew A. Bazar and Mark S. Johnson

US Army Public Health Command (Provisional) [USAPHC (Prov)], Directorate of Toxicology, Aberdeen Proving Ground, MD, 21010, U.S.A.

Sustainable use of training ranges requires the development of compounds that have a minimal impact on the environment when used in a weapon system. Triaminoguanidinium-1-methyl-5-nitriminotetrazolate (TAG-MNT) is a novel, explosive, military compound of interest for application in some weapon systems. Little is known of its toxicologic properties. To ensure the health of potentially exposed personnel and the environment, several initial toxicity investigations were conducted and the results compared with another widely used energetic (hexahydro-1,3,5-trinitro-1,3,5-triazine; RDX). In a novel microplate Ames assay, TAG-MNT was a weak mutagen only at the limit concentration of 2 g/L. However, TAG-MNT was cytotoxic to bacteria and a human liver cell line at 250 mg/L and greater. Unlike RDX, TAG-MNT did not have an affinity for the GABAa receptor convulsant site, and was predicted not to induce seizure. After acute oral dosing in female rats, TAG-MNT had no apparent adverse effect up to the limit dose of 2 g/kg. However, daily oral dosing for 14 days at exposures of 1000 mg/kg-d and above caused reduction in food intake, weight loss, increased kidney weight, leucopenia, and elevated blood urea nitrogen and creatinine levels. Leucopenia, increased liver mass, evidence of liver hepatocyte necrosis and centrilobular hypertrophy were observed at 500 mg/kg-d and above. TAG-MNT was negative in the rat micronucleus assay of blood samples. Based on these data, the 14-day oral No observed adverse effect level (NOAEL) and the lowest observed adverse effect level (LOAEL) is 250 and 500 mg kg^{-1} day^{-1}, respectively.

Key words: RDX, Ames assay, cytotoxicity, oral toxicity, rat micronucleus assay

INTRODUCTION

Current land use patterns and expanding suburban populations present encroachment issues for many military installations. Although there have been efforts to close and realign many installations, wise stewardship of remaining active training sites is a priority of the Army. Therefore, the development of important environmental and occupational toxicity data are important for new compounds; early in development, decisions can be made regarding further testing and implementation (American Society for Testing and Materials (ASTM), 2008).

One novel compound, triaminoguanidinium-1-methyl-5-nitriminotetrazolate (TAG-MNT) is being developed for use as an energetic (Hammer et al., 2005; Klapötke et al., 2008; Klapötke et al., 2008). There is essentially no information on the toxicologic properties of nitriminotetrazolates and TAG-MNT in particular. Several studies were initiated to evaluate the toxicity of this compound. These studies included: determination of

*Corresponding author. E-mail: Larry.Williams45@us.army.mil.

TAG-MNT mutagenicity using a microplate Ames assay; evaluation of its *in vitro* cytotoxicity in bacteria by measuring luminescence (e.g., ATP assay and Microtox® test) and in a human liver cell line by neutral red uptake (NRU) assay; and evaluation of its *in vivo* acute and sub-acute oral toxicity in female, Sprague-Dawley rats. An *in vivo* genotoxicty assay was also included as part of the oral 14-days exposures (that is, rat micronucleus assay).

MATERIALS AND METHODS

Test compound

The TAG-MNT was synthesized and obtained from the Army Research Development and Engineering Center (ARDEC, Picatinny Arsenal, NJ). The lot number was identified as RDD09A004E001 with a purity of 98 to 99% by weight as measured by NMR spectral analysis (R. Damavarapu, personal communication). The Chromatographic Analysis Division (Explosives Team), Directorate of Laboratory Sciences (DLS; USAPHC (Prov), Aberdeen, MD) developed an analytical method for TAG-MNT using high performance liquid chromatography with ultraviolet detection at 313 nm. The system employed separation on a reverse phase C-18 column (150 × 4.6 mm) with a mobile phase of 75/25 acetonitrile and water at a flow rate of 1.0 ml/min. The concentration ranged from 1 to 100 µg/ml in an injection volume of 100 µl. TAG-MNT had a retention time of 1.52 min.

TAG-MNT is readily soluble in water up to 98.3 mg/ml at 22°C (R. Pesce-Rodriguez, personal communication). For the *in vitro* microplate Ames assay, a 25-X stock solution of TAG-MNT was prepared (50 mg/ml in sterile water). For the *in vivo* oral toxicity studies, intended doses of 2000 mg/kg required the use of supersaturated suspensions of TAG-MNT as the maximum daily oral dose volume could not exceed 10 ml/kg. For the *in vivo* sub-acute experiments, 1% methylcellulose, 0.2% Tween 80, in tap water was used as the vehicle; the methylcellulose (CAS # 9004-67-5, lot number 037690), and the Tween 80 (CAS # 9005-65-6 lot number 032097) were purchased from Fisher Scientific (Fairlawn, NJ). Suspensions and solutions of TAG-MNT were verified to be homogenous and stable at 22°C up to 3 weeks.

GABAa receptor binding assay

TAG-MNT was tested at a single concentration of 33 µM for affinity to the GABAa convulsant site. The [35S]-TBPS convulsant site assay, based on the method of Maksay (1993) was performed by MDS Pharma Services, (King of Prussia, PA) using picrotoxin as a standard.

In vitro microplate Ames assay

A microplate Ames assay (Xenometrics MPF™ Ames Assay, AG, Switzerland) was used that provided a convenient, high throughput capability for mutagenicity testing (Xenometrix MPF™ is a trademark of Xenometrix AG, Switzerland). The test uses the *Salmonella typhimurium* strains TA98, TA100, TA1535 and TA1537, and *Escherichia coli* strains WP2 uvrA and WP2 [pKM101] of bacteria (Kamber et al., 2009; Umbuzeiro et al., 2010).

Briefly, bacterial suspensions were diluted with incubation medium and aliquoted to tissue culture wells containing serial dilutions of the 50 mg/ml, 25-X concentrate of TAG-MNT. The bacterial suspensions were incubated with shaking for 90 min at 37°C with and without the inclusion of Aroclor-induced, S9 liver extract (±S9). At the end of the incubation, each well was diluted 11-fold with purple indicator medium and 50 µl aliquots were distributed appropriately into 384-well plates. These plates were incubated anaerobically for 2 days at 37°C and then scored for presence of revertant colonies, that is, evidence of a positive mutagenic event. The colormetric indicator reacts to changes in pH resulting from metabolism by living, mutated bacteria. Positive wells change from purple to yellow and are scored as a mutagenic event.

A positive control appropriate for each strain was run coincidently with TAG-MNT to assure the assay was valid for each trial: 2 µg/ml 2-nitrofluorene TA98 –S9; 0.1 µg/ml 4-nitroquinoline-N-oxide TA100 –S9; 100 µg/ml N4-aminocytidine TA1535 –S9; 15 µg/ml 9-aminoacridine TA1537 –S9; 1 µg/ml 4-nitroquinoline-N-oxide *E. coli* -S9; 5 µg/ml 2-aminoanthracene TA98, TA100, TA1535 and TA1537 +S9; and 50 µg/ml 2-aminoanthracene *E. coli* +S9. The assay as a whole is determined to be valid if the number of control background reversions and the number of positive control reversions is within prescribed limits (Xenometrix 2009). A positive result was indicated if the number of reversions induced by the test compound was at least 2-fold above the background control, and there is a concentration-related increase over the range tested and/or a reproducible increase at one or more concentrations in the number of revertant colonies per plate in at least one strain with or without metabolic activation system (Xenometrix, 2009).

In vitro cytotoxicity assays

ATP assay

Coincident with the mutagenic incubation, a duplicate plate was prepared for determination of cytotoxicity of TAG-MNT using ATP luminescence. After the 90 min incubation at 37°C, samples from the incubation plate were aliquoted to a 96-well plate. An equal volume of luminescent reagent was added to each well according the method described for the BacTiter-Glo™ microbial cell viability assay (BacTiter-Glo™ is a trademark of Promega Corporation Madison, WI). Luminescence was measured using a synergy TM HT multi-detection microplate reader (Model SIAFRTD) and Gen5TM software (BioTek Instruments Inc., Winooski, VT). Cytotoxicity of TAG-MNT in the BacTiter-Glo assay was indicated when the luminosity of ATP in compound-treated cultures was decreased below the levels in vehicle-treated cultures; the level of ATP generated luminosity correlates with the number of living bacteria. Data are expressed as a percentage of the level of luminosity generated by the control, vehicle-treated bacteria.

Microtox® acute toxicity test

Cytotoxicity of TAG-MNT to the marine bacterium, *Vibrio fischeri* was evaluated using a Microtox Model 500 Analyzer (SDIX Strategic Diagnostic, Inc.; (Choi and Meier, 2001). *V. fischeri* NRRL-B-11177, BSL-1 (AZF686018A) was purchased from SDIX. Loss of luminescence, indicating a loss or decrease in cell viability and cell death, was recorded at 5, 15 and 30 min; EC50 values at 5, 15, and 30 min were determined by the MicrotoxOmni™ software.

Neutral red uptake assay

The neutral red uptake assay was run in accordance with National Institute of Environmental Health Sciences guidelines (Interagency Coordinating Committee on the Validation of Alternative Methods (ICCVAM) et al., 2006). Cells of human liver origin (Chang liver CCL-13, ATCC), were seeded into 96-well plates at 5.0×10^3 cells/well/0.1 ml and maintained in culture for 24 h to form a semiconfluent monolayer. The cells were treated with TAG-MNT over a range of 8 concentrations for 48 ± 1 h in 37°C, 5% CO_2 incubator. Treatment medium was removed and the cultures were washed once with phosphate buffered saline (PBS). Neutral red medium (NRM containing 33 µg dye/ml) was added to each well (0.2 ml/ml/well). After 3 h incubation, NRM was discarded and the cultures were washed once with PBS and received 0.1 ml of NR desorbing fixative per well. The plates were placed on a shaker for 20 min at room temperature (24 ± 2°C). NR absorption was detected at optical density (OD) 540 nm using a synergy[TM] HT multi-detection microplate reader (Model SIAFRTD) and Gen5[TM] software (BioTek Instruments Inc., Winooski, VT).

Animals

Oral toxicity studies were conducted using young adult female Sprague-Dawley rats obtained from Charles River Laboratories (Wilmington, MA). Rats were 8 weeks old and six weeks old for the acute and 14-day subacute toxicity studies, respectively. The attending veterinarian examined the animals and found them to be in acceptable health. The animals were quarantined for a minimum of 5 days after arrival. All rats were maintained in a temperature-, relative humidity- and light-controlled room. The conditions were 64 to 79°F, 30 to 70% relative humidity with a 12-h light/dark cycle. A certified pesticide-free rodent chow (Harlan Teklad®, 8728C Certified Rodent Diet) and drinking quality water were available ad libitum (® Teklad Certified Rat Diet is a registered trademark of Harlan, Teklad, Madison, Wisconsin). Rats were housed individually in suspended polycarbonate boxes with Harlan Sani-Chip® bedding (® Harlan Sani-Chip is a registered trademark with P. J. Murphy Forest Products Corporation, Montville, New Jersey). Each rat was uniquely identified by number using cage cards only for the acute study and both cage cards and microchip implants (BioMedic Data Systems, Inc., Seaford, DE) for the 14-day study.

Acute study (Sequential Stage-Wise Probit, SSWP)

The objective of this study was to determine the acute oral LD50 of TAG-MNT in the female Sprague-Dawley rat using the SSWP (Feder et al., 1991; Feder et al., 1991), and to guide oral exposures for the subacute (14-day) study. The general procedures of this acute study followed the USEPA Health Effects Test Guidelines for Acute Oral Toxicity (OPPTS 870.1100) (USEPA, 1998). Tests were performed using two separate stages of dosing.

All animals were fasted overnight prior to dosing and for up to 4 h post-dosing. Doses for the first stage of the acute tests were 180, 270, 400, 600, 900, 1350 and 2000 mg/kg. All doses were calculated based on body weights taken immediately prior to dosing. The amount of TAG-MNT appropriate for each rat was weighed individually in a weigh pan, suspended in corn oil and administered by oral gavage using a 16 gauge × 2-inch stainless steel gavage needle; maximum volume did not exceed 10 ml/kg. In the second stage, two groups of 3 female rats received 2000 mg TAG-MNT/kg body mass. During the course of the acute study, it was determined that the preferred vehicle for the 14-day, sub-acute study would be 1% methylcellulose, 0.2% Tween 80, in tap water; suspension of TAG-MNT was prolonged and the TAG-MNT more soluble in the methylcellulose/Tween 80/ tap water vehicle. To rule out an effect of vehicle on possible toxicity of TAG-MNT, the corn oil vehicle was repeated coincidently with a second group dosed with TAG-MNT suspended in methylcellulose.

Following the administration of the test compound for each stage of the acute test, the rats were observed for 14 days. All clinical signs or incidences of death were recorded on a daily basis. Individual body weights were recorded daily (5 days a week) throughout the -14-day observation period to determine recovery.

Surviving animals were euthanized on day 14 and necropsied for gross pathological examination.

14-Day oral repeated dose toxicity study

Seventy female Sprague-Dawley rats were randomly distributed into seven treatment groups consisting of 10 rats each. The animals were then divided into three evenly distributed experimental groups; the start dates for each group were staggered over a period of three days to facilitate scheduling of necropsies. On the morning of each day, each rat received either 0 (methyl cellulose and water vehicle control), 62.5, 125, 250, 500, 1000, or 2000 mg TAG-MNT/kg body mass-day via gavage using a 16 gauge × 2-inch stainless steel gavage needle. Similar volumes of dosing solutions were administered to all animals using three dosing solutions at concentrations of 2000, 500 and 125 mg/L which resulted in dosing volumes of either 10 or 5 ml/kg. The control animals received the same volume per body weight as the highest dosage group, that is, 10 ml/kg. The doses were administered daily, 7 days per week (total of 14 doses) for the 14-day study. The solution/suspensions were sampled and analyzed to verify concentrations and stability prior to the first day of exposure.

Body weights and feeder weights were recorded on days 0, 1, 3, 7 and 14. Animals were observed daily for toxic signs and morbidity. Water consumption was not monitored during this study.

Following the 14-day study period, the rats were anesthetized with isoflurane gas. Blood was collected by intracardiac puncture and the rats were euthanized using carbon dioxide. Clinical chemistry and hematology values were determined from all valid samples. The adrenals, brain, heart, kidneys, liver, ovaries, spleen, thymus, and uterus were removed and weighed for absolute organ weights, organ-to-body weight ratios, and organ-to-brain weight ratios. Gross necropsies were completed on all terminal animals. The following parameters, by test group, were analyzed and compared to the controls: Body weights; weight gains; food consumption; absolute organ weights; organ-to-body weight ratios; and organ-to-brain weight ratios.

Hematology on blood samples was accomplished using a Cell-Dyn 3700 Hematology Analyzer (Abbott Laboratories, Abbott Park, Illinois). Parameters measured included: White blood cell count (WBC), WBC differential (% neutrophils (NEU, %N), % lymphocytes (LYM, %L), % monocytes (MONO, %M), % eosinophils (EOS, %E), % basophils (BASO, %B)), red blood cell count (RBC), hemoglobin (HGB), hematocrit (HCT), mean cell volume (MCV), mean cell hemoglobin (MCH), mean cell hemoglobin concentration (MCHC), red blood cell distribution width (RDW), platelets (PLT), and mean platelet volume (MPV).

Clinical chemistry was accomplished using a VetTest 8008 chemistry analyzer and VetLyte Na, K, Cl analyzer (IDEXX Laboratories, Inc., One IDEXX Drive, Westbrook, ME). Parameters measured included albumin (ALB), alkaline phosphatase (ALK P), alanine aminotransferase (ALT), blood urea nitrogen (BUN), calcium (Ca), cholesterol (CHOL), creatinine (CREA), glucose (non-fasting) (GLU), globulin (GLOB), lactate dehydrogenase (LDH), phosphorus (PHOS), total bilirubin (TBIL), total protein (TP), sodium (Na), potassium (K), and chlorine (Cl).

Genotoxicity to rat peripheral blood was evaluated using a micronucleus assay. Following the 14-day study period, blood was collected from the lateral saphenous vein of animals from the top three surviving dose groups (250, 500, 1000 mgkg[-1]day[-1]), vehicle control, untreated control, and positive control groups. A positive control group (n = 10) was given three oral doses (48, 24, and 4 h prior to sacrifice) of ethylmethane sulfonate at 200 mg/kg. The micronucleus assay was conducted on peripheral blood using the MicroFlow Plus Kit® (Litron Laboratories) following the manufacturer's instructions. Briefly, blood was placed in anticoagulant, fixed in ultracold methanol, and store at -75 to -85°C

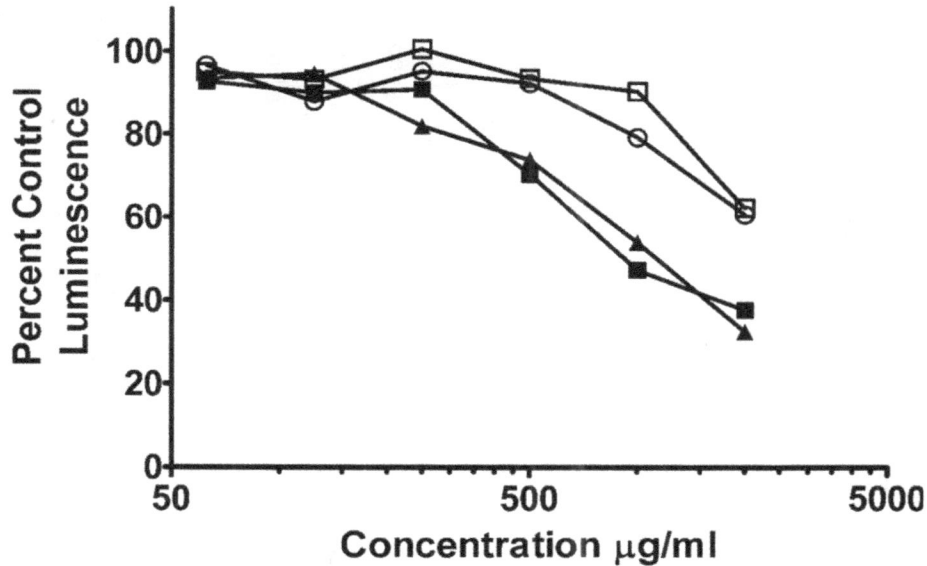

Figure 1. Effect of TAG-MNT on viability of bacteria. ATP luminescence was measured after 90 min incubation at 37°C with increasing concentrations of TAG-MNT. Decreased luminescence from the level expressed in control cultures is evidence of cytotoxicity. The LD_{50} is estimated to range from 1 to 2 g/L depending on bacterial strain. ■ – TA98, ▲ – TA 100, ○ – TA1535, □ – *E. coli*. Cytotoxicity in TA1537 strain was not evaluated.

until analysis. On the day of analysis by flow cytometry, the blood samples were washed with PBS to remove fixative, treated with RNase A, labeled with anti-CD71-FITC and anti-CD61-PE, and then stained with DNA staining solution (propidium iodide, PI). Prior to analysis of samples, the flow cytometer (Beckman Coulter EPICS XL) was calibrated using the kit provided standards. Anti-CD71, anti-platelet–PE and PI fluorescence signals were detected in the FL1, FL2 and FL3 channels, respectively. Micronucleated reticulocytes (MN-RET) were identified as those that show both CD71 and PI-associated fluorescence. A total of 20,000 MN-RETs were analyzed per sample. The data collected from the micronucleus assay were expressed as the percentage of reticulocytes with micronuclei (%MN-RET) and the percentage of red blood cells that were reticulocytes (%RET).

Statistical analyses

Data were analyzed using a one-way ANOVA. If significant, a post hoc Dunnett's multiple comparison test was used to compare dose groups to the control group. Statistical significance was defined at the $p < 0.05$ level. To allow for a consistent characterization of the results, means and standard deviations are presented for all data. Tests were conducted using Prism 4 (GraphPad Software, La Jolla, CA).

For the micronucleus assay, a one-way analysis of variance (ANOVA) was used to test for significant differences in %MN-RET and %RET. The Tukey multiple comparison test was used to evaluate the differences between groups. The results were considered to be statistically significant at $p < 0.05$. SPSS® version 16.0 (SPSS Inc., Chicago, IL) was used for all analyses.

Histopathology

Selected samples of liver and kidney from each dose group (n = 5

per group) were collected, trimmed, fixed in formalin, and embedded in paraffin. These tissues were then sectioned at 6 microns, stained with hematoxylin and eosin, and examined via light microscopy.

This study was conducted consistent with the standards found in Title 40 Code of Federal Regulations (CFR), Part 792, Good Laboratory Practices. The investigators and technicians adhered to the following guidelines: The Public Health Service Policy on Humane Care and Use of Laboratory Animals, "U.S. Government Principles for the Utilization and Care of Vertebrate Animals Used in Testing, Research, and Training", and the Animal Welfare Act. The studies were performed in animal facilities fully accredited by the Association for Assessment and Accreditation of Laboratory Animal Care.

RESULTS

GABAa receptor binding assay

TAG-MNT was tested at a single concentration of 33 µM for affinity to the GABAa convulsant site. At this concentration, TAG-MNT did not displace [^{35}S]-TBPS, that is, had no affinity for the receptor convulsant site (data not shown).

In vitro cytotoxicity of TAG-MNT

Figure 1 illustrates the levels of toxicity demonstrated by TAG-MNT in *S. typhimurium* and *E. coli*. The four bacterial strains tested demonstrated differential sensitivity to TAG-MNT. TAG-MNT was cytotoxic at 1000 µg/ml in the TA1535 and *E. coli*. strains; cytotoxic at 500

Table 1. Results of TAG-MNT Ames assay.

Bacteria strain	-S9			+S9			Level of mutagenicity		Cytotoxicity (µg/ml)
	Control (# Revertants)	TAG-MNT (fold increase)	Positive control (fold increase)	Base line (# Revertants)	TAG-MNT (fold increase)	Positive control (fold increase)	-S9	+S9	
TA98	1.0	3.0*	41.0	1.8	3.3*	26.7*	+ 2 mg/ml	+ 2 mg/ml	500
TA100	6.4	3.5*	7.5*	3.7	0.4	10.5*	+ 1 mg/ml	negative	250
TA1535	2.4	2.7*	20*	1.6	3.4*	16	+ 2 mg/ml	+ 2 mg/ml	1000
TA1537	1	0.3	48*	2.9	0.3	8.1*	negative	negative	1000
E. coli	4.7	0.2	4.7*	7.4	0.7	1.8	negative	negative	2000
V. fischeri									864
Chang cells									316

* indicated significantly different from controls at ρ ≤ 0.05.

µg/ml in the TA98 strain and at 250 µg/ml in the TA 100 strain. However, compounds can still be mutagenic even at cytotoxic concentrations (Xenometrix, 2009). The LC50 was estimated to range from 1 to 2 g/L depending on bacterial strain.

TAG-MNT appeared to be more toxic to *V. fischeri* and Chang liver cells in the microtox and neutral red uptake assays, respectively. The calculated EC50 15 min of TAG-MNT to *V. fischeri* was 864 µg/ml and the calculated IC 50 48 h to Chang liver cells was 316 µg/ml ((Table 1).

Mutagenicity of TAG-MNT

The results for the mutagenicity tests are provided in Table 1. TAG-MNT was mutagenic at the limit dose of 2 g/L in three of the five strains, with or without S9 incubation. TAG-MNT was not mutagenic in the TA1537 or *E. coli* strains.

In all assays, the background control and the positive controls for incubations ± S9 were within the limits specified (Xenometrix 2009), that is, all assays met the criteria for validity. In several strains (Table 1), TAG-MNT was mildly mutagenic

resulting in a two-fold increase only at the limit dose of 2 mg/ml, and showed a dose response only in the TA100 strain –S9.

Acute toxicity

After a single oral dose with TAG-MNT, none of the animals showed any signs of distress during the 14 day period of observation. Even at the limit dose of 2000 mg/kg, TAG-MNT showed no indication of toxicity and no sign of seizure. In the second stage of dosing, six animals received the limit dose of 2000 mg/kg; three were administered TAG-MNT in corn oil and three using the 1% methylcellulose/water vehicle. All animals in both Stages 1 and 2 survived the 14 day observation period and were then euthanized. Gross pathology observations in these animals were unremarkable.

14-Day oral repeated dose toxicity study

Clinical signs of toxicity were observed in the 2000 mg/kg[-1]day[-1] dose group. These signs

included, rough coat, piloerection and lethargy at the end of the first week of dosing progressing to stained hair coat, a hunched or crouched gait, and forelimb impairment accompanied with weight loss. Five of the rats from this limit dose group lost greater than 20% of their starting body weight, became moribund, and were removed from the study and euthanized according to protocol.

The net body weight change of the animals increased similarly with time for all dose groups except for animals in the 1000 and 2000 mg/kg[-1]day[-1] treatments. Weight gain was reduced in these two groups and was evident on the first few days of treatment (Figure 2). However, changes in weight were observed between treated and control animals only at Day 7 in the 2000 mg/kg group; five rats were severely affected (by TAG-MNT during the second week and were removed from the study. The weights of the five remaining rats recovered somewhat during the second week and the weight change was not significantly different from control (Figure 2). The net food consumption during the first week was reduced in the 1000 and 2000 mg/kg groups (data not shown).

Differences were observed in mean absolute organ weights and organ to body weight

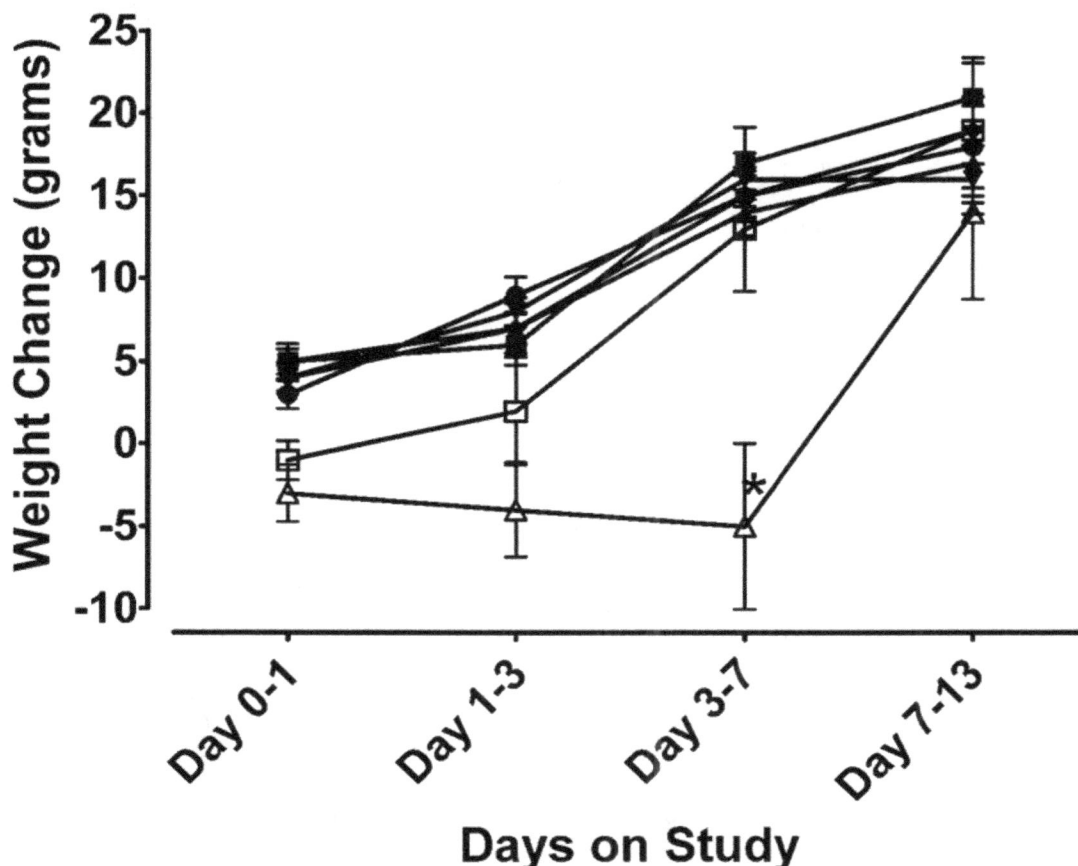

Figure 2. Body weight changes during 14-day TAG-MNT dosing study. The net body weight change of the animals increased similarly with time for all dose groups except for the two highest dose groups, 1000 and 2000 mg/kg. Weight gain was reduced in these two groups. ■ – vehicle, ▲ – 62.5 mg/kg, ▼- 125 mg/kg, ♦ - 250 mg/kg, ● – 500 mg/kg, □ – 1000 mg/kg, ○ – 2000 mg/kg. *$p \leq 0.05$.

ratios for the kidneys, liver, adrenal glands, and spleen between the higher dose groups and the controls (Tables 2 and 3, respectively). The 500, 1000 and 2000 mg/kg^{-1}day^{-1} dose groups had elevated kidney and liver to organ/body weight ratios when compared to control animals. The adrenals and spleen had decreased organ/body weight ratios when compared to controls.

Analysis of the clinical chemistry results revealed treatment-related differences for ALB, ALT, BUN, CREA, GLU, and PHOS analytes when compared to those from samples collected from the vehicle control group (Table 4). ALT, ALB, and GLU levels were decreased relative to control values from rats in the 500, 1000, and 2000 mg/kg^{-1}day^{-1} dose groups whereas BUN, CREA, and Phos were increased (Table 4).

Treatment-related differences in the concentrations of WBC, NEU, BASO, EOS, and LYM were found when compared to the controls; however, there was no difference in the percentage of these cells in the total population (Table 5). There was no decrease in the concentration of monocytes in the WBC population and

thus the percentage MONOs significantly increased. There were no treatment-related differences in RBCs, hematocrit, or other hematologic parameters (Table 5).

In the micronucleus assay, the %MN-RET in the vehicle control group was 0.28% and ranged from 0.27 to 0.34% in female rats treated with 250, 500 and 1000 mg/kg^{-1}day^{-1} of TAG-MNT in methylcellulose (Figure 3). The %MN-RET in the positive control (EMS) group was 1.16%, a marked and statistically significant (p<0.001) increase over that seen in the controls, indicating that the assay system was performing as expected with the known genotoxin.

Treatment with TAG-MNT did not increase %MN-RET relative to the vehicle control (p = 0.672). These results indicate that TAG-MNT does not induce chromosomal aberrations in rat peripheral blood and is not genotoxic in this tissue at the doses tested.

The %RET was 1.36% in the peripheral blood of the rats given the vehicle control and 0.55% in the positive (EMS) control. For rats given 250, 500 and 1000 mg/kg^{-1}day^{-1} of TAG-MNT, the %RET ranged from 1.38% at the

Table 2. Absolute weights (mg) of organs from animals treated 14 days with increasing concentrations of TAG-MNT.

Organ		Vehicle Control	TAG-MNT (mg/kg-day)						
			0	62.5	125	250	500	1000	2000
Adrenal	Mean	0.07	0.08	0.07	0.07	0.07	0.06[*]	0.05[*]
	S.D.	0.01	0.02	0.01	0.01	0.01	0.01	0.01
Brain	Mean	1.81	1.83	1.85	1.81	1.81	1.77	1.81
	S.D.	0.10	0.08	0.07	0.06	0.05	0.10	0.09
Heart	Mean	0.80	0.83	0.82	0.81	0.81	0.69	0.76
	S.D.	0.10	0.09	0.07	0.07	0.10	0.08	0.28
Kidney	Mean	1.65	1.67	1.67	1.70	1.78[*]	1.82[*]	1.75[*]
	S.D.	0.13	0.14	0.12	0.13	0.13	0.19	0.16
Liver	Mean	6.66	6.60	6.68	6.77	8.03	8.75	7.89
	S.D.	0.71	0.62	0.42	0.37	0.92	0.82	0.56
Lungs	Mean	1.22	1.13	1.10	1.17	1.20	1.11	1.14
	S.D.	0.09	0.15	0.11	0.09	0.17	0.15	0.26
Ovaries	Mean	0.11	0.12	0.13	0.12	0.12	0.11	0.08
	S.D.	0.02	0.02	0.02	0.02	0.02	0.03	0.01
Spleen	Mean	0.46	0.42	0.41	0.45	0.44	0.34[*]	0.31[*]
	S.D.	0.13	0.07	0.07	0.08	0.12	0.05	0.06
Thymus	Mean	0.50	0.56	0.48	0.51	0.51	0.38	0.35
	S.D.	0.19	0.16	0.10	0.08	0.12	0.10	0.15
Uterus	Mean	0.48	0.40	0.39	0.53	0.38	0.39	0.30
	S.D.	0.21	0.07	0.05	0.19	0.07	0.10	0.10

*significantly indicated different from controls at $p \leq 0.05$.

Table 3. Normalized organ to body weight ratios from animals treated 14 days with increasing concentrations of TAG-MNT.

Organ		Vehicle Control	TAG-MNT (mg/kg-day-day)						
			0	62.5	125	250	500	1000	2000
Adrenal	Mean		3.82	4.06	3.88	3.78	3.47	3.13[*]	3.33
	S.D.		0.45	0.69	0.62	0.54	0.33	0.32	0.68
Brain	Mean		9.60	9.95	9.92	9.65	9.70	10.17	11.14[*]
	S.D.		0.71	0.40	0.44	0.62	0.69	0.57	0.71
Heart	Mean		4.26	4.49	4.40	4.31	4.33	3.97	4.64
	S.D.		0.47	0.36	0.31	0.35	0.32	0.27	1.51
Kidney	Mean		8.73	9.05	8.90	9.06	9.55[*]	10.44[*]	10.79[*]
	S.D.		0.69	0.43	0.43	0.55	0.64	0.54	0.98
Liver	Mean	3.52	3.59	3.57	3.60	4.28[*]	5.00[*]	4.84[*]
	S.D.	0.25	0.28	0.18	0.13	0.24	0.29	0.28

Table 3. Contd

Lungs	Mean	6.46	6.13	5.90	6.21	6.38	6.32	6.96
	S.D.	0.41	0.61	0.49	0.42	0.53	0.44	1.37
Ovaries	Mean	5.99	6.71	7.01	6.43	6.39	6.20	5.08
	S.D.	1.20	1.17	1.03	0.92	1.07	1.58	1.08
Spleen	Mean	2.43	2.29	2.20	2.37	2.35	1.94*	1.89
	S.D.	0.56	0.30	0.30	0.38	0.48	0.17	0.39
Thymus	Mean	2.65	3.04	2.56	2.73	2.74	2.17	2.14
	S.D.	0.95	0.75	0.57	0.36	0.63	0.70	0.83
Uterus	Mean	2.60	2.16	2.09	2.81	2.03	2.18	1.81
	S.D.	1.25	0.37	0.26	1.06	0.42	0.48	0.64

* indicated significantly different from controls at $p \leq 0.05$.

Table 4. Clinical plasma chemistry results for female rats exposure orally to TAG-MNT for 14-days.

Organ	Vehicle	TAG-MNT (mg/kg-day)						
	Control	0	62.5	125	250	500	1000	2000
ALB	Mean	3.1	3.2	3.2	3.2	3.2	3.1	2.6
(g/dl)	S.D.	0.2	0.2	0.1	0.2	0.3	0.2	0.3
ALKP	Mean	209.6	216.9	215.6	213.2	223.2	168.0	204.4
(U/L)	S.D.	42.7	47.4	35.5	26.8	41.5	52.4	60.6
ALT	Mean	50.2	48.8	43.7	39.1*	43.0	31.4*	22.6*
(U/L)	S.D.	5.4	6.1	4.2	5.3	6.1	8.5	5.0
BUN	Mean	17.3	18.1	18.8	19.0	21.0	24.4*	25.0*
(mg/dl)	S.D.	2.9	1.7	3.4	3.3	4.6	1.8	4.8
Ca	Mean	10.4	10.1	10.4	10.4	10.4	10.3	10.1
(mg/dl)	S.D.	0.2	0.4	0.3	0.4	0.3	0.3	0.4
CHOL	Mean	44.0	49.3	42.4	48.3	48.2	37.3	33.6
(mg/dl)	S.D.	16.0	12.5	14.3	8.5	21.8	15.5	5.8
CREA	Mean	0.5	0.5	0.5	0.6	0.6	0.6*	0.8*
(mg/dl)	S.D.	0.1	0.1	0.1	0.1	0.1	0.1	0.1
GLOB	Mean	2.6	2.7	2.7	2.7	2.7	2.5	2.5
(g/dl)	S.D.	0.1	0.2	0.1	0.2	0.1	0.2	0.2
GLU	Mean	146.3	136.2	134.0	133.5	119.2*	119.7*	118.0*
(mg/dl)	S.D.	13.9	17.6	17.8	13.6	20.9	10.9	5.0
LDH	Mean	2038.9	3209.9	2423.2	1667.0	2717.1	2889.2	3565.8
(U/L)	S.D.	1038.7	1647.1	1493.2	792.5	1722.1	1508.6	2306.3

Table 4. Contd

PHOS	Mean	7.7	7.5	7.6	8.1	8.4	9.5*	10.2*
(mg/dl)	S.D.	0.4	0.5	0.4	0.9	1.0	1.4	1.7
TBIL	Mean	0.1	0.3	0.2	0.1	0.2	0.3	0.3
(mg/dl)	S.D.	0.0	0.3	0.1	0.0	0.1	0.2	0.5
TP	Mean	5.7	5.9	5.9	5.9	5.9	5.6	5.1
(g/dl)	S.D.	0.2	0.3	0.2	0.3	0.3	0.3	0.4
Na	Mean	139.3	138.9	139.4	140.4	138.7	137.1	134.4
(mmol/l)	S.D.	1.3	1.8	2.3	2.4	2.2	3.0	1.1
K	Mean	4.5	4.8	4.6	4.6	4.4	4.6	5.5
(mmol/L)	S.D.	0.5	0.4	0.5	0.8	0.6	0.8	1.0
Cl	Mean	105.4	105.4	104.9	104.6	103.2	101.3	101.2
(mmol/L)	S.D.	2.2	1.0	2.0	2.0	2.5	1.6	2.7

* indicated significantly different from controls at $p \leq 0.05$.

Table 5. Changes in red blood cell count, hematocrit, hemoglobin, white blood cell count, and proportion of white blood cell types as a function of oral TAG-MNT exposure in female rats.

	Vehicle		TAG-MNT (mg/kg-day)					
	Control	0	62.5	125	250	500	1000	2000
WBC	Mean	16.1	11.2*	9.3*	12.6	10.2*	7.3*	6.0*
(K/ΟL)	S.D.	3.2	4.4	1.8	3.2	4.3	3.2	5.7
NEU	Mean	1.3	0.8	0.8	0.7	0.8	0.5*	0.2*
(K/ΟL)	S.D.	1.0	0.4	0.4	0.2	0.4	0.3	0.2
(%N)	Mean	8.3	8.2	8.7	5.6	8.1	7.6	5.4
	S.D.	6.4	6.1	3.3	2.1	2.1	4.1	1.9
LYM	Mean	13.9	9.8	7.2*	11.2	8.7*	6.2*	5.2*
(K/ΟL)	S.D.	3.0	4.2	2.7	3.0	3.6	2.8	5.3
(%L)	Mean	85.9	86.0	84.8	88.6	85.3	85.1	82.3
	S.D.	7.4	6.5	4.3	2.6	3.5	6.0	8.8
MONO	Mean	0.5	0.3	0.3	0.4	0.4	0.3	0.4
(K/ΟL)	S.D.	0.1	0.1	0.1	0.2	0.2	0.2	0.3
(%M)	Mean	2.9	2.7	3.2	3.1	3.3	4.4*	8.2*
	S.D.	0.7	0.7	0.7	0.7	1.4	1.8	3.5
EOS	Mean	0.3	0.1	0.1	0.1	0.1	0.05*	0.04
(K/ΟL)	S.D.	0.5	0.0	0.1	0.0	0.0	0.0	0.0
(%E)	Mean	1.1	1.0	1.2	0.9	1.1	0.8	0.9
	S.D.	0.3	0.3	0.4	0.4	0.4	0.4	0.5

Table 5. Contd

BASO	Mean	0.3	0.2	0.2	0.2	0.2	0.16*	0.1
(K/⊙L)	S.D.	0.1	0.2	0.1	0.1	0.1	0.1	0.0
(%B)	Mean	1.9	2.1	2.1	1.8	2.2	2.1	3.2
	S.D.	0.5	0.6	0.6	0.4	0.7	0.8	2.9
RBC	Mean	7.2	7.1	7.2	7.1	7.0	7.2	7.7
(M/⊙L)	S.D.	0.3	0.2	0.2	0.5	0.5	0.3	0.2
HGB	Mean	14.1	13.6	14.0	13.9	13.7	13.5	14.0
(g/dL)	S.D.	0.5	0.4	0.5	1.0	0.5	0.6	0.1
HCT	Mean	41.5	40.0	41.1	41.0	40.0	40.1	41.4
(%)	S.D.	1.6	1.2	1.4	2.8	2.6	1.6	1.0
MCV	Mean	57.7	56.4	56.9	57.5	57.4	55.7	54.1
(fL)	S.D.	1.1	1.7	1.9	1.6	2.4	1.8	3.0
MCH	Mean	19.6	19.2	19.3	19.4	19.8	18.7	18.3
(pg)	S.D.	0.3	0.8	0.6	0.6	1.4	0.7	0.4
MCHC	Mean	33.9	34.0	33.9	33.8	34.4	33.6	33.9
(g/dL)	S.D.	0.3	0.5	0.2	0.2	1.8	0.5	1.1
RDW	Mean	14.4	15.1	14.2	14.9	14.2	15.3	17.4
(%)	S.D.	0.6	0.7	0.5	0.7	0.8	0.7	0.0
PLT	Mean	1073.6	1071.1	1062.6	1035.8	1084.0	848.1	1086.5
(K/⊙L)	S.D.	208.6	107.4	180.0	316.7	216.8	426.6	88.4
MPV	Mean	5.4	5.8	5.4	5.2	5.2	5.4	5.3
(fL)	S.D.	0.5	0.4	0.2	0.4	0.2	0.4	0.1

*Significantly different from Vehicle-treated animals.

low dose to 1.09% at the high dose (Figure 4). There were no differences in the frequency of reticulocytes in the peripheral blood among treated and vehicle control groups (p = 0.401). These results suggest that erythropoiesis in the bone marrow was not affected by treatment with TAG-MNT.

Histopathology was performed on liver and kidney tissues in all dose groups (n = 5) with preference to animals showing signs of organ weight changes or organ discoloration. No adverse events were observed in kidney. However, dose-related adverse events were observed in the liver tissues (Figure 5). Observed lesions included: Single-cell necrosis, macrocytic and microcytic cytoplasmic vacuolization of hepatocytes, inflammation, hepatocyte degeneration, centrilobular hepatocyte hypertrophy, increased mitoses, megakaryocytic hepatocytes, and bile duct hyperplasia. There was a dose-relationship of single-cell necrosis with increasing incidence and severity at the three highest doses, 500, 1000 and 2000 mg/kg^{-1}day^{-1} (Table 6). Lymphocytic and acute inflammation was noted in all treated groups, most evidently in the three highest dose groups, but not in the control group.

Hepatocyte degeneration was observed in one animal in each of the two highest-dose groups. Observations of centrilobular hypertrophy was also dose-related, with the incidence and severity consistent with increasing dose (periportal hypertrophy was present only at the highest dose). Megakaryocytic hepatocytes and increased heaptocyte mitoses were noted in several animals, indicative of a response to injury. Granulomatous inflammation or infiltration present in the control and treated groups was an incidental finding not considered to be related to treatment.

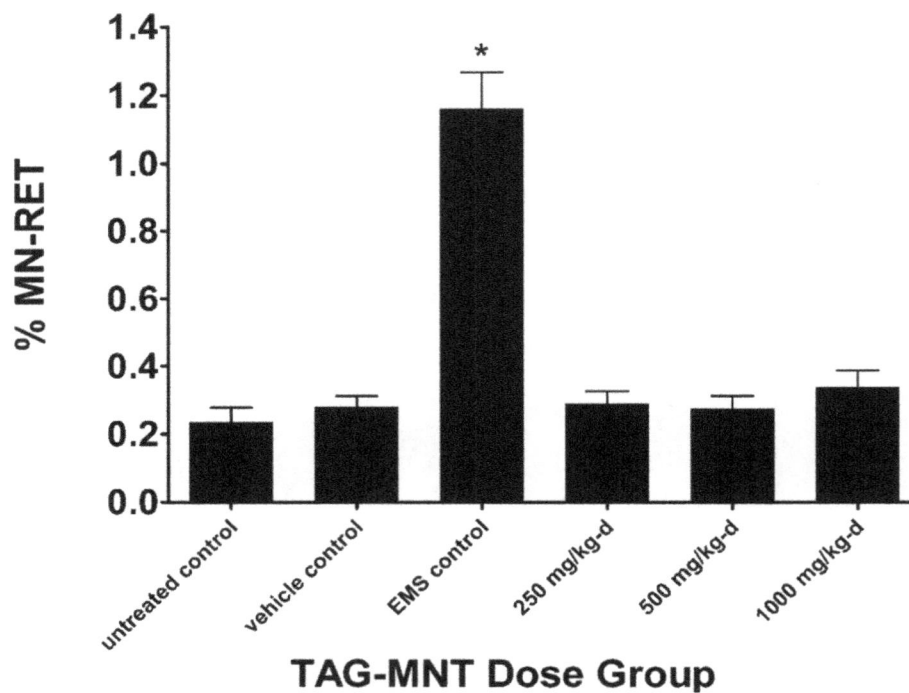

Figure 3. Percent micronucleated reticulocytes in female rats orally dosed with TAG-MNT for 14-days. There were no differences in the frequency of micronucleated reticulocytes between the vehicle control group and TAG-MNT dose groups. Treatment with the positive control (EMS) increased the frequency of micronucleated reticulocytes.

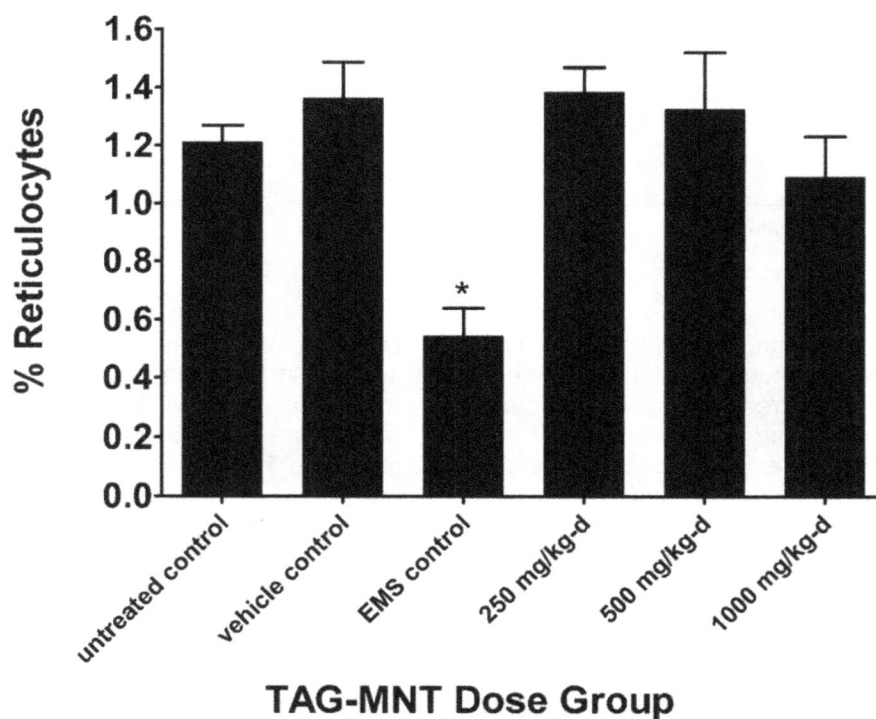

Figure 4. Percent reticulocytes in female rats orally dosed with TAG-MNT for 14-days. The frequency of reticulocytes did not differ between the vehicle control group and TAG-MNT dose groups. Treatment with the positive control (EMS) reduced the frequency of reticulocytes.

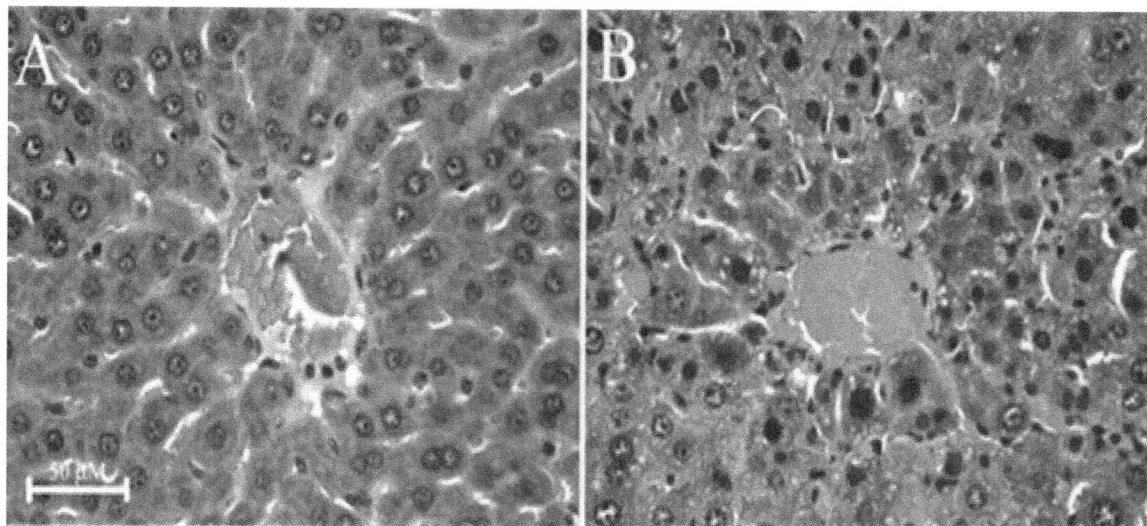

Figure 5. Histopathology of Liver. Micrograph of a central vein from a vehicle-treated animal (control, A) and from a high dose rat receiving 2000 mg/kg-d (B) for 14 days. Hepatocyte degeneration and necrosis is shown along with megakaryocytoses, cytoplasmic vacuolization, inflammation, and centrilobular hepatocyte hypertrophy. Scale bar equals 50 μm.

Table 6. Histological hepatic observations of oral TAG-MNT exposure in female rats.

Observation	TAG-MNT (mg/kg)						
	0	62.5	125	250	500	1000	2000
Necrosis, single cell	1/5	2/5	0/5	2/6	3/5	2/5	5/5
Trace	1/5	1/5	0/5	2/6	1/5	0/5	0/5
Mild	0/5	1/5	0/5	0/6	2/5	2/5	3/5
Moderate	0/5	0/5	0/5	0/6	0/5	1/5	2/5
Centrobular hypertrophy	0/5	0/5	0/5	0/6	1/5	2/5	3/5
Mild	0/5	0/5	0/5	0/6	1/5	2/5	2/5
Moderate	0/5	0/5	0/5	0/5	0/5	0/5	1/5
Vaculolar change	3/5	4/5	2/5	4/6	6/5	5/5	4/5
Trace	3/5	2/5	2/5	0/6	0/5	4/5	1/5
Mild	0/5	2/5	0/5	4/6	4/5	2/5	2/5
Moderate	0/5	0/5	0/5	0/5	2/5	1/5	2/5
Severe	0/5	0/5	0/5	0/5	0/5	0/5	1/5

DISCUSSION

Testing new compounds requires an integrative, holistic approach that is phased proportional with the level of investment or value of the new substance and how it will be used (American Society for Testing and Materials (ASTM), 2008). Many assays are available that provide value in assessing the risk from substance exposure. However, all assays have drawbacks, shortcomings, and uncertainties when applying assay results in a risk assessment paradigm. Therefore, the results of many tests can help provide a more complete evaluation of relative impact and serve to reduce levels of certainty while maintaining costs consistent with the level of efforts devoted to the system or platform. Here we compare there sults of these tests on TAG-MNT relative to a compound with a similar function, that is, RDX.

Cytotoxicity of TAG-MNT was observed in *S. typhimurium* and *E. coli* after a 90 min exposure of concentrations greater than 250 mg/L with an estimated

LD50 of 1 g/L. Data from *V. fischeri* exposures to TAG-MNT suggest an increased sensitivity than the other bacteria species tested (EC50 15 min = 864 µg/ml). However, using US Fish and Wildlife Service aquatic toxicity criteria, TAG-MNT is considered "Practically Nontoxic" as the EC50 is greater than 100 µg/ml (US Fish and Wildlife Service (USF&WS) 1984). *V. fischeri* are frequently more sensitive to xenobiotics than other aquatic organisms and thus, the results from Microtox® tests are often useful screens in the assessment of relative toxicity to aquatic organisms (Dutka and Kwan, 1981; Argese et al., 1998; Codina et al., 1993; McFeters et al., 1983; Choi et al., 2004). Thus, TAG-MNT is predicted to have a low probability for inducing aquatic toxicity.

Change in uptake of neutral red in Chang liver cells was also observed as a result of TAG-MNT exposure. Sufficient data have been obtained from the NRU studies that it can be used as a predictor of *in vivo* toxicity (Interagency Coordinating Committee on the Validation of Alternative Methods (ICCVAM) et al., 2006). The results of TAG-MNT in the NRU assay (IC 50 48 h = 316 µg/ml) predicted that TAG-MNT would have an oral LD50 of 900 mg/kg in rat. However, the present *in vivo* study found that acute oral exposures of up to 2000 mg TAG-MNT/kg failed to produce death or other overt toxicity in Sprague-Dawley female rats.

TAG-MNT showed a weak positive mutagenic response both with and without S9 metabolism at the highest, limit dose. Although meeting the criteria for a positive response for mutagenicity, the number of mutant reversions was less than 3.5-fold above background; the positive controls for the *S. typhimurium* strains were greater than 10-fold above background. However, OECD guidelines indicate that mutagenicity testing is valid only above cytotoxic concentrations (OECD, 1997). Thus, TAG-MNT is not considered mutagenic as signs of mutagenicity only occur at concentrations 2 to 4 times above the cytotoxic concentration (Table 1).

Unlike other tetrazole derivatives (Luttjohann et al., 2009; Sharma et al., 1994; Squires et al., 1984), TAG-MNT did not have affinity for the GABAa receptor in the *in vitro* binding assay. This result predicted that TAG-MNT would probably not induce convulsions like RDX when administered to rats. In fact, oral administration of TAG-MNT did not cause seizure or any other sign of acute toxicity up to the limit dose of 2000 mg/kg. However, daily administration for 14-days resulted in effects to other targets. TAG-MNT primarily affected the liver causing hepatocyte hypertrophy, inflammation, and increased liver weight associated with weight loss and animal morbidity. The histopathologic observations combined with other evidence suggests a No observed adverse effect level (NOAEL) and a lowest observed adverse effect level (LOAEL) of 250 and 500 mg/kg^{-1}day^{-1}, respectively.

The statistically significant increases in BUN and CREA in the high dose groups are consistent with dysfunction of the kidneys. However, this was not supported by evidence of histopathology in the kidney tissues examined. The significant decrease in ALT is counterintuitive as a marker of liver dysfunction as increases in free (systemic) ALT concentrations are usually attributed to increased liver parenchymal cell necrosis. Potential causes of decreased serum activities of ALT are reported to include: Decreased hepatocellular production or release of the enzymes; inhibition or reduction of the enzyme activity; or interference with the enzyme assay (PSD Guidance Document, 2007). Decreases in ALT have previously been reported in other chemical-induced, liver pathology models (Waner and Nyska, 1991). However, the most widely recognized cause is considered to be a negative effect, directly or indirectly, on pyridoxal 5'-phosphate (ECETOC 2002; PSD Guidance Document, 2007).

There was a reduction in white blood cell concentrations in rats receiving TAG-MNT oral doses of 500 mg/kg^{-1}day^{-1} and above. This included a general decrease in number of all subpopulations of WBCs, particularly neutrophils and lymphocytes, and thus no change in the overall percentage of these cells. Typically, these reductions are a result of increased levels of circulating stress proteins, e.g. glucocorticoids, an observation common in stressed animals (Burns-Nass et al., 2001).

The current USEPA regulatory values for RDX are based on prostatic inflammation in rodents as a result of chronic oral exposures (Levine et al., 1983), and hepatocellular carcinomas in only female mice (Lish et al., 1984; McLellan et al., 1992; Parker et al., 2006). However, the primary toxicity of RDX is its induction of epileptiform seizure following oral ingestion (Burdette et al., 1988; Kasuske et al., 2009; Stone et al., 1969). The mechanism of seizure induction has recently been shown to involve blockage of chloride flux through the GABAa receptor ligand-gated channel as a result of RDX binding to the convulsant site of the receptor (Williams et al., 2010). TAG-MNT did not have affinity for the GABAa receptor when tested in an *in vitro* binding assay, and oral administration of TAG-MNT to female rats did not cause seizure or any other sign of acute toxicity up to the limit dose of 2000 mg/kg. In contrast, the LD50 of RDX in rat is 60 mg/kg where death always follows induction of seizure (USACHPPM. 2006). Thus, the acute oral toxicity of TAG-MNT is orders of magnitude less than RDX.

As a group, relatively few toxicity data are available for tetrazoles. Tetrazoles have been associated with convulsant activity similar to that found with RDX where the GABA-chloride ionophore receptor complex is inhibited (Sharma et al., 1994; Squires et al., 1984; Williams et al., 2010; Luttjohann et al., 2009). The present study found daily administration of TAG-MNT for 14-days resulted in adverse effects primarily in the liver resulting in hepatocyte hypertrophy, inflammation, increased liver weight and morbidity; no evidence of neurotoxcity was observed nor was there activity associated with the GABA picrotoxin convulsant site. These data suggest that TAG-MNT primarily affects blood conditioning organs; though additional study is needed to further understand the complete spectrum of adverse effects from those exposed through manufacture and use.

ACKNOWLEDGMENTS

This work was supported by the US Army Environmental Quality Technology Program through the U.S. Army Research, Development and Engineering Command, Environmental Acquisition and Logistics Sustainment Program.

The authors would like to acknowledge the support and encouragement of Erik Hangeland of the U.S. Army Research, Development and Engineering Command, Environmental Acquisition and Logistics Sustainment Program. We also thank Pat Beall, Anne MacLarty, Will McCain, Valerie Adams, Terry Hanna, John Houpt, and Mark Way for providing technical support. Particular acknowledgment goes to Martha Thompson for collecting LABCAT® data and generating the appropriate reports. We also thank Desmond Bannon and Cindy Landgren for their critical review of the manuscript.

REFERENCES

American Society for Testing and Materials (ASTM) (2008). Standard Guide for Assessing the Environmental and Human Heath Impacts of New Energetic Materials. In Water and Environmental Technology, Biological Effects and Environmental Fate, Biotechnology. ASTM International, Conshohocken, PA, USA., pp. 1-15

Argese E, Bettiol C, Ghirardini AV, Fasolo A, Giurin G, Ghetti PF (1998). Comparison of in vitro Submitochodrial Particle and microtox® Assays for Determining the Toxicity of Organotin Compounds. Environ. Toxicol. Chem., 17(6): 1005-1012.

Burdette LJ, Cook LL, Dyer RS (1988). Convulsant properties of cyclotrimethylenetrinitramine (RDX): spontaneous audiogenic, and amygdaloid kindled seizure activity. Toxicol. Appl. Pharmacol., 92(3): 436-442.

Burns-Nass LA, Meade BJ, Munson AE (2001). Toxic Responses of the Immune System. In Casarett and Doull's Toxicology edited by CD Klaassen: McGraw-Hill, New York, pp. 335 - 402

Choi K, Meier PG (2001). Toxicity evaluation of metal plating wastewater employing the Microtox assay: a comparison with cladocerans and fish. Environ. Toxicol., 16(2): 136-141.

Choi K, Sweet LI, Meier PG, Kim PG (2004). Aquatic toxicity of four alkylphenols (3-tert-butylphenol, 2-isopropylphenol, 3-isopropylphenol, and 4-isopropylphenol) and their binary mixtures to microbes, invertebrates, and fish. Environ. Toxicol., 19(1): 45-50.

Codina JC, Perez-Garcia A, Romero P, de Vicente A (1993). A comparison of microbial bioassays for the detection of metal toxicity. Arch. Environ. Contam. Toxicol., 25(2): 250-254.

Dutka BJ, Kwan KK (1981). Comparison of three microbial toxicity screening tests with the Microtox test. Bul. Environ. Contam. Toxicol., 27(6): 753-7.

ECETOC (2002). Recognition of, and differentiation between, adverse and non adverse effects in toxicology studies. edited by European Centre for Ecotoxicology and Toxicology of Chemicals, Technical Report, No. 85.

Feder PI, Hobson DW, Olson CT, Joiner RL, Matthews MC (1991). Stagewise, adaptive dose allocation for quantal response dose-response studies. Neurosci. Biobehav. Rev., 15(1): 109-114.

Feder PI, Olson CT, Hobson DW, Matthews MC, Joiner RL (1991). Stagewise, group sequential experimental designs for quantal responses. one-sample and two-sample comparisons. Neurosci. Biobehav. Rev., 15(1): 129-133.

Hammer A, Hiskey MA, Holl G, Klapötke TM, Polborn K, Stierstorfer J, Weigand, JJ (2005). Azidoformamidinium and Guanidinium 5,5'-Azotetrazolate Salts. Chem. Mater., 17(14): 3784-3793.

Interagency Coordinating Committee on the Validation of Alternative Methods (ICCVAM), and National Toxicology Program (NTP), Interagency Center for the Evaluation of Alternative Toxicological Methods (NICEATM), and of the National Institute of Environmental Health Sciences (NIEHS), National Institutes of Health (NIH) (2006). "In vitro Cytotoxicity Test Methods for Estimating Acute Oral Systemic Toxicity", 1: 2.

Kamber M, Fluckiger-Isler S, Engelhardt G, Jaeckh R, Zeiger E (2009). Comparison of the Ames II and traditional Ames test responses with respect to mutagenicity, strain specificities, need for metabolism and correlation with rodent carcinogenicity. Mutagenesis, 24(4): 359-366.

Kasuske L, Schofer JM, Hasegawa K (2009). Two marines with generalized seizure activity. J. Emerg. Nurs., 35(6): 542-543.

Klapötke TM, Laub HA, Stierstorfer J (2008). Synthesis and characterization of a new class of energetic compounds - ammonium nitriminotetrazolates. Propellants Explos. Pyrotech. 33(23):421-430.

Klapötke TM, Stierstorfer J, Wallek AU (2008). Nitrogen-rich salts of 1-methyl-5-nitriminotetrazolate: An auspicious class of thermally stable energetic materials. Chem. Mater., 20: 4519-4530.

Levine BS, Furedi-Machacek EM, Rac VS, Gordon DE, Lish PM (1983). Determination of the chronic mammalian toxicological effects of RDX: twenty-four month chronic toxicity/carcinogenicity study of hexahydro-1,3,5-trinitro-1,3, 5-traizine (RDX) in the Fischer 344 rat. Phase V.: Contract No. DAMD 17-79-C-9161. AD A160744, US Army Medical Research and Development Command. pp. 1-354

Lish PM, Levine BS, Marianna-Furedi EM, Sagartz JM, Rac VS (1984). Twenty-Four Month Chronic Toxicity/Carcinogenicity Study of Hexahydro-1,3,5-Trinitro-1,3,5-Triazine (RDX) in the B6C3F1 Hybrid Mouse, ADA No. 160774: U.S. Army Medical Research and Development Command. pp. 1-364

Luttjohann A, Fabene PF, van Luijtelaar G (2009). A revised Racine's scale for PTZ-induced seizures in rats. Physiol. Behav. 98(5): 579-86.

Maksay G (1993). Partial and full agonists/inverse agonists affect [35S]TBPS binding at different occupancies of central benzodiazepine receptors. Eur. J. Pharmacol., 246(3): 255-260.

McFeters GA, Bond PJ, Olson SB, Tchan YT (1983). A comparison of microbial bioassays for the detection of aquatic toxicants. Water Res., 17(12): 1757-1762.

McLellan WL, Hartley WR, Brower ME (1992). Hexahydro-1,3,5-trinitro-1,3,5-triazine (RDX). In Drinking Water Health Advisory: Munitions, United States Environmental Protection Agency, Office of Drinking Water Health Advisories, edited by WC Roberts and WR Hartley: Lewis Publishers, pp. 130-188

OECD (1997). OECD guideline for testing of chemicals: Baceterial reverse mutation assay: Organisation for Economic Co-operation and Development. pp. 1-11

Parker GA, Reddy G, Major MA (2006). Reevaluation of a twenty-four-month chronic toxicity/carcinogenicity study of hexahydro-1,3,5-trinitro-1,3,5-triazine (RDX) in the B6C3F1 hybrid mouse. Int. J. Toxicol., 25(5): 373-378.

PSD Guidance Document (2007). Toxicological significance of reduced levels of serum ALT and/or AST in animal studies, edited by Health and Safety Executive Chemical Regualtion Directorate, pp. 1-17

Sharma SK, Bolster B, Dakshinamurti K (1994). Picrotoxin and pentylene tetrazole induced seizure activity in pyridoxine-deficient rats. J. Neurol. Sci., 121(1): 1-9.

Squires RF, Saederup E, Crawley JN, Skolnick P, Paul SM (1984). Convulsant potencies of tetrazoles are highly correlated with actions on GABA/benzodiazepine/picrotoxin receptor complexes in brain. Life Sci., 35(14): 1439-1444.

Stone WJ, Paletta TL, Heiman EM, Bruce JI, Knepshield JH (1969). Toxic effects following ingestion of C-4 plastic explosive. Arch. Intern. Med., 124(6): 726-730.

Umbuzeiro Gde A, Rech CM, Correia S, Bergamasco AM, Cardenette GH, Fluckiger-Isler S, Kamber M (2010). Comparison of the

Salmonella/microsome microsuspension assay with the new microplate fluctuation protocol for testing the mutagenicity of environmental samples. Environ. Mol. Mutagen. 51(1):31-38.

US Fish and Wildlife Service (USF&WS) (1984). Acute-toxicity rating scales. Research Information Bulletin No. 84-78. Department of the Interior, Washington D.C., pp. 1-23

USACHPPM (2006). Toxicology Study No. 85-XC-5131-03, Subchronic Oral Toxicity of RDX in Rats edited by LCB Crouse, MW Michie, MA Major, MS Johnson, RB Lee and HI Paulus. US Army Center for Health Promotion and Preventive Medicine, Aberdeen Providng Ground, MD 21010-5403. pp. 1-218

USEPA (1998). Health Effects Test Guidelines OPPTS 870.1100, Acute Oral Toxicity Study in Rodents. EPA 712-C-98-199, edited by United States Environmental Protection Agency. Washington D.C. pp. 1-11.

Waner T, Nyska A (1991). The toxicological significance of decreased activities of blood alanine and aspartate aminotransferase. Vet. Res. Commun., 15(1): 73-78.

Williams LR, Aroniadou-Anderjaska V, Qashu F, Finne H, Pidoplichko V, Bannon DI, Braga MFM (2010). RDX Binds to the GABAA Receptor-Convulsant Site and Blocks GABAA Receptor-Mediated Currents in the Amygdala: a Mechanism for RDX-Induced Seizures. Env. Health Perspect. (In Press).

Xenometrix (2009.) Ames MPF™ Penta I Microplate Format Mutagenicity Assay Instructions for use Available from http://www.aniara.com/pdf/literature/ICT-ANIARA-MPF-PENTA-AD.pdf. pp. 1-28.

8

Effects of parathion on cardiac rate, ventilatory frequency and hemolymphatic gas levels in the estuarine crab *Neohelice granulata* (Decapoda, Brachyura)

Daniel A. Medesani[1], Claudio O. Cervino[2], Martín Ansaldo[3] and Enrique M. Rodríguez[1]*

[1]Department of Biodiversity and Experimental Biology, FCEyN, Pab. II, University of Buenos Aires, Ciudad Universitaria, C1428EHA Buenos Aires, Argentina.
[2]School of Medicine, University of Morón, Machado 914, (1708) Morón, Pcia, Buenos Aires, Argentina.
[3]Argentine Antartic Institute, Cerrito 1248, 1010 Buenos Aires, Argentina.

Adult crabs *Neohelice granulata* were exposed for 96 h to 0.125 mg/L of parathion under complete submersion. Crabs were then exposed to air for 6 h (emersion period) to be finally re-submerged. Cardiac rate and ventilatory frequency, as well as hemolymphatic pH, pCO_2 and pO_2 were measured at different times throughout the experiment. Cardiac rate was significantly lower in crabs exposed to parathion, but only during the submersion periods, when cardiac activity is normally high. The insecticide could be affecting the neurological regulation of heart, avoiding the normal increase of cardiac activity. Ventilatory frequency, hemolymphatic, pH and gas parameters remained unaltered by exposure to 0.125 mg/L parathion, both under submersion and emersion.

Key words: Parathion, cardiac rate, crabs, submersion, emersion.

INTRODUCTION

The South American crab *Neohelice* (formerly *Chasmagnathus*) *granulata* (Decapoda, Brachyura, Varunidae) is an intertidal, semiterresterial species, widely distributed along the coast of Argentine and Brazil. As stated by Spivak (2010), this species has been extensively studied during the last 25 years, becoming a reference model for both physiological and toxicological research.

Intertidal crabs are daily exposed to the tidal flux, therefore being alternatively exposed to air and water. *N. granulata* actually breathes in both media; thus, several interacting mechanisms are involved for gas exchange and distribution, as in other crab species (Burnett, 1988;

Truchot, 1990; McMahon, 2001). Besides, gas exchange, circulation, ionic regulation and acid-base balance are processes strongly inter-related in crustaceans, especially in intertidal crabs (Böttcher and Siebers, 1993; Burnett, 1988; Bianchini et al., 2008). Since heart rate is influenced by several physiological processes, such as ventilation, excretion and osmoregulation, it has a direct relevance for both the organism performance and population health (Depledge et al., 1995). In several previous studies made on crustaceans, heart rate has been used as a physiological biomarker of pollution (Depledge et al., 1995; Lundebye et al., 1997; Brown et al., 2004). An increased heart rate was observed in the crab *Carcinus maenas,* exposed to copper (Bamber and Depledge, 1997; Camus et al., 2004). On the contrary, cardiac frequency decreased in *C. maenas* exposed to the organophosphate insecticide dimethoate (Lundebye et al., 1997).

*Corresponding author. E-mail: enrique@bg.fcen.uba.ar.

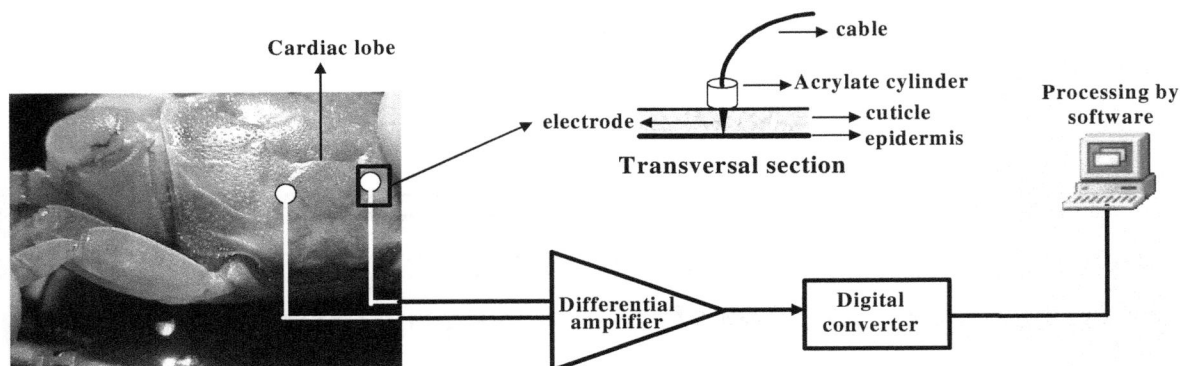

Figure 1. Diagram of the electrodes used for measuring cardiac rate. Data acquisition and processing are also indicated.

Parathion is an organophosphate insecticide extensively used in Argentina since several decades ago. It was detected above the permissible levels in the "Río de La Plata" estuary and its main affluent is the Paraná River (Administrative Commission of Río de La Plata, 1990; Lenardon et al., 1984). The mode of action of this highly toxic pesticide is based on the inhibition of the enzyme acetylcholinesterase (Verhaar et al., 1992). The toxicity of parathion and other organophosphate pesticides have been studied in several species of fish and crustaceans (Van der Oost et al., 2003; Hanazato, 2001; Rodríguez et al., 2007). In the studied species, the lethal toxicity of parathion, as well as several of their sublethal effects, especially on reproduction, has been previously reported (Rodríguez and Lombardo, 1991; Rodríguez et al., 1992; Rodríguez and Pisanó, 1993; Rodríguez et al., 1994; Bianchini and Monserrat, 2007). However, no studies concerning the effects of pesticides on the respiratory gas exchange or circulatory physiology have been made in the studied species, although some studies on the effects of heavy metals on several hemolymphatic parameters have been made (Rodríguez et al., 2001).

The current study was aimed at evaluating the effects of parathion on the ventilatory and cardiac frequencies of adult crabs *N. granulata* alternatively submerged and emerged. Some relevant hemolymphatic parameters, such as hemolymphatic pH, pCO_2 and pO_2, were also measured.

MATERIALS AND METHODS

Adult male crabs (mean wet weight = 13.85 ± 2.07 g, n = 100) were collected at Faro San Antonio beach, located at the southern end of the Samborombón Bay (36°18'S, 56°48'W). After carefully transporting the animals to the laboratory, they were maintained for two weeks at the same environmental conditions to be used later for bioassays: 12L:12D photoperiod (fluorescent light), temperature 20 ± 1 °C, pH 7.4 ± 1 and salinity 12 g/L prepared from dechlorinated tap water (total hardness: 80 mg/L as $CaCO_3$ equivalents) and salts for preparing artificial marine water (HW Marine Mix®). During the 2-week acclimation period, crabs were fed twice a week with commercially available pellets of rabbit food (Cervino et al., 1996).

Parathion stock solution at 1 g/L was made from pesticide technical grade (purity 99.9%, Compañia Química, Argentina). Pentaethylene nonylphenolate was used as solvent carrier, in equal proportion to the pesticide, before adding distilled water. This solvent has been used in previous studies, showing no lethal or sublethal effects on the studied species, at concentrations similar or even higher than that used in the current study (Rodríguez and Lombardo, 1991; Rodríguez et al., 1992, 1994; Rodríguez and Pisanó, 1993). A small aliquot of the stock solution was then diluted in 12 g/L saline water to achieve a final pesticide concentration of 0.125 mg/L, this concentration representing 25% of the parathion 96 h-LC50 for *N. granulata* adults, which were determined by Rodríguez and Lombardo (1991) as the incipient lethal threshold concentration for the studied species. A water dilution control was run, containing only the artificial saline water previously specified. A solvent control was also run, containing the solvent concentration presented in the parathion concentration. Ten crabs were randomly assigned to each treatment (parathion or controls), placing them in a glass aquarium of 24-L capacity filled with 3 L of 12 g/L saline water, prepared as previously specified. All solutions were renewed every 48 h. Crabs were exposed for 96 h to all solutions tested, following the specific procedure previously outlined by Rodríguez and Lombardo (1991).

After the 96-h exposure to parathion, every crab from each treatment was placed in the experimental device. It consisted of several single plastic containers with a plastic mesh screen floor separating the crabs from the bottom. In each container, a single crab was submerged in 500 ml of test solution (water dilution, solvent or parathion 0.125 mg/L), under continuous aeration. The bottom of each recipient was connected to a water reservoir by means of a plastic tube. By changing the relative position of the reservoir, the crab-recipient was slowly filled or emptied out, avoiding stress of animals.

The heart beats were recorded from the bioelectrical activity of the cardiac muscles fibbers, by means of two electrodes implanted in the carapace (Figure 1). This procedure allowed recording the electrocardiogram (EKG) in both submerged and emerged crabs. Each electrode was built with stainless steel wire (diameter 0.5 mm) included in a dental acrylate cylinder (diameter 5 mm, height 5 mm) leaving 1 mm of free-tip, and soldered to an electrically isolated copper cable. To fix the electrodes, crabs were cold-anesthetized and the pericardial membrane exposed, but not damaged, by carefully drilling through the dorsal carapace. The two active electrodes were adhered with cyanoacrylate glue at each side of cardiac lobe, while a third electrode was fixed in the gastric lobe and taken as ground reference.

The ventilatory frequency was determined by the movements of the scaphognathite (SGM), recorded by means of two copper wires (0.3 mm wide), electrically isolated with exception of the tip. The wires were introduced in the left branchial chamber, through the exhaling channel, until the electrode's tips touched the base of the scaphognathite.

For both cardiac rate and ventilatory frequency recordings, the free ends of the cables were connected to a differential amplifier, AC coupled (Figure 1). The transducer signals were amplified and filtered (bandpass 1 to 40 Hz). Data were acquired and digitized with a sample frequency of 64 Hz, and later processed by Rhythms (Stellate Systems®) software.

All the measurements were performed in a soundproof room, built as a Faraday cage and maintained at $20 \pm 1°C$. All recordings started at 10 AM. Crabs were kept under complete submersion in the apparatus during 1 h before starting the records, to acclimate them to the experimental device, according to previous studies (Cervino et al., 1995). A first recording of EKG and SGM was made at this initial submersion (SUB1). Next, the test solution was drawn off and the crabs were exposed to air for 6 h, the expected duration of low tide in the natural environment of the studied species. During this emersion period (EM), records were taken at 15, 30, 45, 60, 90, 120, 180, 240, 300 and 360 min. Finally, the recipients of crabs were completely re-filled with the corresponding test solution, and new records were taken at 15 and 30 min after re-submersion (SUB2). In all cases, the duration of each record was 5 min. The cardiac and ventilatory frequencies were determined by counting the wave-spike complex of EKG and waves of SGM in a minute.

A replicate of all treatments was assigned for determination of hemolymphatic pH and partial pressures of gases (pCO_2 and pO_2). Hemolymph samples were taken from 10 different crabs at the base of the fourth pereiopods by means of a capillary tube, at the same times previously specified; pH, pCO_2 and pO_2 were analyzed by means of a Radiometer BMS3 Mk2 Blood micro system held at $20°C$.

Two-way ANOVA (treatment, considered as independent factor, and period, considered as a repeated measure factor) was applied to compare mean values. Tukey's test was used for multiple comparisons of means (Sokal and Rohlf, 1981). A 5% confidence level was always considered.

RESULTS

No mortality was recorded in any treatment, during the 96-h exposure period. Mean values of both cardiac (CF) and ventilatory (VF) frequencies for each considered period (SUB1, EM and SUB2) are shown in Figure 2A and B. Due to their similarity, data from 0 to 60 min of the emersion period were averaged (emersion 1 period: EM1), while the same calculation was applied to data from 90 to 360 min (emersion 2 period: EM2).

Crabs from both water dilution control (WDC) and solvent control (SC) showed, during the EM2 period, a CF significant ($p < 0.05$) lower than during both submersion periods. This pattern was considered as the normal response. On the contrary, CF of crabs exposed to

parathion remained constant throughout the experiment. During both the SUB1 and SUB2 periods, a significantly ($p < 0.05$) lower CF was detected in parathion-exposed crabs compared with either WDC or SC crabs, while no differences ($p > 0.05$) were observed between parathion and either control during both emersion periods (Figure 2A).

No significant differences ($p > 0.05$) between either control or parathion-exposed crabs were noted in the VF throughout the experiment (Figure 2B). For all the treatments, a significant ($p < 0.05$) decrease was observed at the beginning of the EM1 period (Figure 2B). The VF values of the SUB2 period were similar ($p > 0.05$) to those of the SUB1 period. No significant differences ($p > 0.05$) were noted between treatments or between exposure periods, in the hemolymphatic pH, pCO_2 or pO_2, at any considered time (Figures 2C to E).

DISCUSSION

The value of cardiac rate for *N. granulata* control crabs during submersion (about 130 beats/min) was similar to that of the aquatic crab *Callinectes sapidus*, measured at 22 to $25°C$ (DeFur and Mangum, 1979). Besides, a significant bradycardia developed in *N. granulata* control crabs exposed to air, but just after 60 min of exposure. This later result was in accordance with previous reports made for crabs that typically live in aquatic and intertidal environments (Burggren and McMahon, 1988). The bradycardia observed in *N. granulata* was also in accordance with the metabolic depression suggested for the same species during emersion, concomitant with a reduction in both O_2 consumption and locomotive activity (Schmitt and Santos, 1993).

A significant decrease in cardiac frequency was noted in submerged crabs exposed for 96 h to 0.125 mg/L of parathion, compared to controls. Moreover, this decrement was not associated with changes in ventilatory frequency, pH or partial pressures of gases. Therefore, a direct effect of parathion on cardiac physiology seems to be plausible. A correlation between the inhibition of acetylcholinesterase activity and the reduction of heart rate was observed in the crab *C. maenas* exposed to the organophosphorous pesticide dimethoate, suggesting that a disruption in the normal nervous control of heart is taking place (Lundebye et al., 1997).

In crustaceans, the nervous regulation of cardiac rate involves several pathways. First, the neurons of cardiac ganglion, attached to the neurogenic heart, are acting as pacemakers, some of them being cholinergic (Lundebye et al., 1997). Second, both excitatory and inhibitory nerves have been described to innervate the heart (McMahon, 2001), and third, a cholinergic pathway originated from thoracic ganglion regulates the secretory activity of the pericardial organ (Atwood, 1982). Several

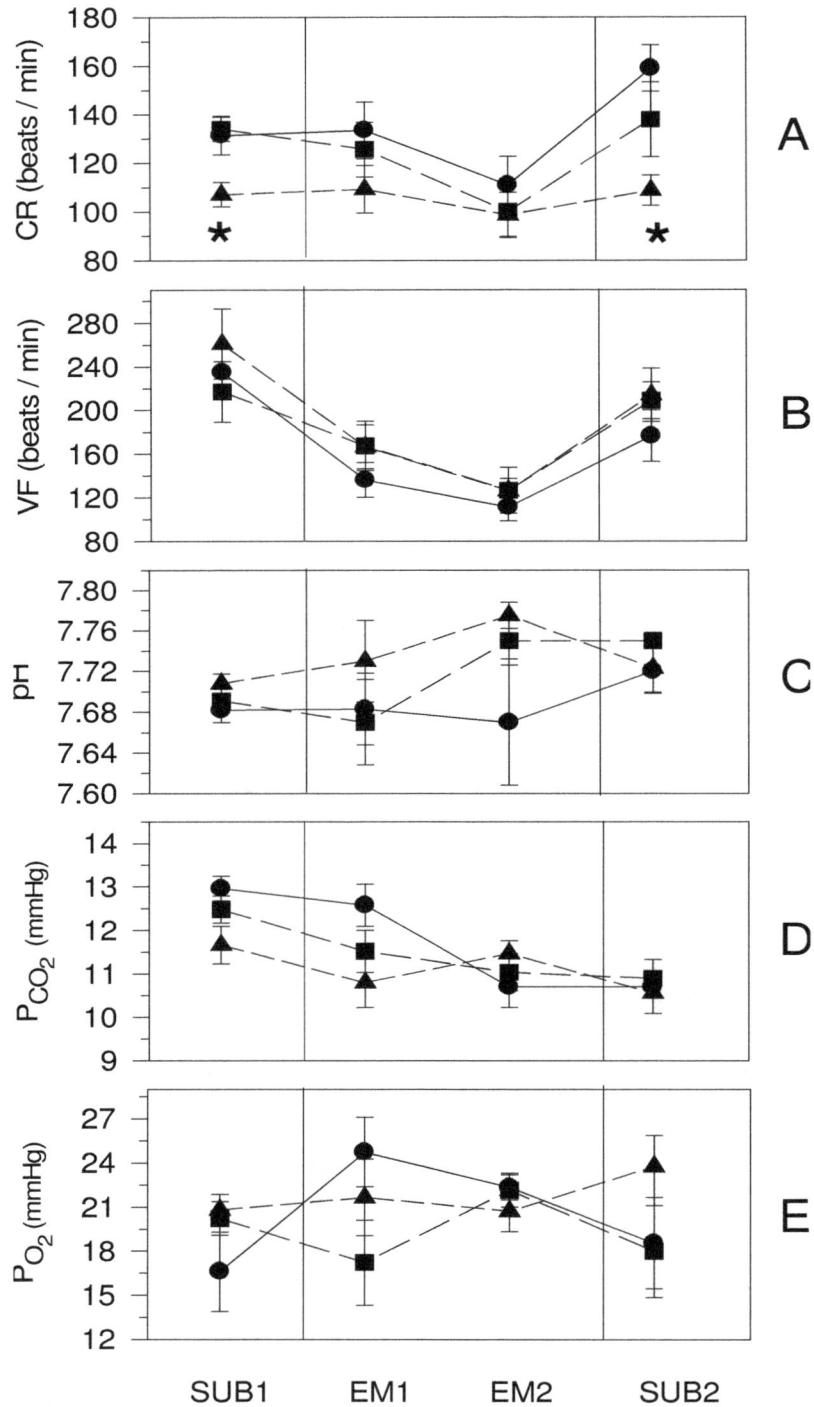

Figure 2. Mean values (± standard error) of (A) Cardiac rate (CR), (B) Ventilatory frequency (VF), (C) pH, (D) PCO_2 and (E) PO_2 during submersion and emersion periods. Each mean represents ten crabs. SUB1: First (initial) period of submersion; EM1: First period of emersion (0 to 60 min); EM2: Second period of emersion (90 to 360 min); SUB2: Second (final) period of submersion. Circles: Water dilution control; squares: Solvent control; triangles: Parathion at 0.125 mg/L. Asterisks indicate significant differences (p < 0.05) between parathion and either water dilution or solvent control.

cardioexcitatory amines are secreted by the pericardial organ of crabs, such as serotonin, octopamine and also the crustacean cardioactive peptide, all of them with a stimulatory effect on cardiac rate (Atwood, 1982; Wilkens, 1987; McMahon, 2001). One or several of these pathways could be affected by the anticholinesterase effect of parathion, producing an impairment of the normal heart rate, which becomes evident when it is to rise during submersion. The effect of parathion of reducing CF would therefore be relevant when, for instance, the mentioned biogenic amines are being actively secreted, that is, during the submersion periods, but not during emersion, when CF normally decreases.

Depledge et al. (1995) have suggested that physiological variables measured on crabs, such as heart rate and ventilatory frequency, can be potentially used as biomarkers of pollution, since they are integrated responses which reflect the overall performance of the animal, also showing to be sensitive under the exposure to several pollutants. As instance, Bamber and Depledge (1997) have reported a consistent increase in the heart rate of the crab C. maenas exposed to copper, suggesting that such increased rate is related to respiratory stress, rather than to a direct effect of copper on heart. The increase of heart rate caused by copper has been also reported, in the same species, to be enhanced by either low or high temperature (Camus et al., 2004). An increased heart rate during daytime was also reported for both the crab Potamon potamios and the crayfish Astacus astacus exposed to mercury (Styrishave and Depledge, 1996).

Concerning hydrocarbons, no significant effects on heart rate of adult C. maenas have been detected at either relatively high doses of benzo[a]pyrene (Bamber and Depledge, 1997), or pyrene (Dissanayake et al., 2008). As for pesticides, Lundebye et al. (1997) have reported, as mentioned before, a decrease in heart rate of crabs C. maenas exposed to the organophosphorous insecticide dimiethoate, in correlation with the inhibition of acetylcholinesterase. Our results with parathion also associate a decreasing heart rate in submerged crabs with an organophosphate intoxication, supporting the idea that this physiological variable could be a specific biomarker for identifying the exposure to acetylcholinesterase inhibitors. In this respect, since parathion exerts an irreversible inhibition on acethylcholinesterase, further assays to determine the degree and time of reversion for heart rate of crabs transferred to clean water would be relevant.

Gill damage was observed in adult crabs N. granulata exposed during 96 h to 0.25 mg/L of parathion; concomitantly, hemolymphatic pCO_2 and lactate were increased due to the impairment of CO_2 and O_2 exchange at gills, leading to a blood acidosis of both respiratory and metabolic origin (Medesani et al., unpublished). In the current study, though, the hemolymphatic pH and partial pressures of gases remained unaffected from one treatment to another, regardless of the submersion state and the exposure to 0.125 mg/L of parathion. The insecticide did not cause changes in ventilatory frequency in any case, in good agreement with absent of effect observed on the hemolymphatic parameters. Therefore, the current study has evaluated the effect of parathion on heart rate and ventilation, at a concentration that allowed discarding the possible modifying effect that hemolymphatic gas level and/or pH could have on those physiological variables.

ACKNOWLEDGMENTS

This work was supported by grants from the University of Buenos Aires, CONICET and ANPCyT.

REFERENCES

Administrative Commission of Río de La Plata (1990). Study for the evaluation of pollution in the Río de La Plata. Advanced Report. Argentine Hydrographic Navy Service, Buenos Aires, pp. 317-382.

Atwood HL (1982). Synapses and neurotransmitters. In: Bliss D (ed.), The Biology of Crustacea. Academic Press, New York, 3: 105-150.

Bamber SD, Depledge MH (1997). Evaluation of changes in the adaptive physiology of shore crabs (Carcinus maenas) as an indicator of pollution in estuarine environments. Mar. Biol., 129: 667-672.

Bianchini A, Machado Lauer M, Maia Nery LE, Colares EP, Monserrat JM, Santos EA (2008). Biochemical and physiological adaptations in the estuarine crab Neohelice granulata during salinity acclimation. Comp. Biochem. Physiol., 151A: 423-436.

Bianchini A, Monserrat JM (2007). Effects of methyl parathion on Chasmagnathus granulatus hepatopancreas: Protective role of Sesamol. Ecotoxicol. Environ. Saf., 67: 100-108.

Böttcher K, Siebers D (1993). Biochemistry, localization and physiology of carbonic anhydrase in the gills of euryhaline crabs. J. Exp. Zool., 265: 397-409.

Brown RJ, Galloway TS, Lowe D, Browne MA, Dissanayake A, Jones MB, Depledge MH (2004). Differential sensitivity of three marine invertebrates to copper assessed using multiple biomarkers. Aquat. Toxicol., 66: 267-278.

Burggren WW, McMahon BR (1988). Circulation. In: Burggren WW, McMahon, BR (eds.), Biology of the land crabs. Cambridge University Press, New York, pp. 298-332.

Burnett LE (1988). Physiological responses to air exposure: Acid-base balance and the role of branchial water stores. Am. Zool., 28: 125-135.

Camus L, Davies, PE, Spicer JI, Jones MB (2004). Temperature-dependent physiological response of Carcinus maenas exposed to copper. Mar. Environ. Res., 58: 781-785.

Cervino CO, Luquet CM, Haut GE, Rodriguez EM (1996). Salinity preferences of the estuarine crab Chasmagnathus granulata DANA, 1851 after long-term acclimation to different salinities. Atlantica, 18: 69-75.

Cervino CO, Medesani DA, Rodriguez EM (1995). Effects of feeding on metabolic rate of the crab Chasmagnathus granulata (Decapoda, Brachyura). Nauplius, 3: 155-162.

DeFur PL, Mangum CP (1979). The effects of environmental variables on the heart rates of invertebrates. Comp. Biochem. Physiol., 62A: 283-294.

Depledge MH, Aagaard A, Gyorkos P (1995). Assessment of trace metal. Toxicity using molecular, physiological and behavioural biomarkers. Mar.Poll. Bull., 31: 19-27.

Dissanayake A, Galloway TS, Jones MB (2008). Physiological responses of juvenile and adult shore crabs *Carcinus maenas* (Crustacea: Decapoda) to pyrene exposure. Mar. Environ. Res., 66: 445-450.

Hanazato T (2001). Pesticide effects on freshwater zooplankton: an ecological perspective. Environ. Poll., 112: 1-10.

Lenardon A, Hevia MIH, Fuse JA, Nochetto, CB, Depetris, PJ (1984). Organochlorine and organophosphorous pesticides in the Parana River (Argentina). Sci. Total Environ., pp. 289-297

Lundebye K, Curtis TM, Braven J, Depledge MH (1997). Effects of the organophosphorous pesticide, dimethoate, on cardiac and acetylcholinesterase (AChE) activity in the shore crab *Carcinus maenas*. Aquat. Toxicol., 40: 23-36.

McMahon BR (2001). Control of cardiovascular function and its evolution in Crustacea. J. Exp. Biol., 204: 923-932.

Rodríguez EM, Lombardo RJ (1991). Acute toxicity of parathion and 2,4-D to estuarine adult crabs. Bull. Environ. Contam. Toxicol., 46: 576-582.

Rodríguez EM, Monserrat JM, Amin OA (1992). Chronic toxicity of ethyl parathion and isobutoxyethanol ester of 2,4-dichlorophenoxyacetic acid to estuarine juvenile and adult crabs. Arch. Environ. Contam. Toxicol., 22: 140-145.

Rodríguez EM, Pisanó A (1993). Effects of parathion and 2,4-D to eggs incubation and larvae hatching in *Chasmagnathus granulata* (Decapoda, Brachyura). Comp. Biochem. Physiol., 104C: 71-78.

Rodríguez EM, Schuldt M, Romano L (1994). Chronic histopathological effects of parathion and 2,4-D on female gonads of *Chasmagnathus granulata* (Decapoda, Brachyura). Food Chem. Toxicol., 32: 811-818.

Rodríguez EM, Bigi R, Medesani DA, Stella VS, López LS, Rodriguez Moreno PA, Monserrat JM, Pellerano GN, Ansaldo M (2001). Acute and chronic effects of cadmium on blood homeostasis of an estuarine crab, *Chasmagnathus granulata*, and the modifying effect of salinity. Braz. J. Med. Biol. Res., 34: 509-518.

Rodríguez EM, Medesani DA, Fingerman M (2007). Endocrine disruption in crustaceans due to pollutants: A review. Comp. Biochem. Physiol., 146A: 661-671.

Schmitt ASC, Santos EA (1993). Lipid and carbohydrate metabolism of the intertidal crab *Chasmagnathus granulata* Dana, 1851 (Crustacea, Decapoda) during emersion. Comp. Biochem. Physiol., 106A: 329-336.

Sokal RR, Rohlf FJ (1981). Biometry, 2nd ed. Freeman, New York, pp. 195-279.

Spivak ED (2010). The crab *Neohelice* (=*Chasmagnathus*) *granulata*: an emergent animal model from emergent countries. Helgol. Mar. Res., 64: 149-154.

Styrishave B, Depledge MH (1996). Evaluation of mercury-induced changes in circadian heart rate rhythms in the freshwater crab, *Potamon potamios* and the crayfish, *Astacus astacus* as an early predictor of mortality. Comp. Biochem. Physiol., 115A: 349-356.

Truchot JP (1990). Respiratory and ionic regulation in invertebrates exposed to both water and air. Annu. Rev. Physiol., 52: 61-76.

Van der Oost R, Beyer J, Vermeulen NPE (2003). Fish bioaccumulation and biomarkers in environmental risk assessment: A review. Environ. Toxicol. Pharmacol., 13: 57-149.

Verhaar HJM, Van Leeuwen CJ, Hermens JLM (1992). Classifying environmental pollutants. Chemosphere, 25: 471-491.

Wilkens JL (1987). Cardiac and circulatory control in decapod Crustacea with comparisons to molluscs. Experientia, 43: 990-994.

9

Toxicity of Aluminum sulphate (alum) to Nubian goats

Medani A. B.[1], El Badwi S. M. A.[2]* and Amin A. E.[2]

[1]Department of Pharmacology and Toxicology, Khartoum College of Medical Sciences, Sudan.
[2]Department of Pharmacology and Toxicology, Faculty of Veterinary Medicine, University of Khartoum, Sudan.

This study was conducted to investigate the toxic effect of Aluminum sulphate (alum). Nine Nubian goat kids were divided into 3 groups, each of three goats. Group 1 animals were the undosed controls. Test groups were given alum at dose rates of 1 and 20% respectively for groups 2 and 3 for a period of 10 weeks after an adaptation period of two weeks during which the animals were X kept under ideal experimental conditions. Clinical signs were closely observed with postmortem and histopathological examinations. Chemical investigations included enzymatic activities of ALP, AST, CK, ALT and LDH and metabolic changes of albumin, urea, total protein, cholesterol, bilirubin, glucose and creatinine were detected. Fluctuations in electrolyte levels of Mg, Fe, Na, K, Ca and P were monitored together with hematological changes in Hb, PCV, RBCs and WBCs. Mortalities occurred to variable degrees irrespective to the dose level. On alum challenge, the test species showed clinical signs of low voice, inappitence, dullness, whitish salivation, watery diarrhea and also recumbency. On atomic absorption only the lungs kept residual alum, while the livers washed out the substance, maybe via bile. Notably oral dosing with alum caused changes consisted of congested liver with white spots, stiff-greenish lungs and inflammed empty intestines. The un-dosed group 1 goats showed a normal picture. On histopathology, alum-dosed group of goats showed necrosis in the cortex and medulla of the kidney in one group member, emphysema in the lungs and necrosis in the hepatocytes and congestion in the liver in all group members. On evaluation of the previous results, alum was considered toxic to Nubian goat kids at all tried dose rates. Practical implications of the results were highlighted as suggestions for future work were put forward.

Key words: Toxicity, alum, Nubian goats, drinking water.

INTRODUCTION

The treatment of waters to make them suitable for subsequent use requires physical, chemical and biological processes such as distillation, gas exchange, coagulation, sedimentation, filtration, adsorption, iron exchange and disinfection (WHO, 1984).

Polymer (polyDADMAC) is one of the synthetic cationic polyelectrolytes which are used widely in drinking water treatment (DWI, 2001, 2002) and historically dirty water is cleaned by treating with alum and lime. Despite the toxicity, alum is known to improve floc formation yielding a large size of fast-sitting rate, improves treated water quality through reducing suspended solids and turbidity, filter-runs, cost effective in use and is widely used to precipitate phosphate in industrial effluent treatment plant (AWWA, 1997).

In Sudan, the first use of chemicals to reduce the turbidity of the Nile water, especially during the flood season, to standardize it for the healthy human consumption with a maximum allowable level (MAL) of 150 mg/L, was at 1925 (Mohammed, 1998). The disposal system to eliminate the outcome of the chemical sludge that happen during treatment of the turbid water, is not adjusted well to the laws of environmental health regulations. Throughout the time alum did not give satisfactory results in the reduction of turbidity of water, especially during the flood season. As Sudan had tried comprehensive drinking water treatment using alum for ages and considering the lack for toxicological data in this area, this experiment was done to explore the risk factor by testing alum in goats for toxicological effects.

*Corresponding author. E-mail: samiaelbadwi@yahoo.com.

MATERIALS AND METHODS

Animals, housing and management

Nine male Nubian goat kids (5 to 7 months old) were purchased from El Sheikh Abu Zeid, a local livestock Market in the vicinity of Omdurman and kept within the premises of the Veterinary Teaching Hospital, Faculty of Veterinary Medicine, University of Khartoum. During the 2 week adaptation period, animals were ear tagged and given prophylactic doses of Oxytetracycline 5% (Bremerpharma, Germany) and Sulphamethazine 33.3% (Norbrook, U.K.) against bacterial infections and coccidiosis respectively and fed *ad libitum* on lucerne and allowed free access to Nile water.

Administration and dose levels of alum

At the end of adaptation period, animals were weight-distributed, and allotted randomly to three groups, each of three goats. Goats of (Group 1) were left un-dosed (controls).

Goats of (Group 2) were treated daily (by the oral route) with alum in concentration of 1% in drinking Nile water while goats of (Group 3) were given alum in a concentration of 20%. Goats of (Group 1) were left un-dosed (controls).

Parameters

Clinical signs and mortality rates were recorded. Blood samples were obtained from the jugular vein before the start of the experimental dosing and thereafter fortnightly for haematological investigations and serum analysis.

Haemoglobin concentration (Hb), packed cell volume (PCV), red blood cell (RBC) and white blood cell (WBC) counts were estimated.

Sera were analyzed for the activities of ALP, AST, CK, GPT and LDH and also for the concentrations of metabolic indicators cholesterol, creatinine, bilirubin, uric acid, urea, albumin, total protein and glucose and also electrolytes calcium, inorganic phosphorus, iron, sodium and potassium.

METHODS

Haematological methods

These methods were described by Schalm (1965). Blood samples from goats were collected into clean dry bottles containing the anti-coagulant heparin from the jugular vein. The concentration of haemoglobin was determined by the cynomethaemoglobin technique by Drabkins solution in g/dl of blood. Fresh blood samples were centrifuged in a micro haematocrit centrifuge to read off the packed cell volume percentage. Blood cells and the white blood cells were counted with an improved Neubauer haemocytometer (Hawksley and Sons Ltd., England).

Histological methods

The specimens were collected immediately after death or slaughter and fixed in 10% formal saline, embeded in paraffin wax, sectioned at 5 μm and stained with haemotoxylin and eosin (H & E) using Mayer's haemalum.

Chemical methods

Blood samples were obtained from the jugular vein before and after dosing with $AlSO_4$. Venous blood samples were centrifuged at 3000 r.p.m. for 5 min and stored at -20°C until analyzed for LDH, AST, ALP, CK, cholesterol, creatinine, total bilirubin, urea, total protein, calcium, albumin, phosphorus, iron and magnesium by a colorimetric method using a commercial kit (Randox Laboratories Ltd., U.K).

Determination of Na^+ and K^+

Samples were diluted in 100 ml of distilled water compared with Na^+ and K^+ standard (Sherwood Scientific Ltd. UK) with concentration of 140 and 50 mmol/l respectively). Concentrations were detected by Flame, using Bio-dynamics Lyteteck Flame Photometer, USA.

Statistical methods

The difference between mean values of data were analysed by the un-paired students- t-test (Snedecor and Cochran, 1989).

RESULTS

Clinical signs

The most obvious signs in the 1% alum-dosed goats included low voice, nervous signs, inappitance, recumbency and death. The 20% alum-dosed goats showed watery involuntary diarrhea followed by dullness, shivering, salivation, inappitence, isolation from the herd and final recumbency, with a 100% mortality rate. Deaths started in the 1% alum-dosed goats in the 9th day up to the 13th, while that of the 20% alum-dosed goats started at the 1st day up to the 5th day of treatment. The control un-dosed goats (group 1) were normal.

Post-mortem changes

In the 1% alum-dosed group of goats, changes consisted of congested liver with white spots, stiff-greenish lungs and inflamed empty intestines. On post-mortem of the 20% alum-dosed group of goats, most livers were congested and were spotted with white foci, lungs were stiff and greenish and intestines were inflamed and empty. The un-dosed group 1 goats showed a normal picture.

Histopathological picture

On histopathology, the 1% alum-dosed group of goats showed necrosis in the cortex and medulla of the kidney in one group member, emphysema in the lungs and necrosis in the hepatocytes and congestion in the liver in all group members (Figure 1).

Lung sections of the 20% alum-dosed group of goats showed emphysema and the intestines were edematous with catarrhal inflammation, hearts slightly necrotic, spleens slightly congested and the liver showed degenerative necrosis and lymphocytic infiltration. No

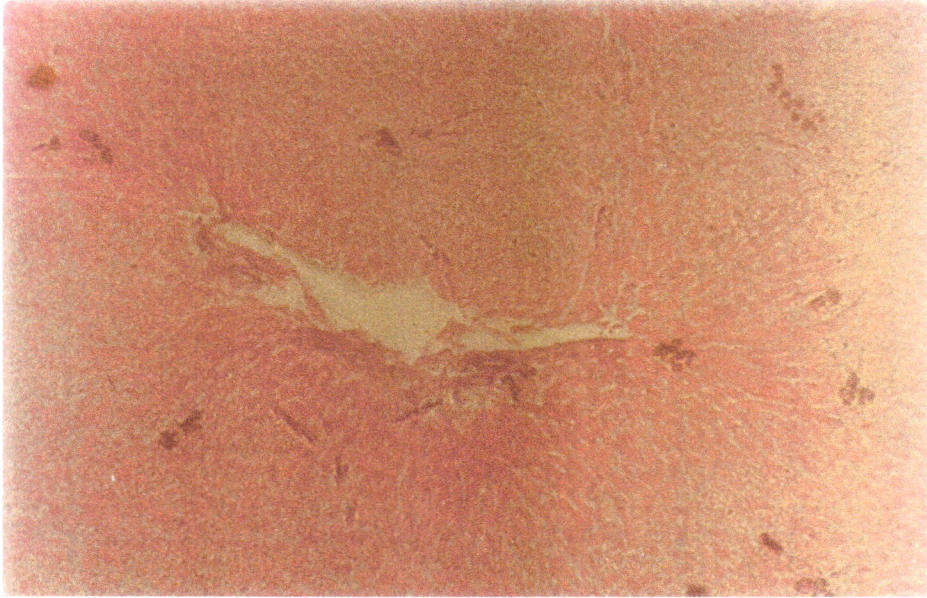

Figure 1. Necrosis in the hepatocytes of goats dosed with 1% alum in drinking water (H & E) X 100.

Table 1. Average activities (mean ± SD) of serum enzymes of the alum-dosed goats.

Group / Dose	ALP (iu / l)	1AST (iu/ l)	CK (iu/ l)	ALP (iu/ l)	LDH (iu/ l)
G₁ un-dose	151.37 ± 11.9	33.814 ± 43	1.90 ± 0.26	18.37 ± 1.62	40.80 ± 0.35
G₂ (1% solution)	101.96 ± 1.73*	24.17 ± 1.77*	3.31 ± 0.09*	24.73 ± 1.77*	67.31 ± 0.62***
G₃ (20% solution)	106.23 ± 7.98*	21.20 ± 0.00*	2.33 ± 0.29*	27.67 ± 2.06*	61.73 ± 0.31***

* denotes P<0.05 *** denotes P<0.001

abnormal histopathological changes were observed in the un-dosed group of goats.

Changes in the activities of serum enzymes

Table 1 shows the changes in serum constituents of goats treated with alum.

Clearly decreased values (P<0.05) were observed in both test groups when their serum activities of ALP and AST were evaluated. Values of CK, and ALT were higher (P<0.05) when compared to the control, and these values were highly (P<0.001) increased for LDH compared to the control. The control group showed normal activities of the serum enzymes.

Serum metabolites values

Table 2 summarizing the changes in serum metabolite in blood of goats treated with alum. Both test groups values

for albumin, total protein, and cholesterol, were found significant in comparison to the un-dosed group, while significant increases (P<0.01 to 0.001) were observed when the urea serum level was evaluated. Marked increases (P<0.05 to 0.01) were detected on evaluating creatinine. Goats of group 2 and 3 showed significant increase and decrease (P<0.05) when investigating both bilirubin and glucose respectively, when compared to the control group 1 which showed values that were within the normal ranges for all the metabolites tested.

Changes in serum electrolytes

The changes in serum electrolytes are summarized in Table 3. Significant decreases (P< 0.01) in magnesium concentrations were seen in both groups. Goats of group 2 and 3 showed significant decreases (P<0.05 to 0.001) in their serum iron, sodium, potassium, and phosphorus, whereas for Ca, group 2 estimates were insignificant (P>0.05) and group 3 estimates were decreasingly

Table 2. Average values (mean ± SD) of serum metabolites of the alum-dosed goats.

Group / Dose	Albumin (g/dl)	Urea (mg/dl)	Total protein (g/dl)	Cholesterol (mg/dl)
G_1 un-dosed	3.01 ± 0.37	35.35 ± 4.37	5.47 ± 0.75	38.10 ± 3.59
G_2 (1% solution)	3.92 ± 0.27[N.S]	76.70 ± 1.42[***]	4.12 ± 1.21[N.S]	40.82 ± 10.78[N.S]
G_3 (20% solution)	3.17 ± 0.40[N.S]	75.47 ± 8.19[**]	3.56 ± 0.60[N.S]	34.57 ± 6.28[N.S]

Group /Dose	Bilirubin (mg/dl)	Glucose (mg/dl)	Creatinine (mg/dl)
G_1 un-dosed	0.18 ± 0.13	31.08 ± 0.87	0.21 ± 0.04
G_2 (1% solution)	0.84 ± 0.15*	5.93 ±10.28*	0.50 ± 0.16**
G_3 (20% solution)	0.89 ± 0.14*	6.11 ± 1.53*	0.44 ± 0.05*

NS = Not significant * denotes P<0.05 ** denotes P<0.01 *** denotes P<0.001.

Table 3. Average values (mean ± SD) of serum electrolytes of the alum -dosed goats.

Group / Dose	Mg (mg/dl)	Iron (µg/dl)	Na (mg/dl)	K (mg/dl)	Ca (mg/dl)	P (mg/dl)
G_1 (un-dosed)	1.44 ± 0.08	228.77 ± 13.73	152.53 ± 0.42	5.69 ± 0.12	6.92 ± 1.44	4.14 ± 0.11
G_2 (1% solution)	0.70 ± 0.01[**]	165.87 ± 0.31*	121.80 ± 0.36[***]	5.03 ± 0.01**	3.70 ± 0.01[N.S]	2.21 ± 0.50**
G_3 (20% solution)	0.27 ± 0.02[**]	62.60 ± 4.68**	30.73 ± 0.46[***]	1.34 ± 0.02***	1.11 ± 0.03*	0.59 ± 0.07***

NS = Not significant · denotes P<0.05 ** denotes P<0.01 *** denotes P<0.001.

Table 4. Average haematological values (mean ± SD) of the alum -dosed goats.

Group / Dose	Hb (g %)	PCV %	RBCs (x10[6])	WBCs (x10[3])
G_1 (un-dosed)	8.13 ± 0.57	26.40 ± 1.97	8.57 ± 0.87	6.36 ± 0.67
G_2 (1% solution)	7.18 ± 1.29[N.S]	22.53 ± 0.70**	4.96 ± 0.01**	7.80 ± 0.01[N.S]
G_3 (20% solution)	7.37 ± 0.66[N.S]	20.11 ± 0.19**	4.67 ± 0.59**	7.67 ± 1.97[N.S]

X NS = Not significant ·· denotes P<0.01.

significant (P<0.05). The serum levels of the afore-mentioned electrolytes in the control group were normal.

Haematological values

Table 4 shows the heamatological values of goats dosed with alum. Concentrations of haemoglobin of both test groups were not deviated (P>0.05) from those of the control group, while those of PCV and RBCS were decreased (P<0.01). WBCS counts significantly decreased (P<0.01) in group 3, while counts of WBCS of group 1 and 2 were not different (P>0.05) compared with the control group goats which were of normal values.

DISCUSSION

Daily routine dosing of alum revealed marked nervous system involvement including frenzies, salivation, shivering involuntary watery diarrhea and finally recumbency (Verstraeten et al., 2008). The whole picture was showing a parasympathetic involvement (Orshoven et al., 2006) whereas the low voice was indicating the local irritant effect of alum on the vocal folds (IDSP, 1980; Barthold, 1996). In addition to the high rate of mortality in ruminant animals, the necrotic intoxicated lung, proved on atomic absorption by the high pulmonary levels of aluminum (Bernardo et al., 2009), marked pulmonary affections were probably contributed to the development of dysnoea which support the hypothesis that alum is harmful specially with regard to lung fibrosis (Parmeggiani, 1983; Sjogren et al., 1983 and Morgan and Dinman, 1989). On atomic absorption of the processed liver - alum values were negligible, showed decreased activity in alkaline phosphatase (ALP) in serum as well as the absence of bilirubinaemia which suggests that alum is not expected to interfere with the excretory ability of the liver cells (Ford, 1963; Aspenstrom-Fagerlund et al., 2009). Necrotic hepatocytes in combination with the rise in some serum activity are suggests hepatocellular damage (Adam et al., 1973; Dafalla and Adam, 1986; Dafalla et al., 1987; Ayed et al., 1991; Kew, 2000). Renal insufficiency was indicated by increase in urea, creatinine, total protein, decrease in albumin concentrations and necrotic, haemorrhagic injured renal tubules (Ahmed and Adam, 1979). Intestinal wall which was spotted with white (probably with alum causing focal enteritis) was greatly affected with the irritant alum and/or its metabolites. When the resin is precipitated by alum in its preparation, commonly the salt intensifies its action and the cream of tartar increases the hydragogue effect (Felter, 1922). This action was very clear on the congested mesenteric blood vessels and symptomatically, by diarrhea and salivation due to nausea. Increased rates of mortality, are proportional with the dose of alum in test goats, may be due to myocardial and nervous system (CNS) involvement (mostly the parasympathetic). Pulmonary haemorrhages, oedema, emphysema, necrosis and adhesions may cause the clinically observed difficulty in respiration which was considered as one of the irritant effects of the toxicant and/or its metabolites. Hepatic damage was manifested in the congested and necrotic central vein, fatty changes, increased level of bilirubin, increased activities of liver enzymes in the serum (Ford et al., 1972; Adam, 1972; Adam et al., 1973). The irritant substance or its metabolites were incriminated for GIT affections manifested as nausea, salivation, vomiting, watery diarrhea and necrosis of muscularis mucosae (Orshoven et al., 2006). The same effects may also be due to increased peristaltic movement, cholinesterase inhibition or both (Michael, 2000). Hypocalcaemia is similar to that described by Ali and Adam (1978), El Dirdiri et al. (1987), Galal and Adam (1988) and Mohamed and Adam (1992) in young ruminants. This was accompanied by a generalized fall in serum electrolytes due to vomiting and diarrhea drain. Myopathy expressed in reluctance to move was sequel to an increased activity of AST (Sousa et al., 2002).

Lymphocyte infiltration probably suggests the development of suppression in the immune system of the poisoned goats. The haemorrhage and necrotic kidneys that were caused by the direct irritant effect of the polymer and/or its metabolites, is indicating renal insufficiency. The additive effect of the increased creatinine, urea AST concentrations in serum indicated damage to renal tubules. The available results and the studies of Mohamed and Adam (1992) and Ahmed and Adam (1979) suggested that serum urea level was of significance in the evaluation of renal toxicity in goats.

Alum is neurotoxic, cardiotoxic and lethal chemical in overdoses, a hepatotoxic, neuro toxic and nephrotoxic chemical even in mild overdoses. It can be deposited in lungs, the dose used routinely + 1.5 mg/l is also toxic.

REFERENCES

Adam SEI (1972). A review of drug hepatotoxicity in animals. Vet. Bull., 42(11): 683-689

Adam SEI, Tartour G, Obeid HM, Idris OF (1973).Effects of Ipomea carnea on the liver and serum enzymes in young ruminants. J. Comp. Pathol., 83(3): 351-356.

Ahmed OM, Adam SE (1979). Effects of Jatropha curcas on calves. Vet. Pathol., 16(4): 476-482.

Ali B, Adam SEI (1978). Effects of Acanthospermum hispidum on goats. J. Comp. Pathol., 88(4): 533-544

Aspenstrom-Fagerlund B, Sundstrom B, Tallkvist J, Ilbak NG, Glynn AW (2009). Fatty acids increase paracellular absorption of aluminium across Caco-2 cell monolayers. Chem. Biol. Interact., 181(2): 272-278. [Medline].

AWWA (1997). History of Alum Use for Water Treatment,AWWA main strem. http:// www.awwa.org, 1: 143

Ayed IAM, Dafalla R, Yagi AI, Adam SEI (1991). Effect of ochratoxin on lohmann-type chicks. Vet. Hum. Toxicol., 33(6): 557-560

Barthold SW (1996). Chronic Lyme Disease: Basic Science and Clinical Approaches:Abstracts of Presentations. 9th Annual International Conference on Lyme Borreliosis and Other Tick-borne Disorders . Lyme Disease Foundation Inc. LDF Conference Abstract - Boston, USA, 74(1): 57–67

Bernardo JF, Edwards MR, Barnett B (2009). Toxicity, Aluminum. Emergency medicine, Medicine Specialists, Toxicology. http://emedicine.medscape.com/article/165315-overview.

Dafalla R, Adam SEI (1986). Effects of various levels of dietary selenium on hypro-type chicks. Vet. Hum. Toxicol., 28(2): 105-108

Dafalla R, Yagi IA, Adam SEI (1987). Experimental aflatoxicosis in Hypro-type chicks: Sequential change in growth and serum constituents and histopathological changes. Vet. Hum. Toxicol., 29(3): 222-226.

DWI (2002). DWI Information Letter: DWI Regulation 25 Letter 1/2001. Department for Enviroment, Food and Rural Affairs (DEFRA), The Drinking Water Inspectorate (DWI). dwi.enquiries@defra.gsi.gov.uk, Ashdown House, London.

El Dirdiri NI, Barakat SEM, Adam SEI (1987). The combined toxicity of Aristolochia bracteata and Cadaba rotundifolia to goats. Vet. Hum. Toxicol., 29(2): 133-137.

Felter HW (1922). Podophyllum. The Eclectic Materia Medica, Pharmacology and Therapeutics. Henriette's Herbal, www.ibiblio.org/herbmed/index.html. 16(2): 106-113.

Ford EJH (1963). In Clinical Biochemistry of Domestic Animals. edit: Cornelius CE, Kaneko J, Academic Press, New York, pp. 256-258.

Ford EJH, Adam SEI, Gopinath C (1972). Hepatic amidopyrine N-demethylase activity in the calf. J. Comp. Pathol., 82(3): 355-364

Galal M, AdamSEI (1988). Experimental Chrozophora plicate poisoning in goats and sheep. Vet. Hum. Toxicol., 30(5): 447-452.

IDSP (1980). Health Effects of Aluminium in the Workplace. Progress Report, Industrial Disease Standards Panel, Worker' Compensation Board of Ontario, Ontario, Canada, pp. 1-24.

Kew M (2000). Serum aminotransferase concentration as evidence of hepatocellular damage. Lancet., 355(9204): 591-592.

Michael W (2000). Rivastigmine, a New-Generation Cholinesterase Inhibitor for the Treatment of Alzheimer's Disease. Pharmacotherapy, 20(1): 1-12.

Mohamed OSA, Adam SEI (1992). Effect of phenobarbitone pretreatment on the toxicity of temik and sumicidin in Nubian goats. Vet. Hum. Toxicol., 34(2): 138-140.

Mohammed SNH (1998). Head Industerial Hygiene and Enviromental Pollution Department. Federal Occupational Health Administration, Personal Communication.

Morgan WKC, Dinman BD (1989). Pulmonary effects of aluminum. In: Gitelman HJ, ed Aluminum and health- A critical review. New York: Marcel Dekker, pp. 203-234.

Orshoven NP van, Andriesse GI, Schelven LJ, Smout VA J, Akkermans LMA, Oey P L (2006).Subtle involvement of the parasympathetic nervous system in patients with irritable bowel syndrome. Clin. Auton. Res., 16(1): 33-39.

Parmeggiani L (1983).Encyclopaedia of occupational health and safety. 3rd (Rev.) ed. Geneva: International Labour Office, pp. 153-154

Schalm OW (1965). Minutes of The Regents of the University of California. http://dynaweb.oac.cdlib.org:8088/dynaweb/uchist/public/regentminutes/regents865/@Generic__BookTextView/, p. 111.

Sjogren B, Lundberg I, Lidums V (1983). Aluminium in the blood and urine of industrially exposed workers. Br. J. ind. Med., 40(3): 301-304.

Snedecor GW, Cochran WG (1989). Statistical Methods. Eighth Edition,

Iowa State University Press, pp. 158-160.

Sousa AB, Soto-Blanco B, Guerra JL, Kimura ET, Górniak SL (2002). Does prolonged oral exposure to cyanide promote hepatotoxicity and nephrotoxicity. Toxicology, 174(2): 87-95.

Verstraeten SV, Aimo L, Oteiza PI(2008). Aluminium and lead: Molecular mechanisms of brain toxicity. Arch. Toxicol., 82(11): 789-802. [Medline].

WHO (1984). Guidelines for Drinking Water Quality.World Health Organization, Geneva. www.who.int/water_ sanitation_ health/dwq/S02.pdf, pp. 1-23.

10

Micronucleus test in post metamorphic *Odontophrynus cordobae* and *Rhinella arenarum* (Amphibia: Anura) for environmental monitoring

Beatriz Bosch[1]*, Fernando Mañas[2,3], Nora Gorla [2,3] and Delia Aiassa[1]

[1]Departamento de Ciencias Naturales, Facultad de Ciencias Exactas Físico-Químicas y Naturales (FCEFQN), Universidad Nacional de Río Cuarto (UNRC), Argentina.
[2]Facultad de Agronomía y Veterinaria (FAV) Universidad Nacional de Río Cuarto (UNRC), Argentina.
[3]CONICET, Argentina.

The genotoxic effect of cyclophosphamide and glyphosate in a commercial formulation were determined using the micronucleus test in peripheral blood erythrocytes of *Odontophrynus cordobae* and *Rhinella arenarum*, amphibians widely distributed in the Province of Córdoba, Argentina. For this, the basal frequency of the micronucleated erythrocytes (MNE) was determined by: 0.40 ± 0.18 MNE/1000 erythrocytes in *Odontophrynus cordobae* and 0.30 ± 0.09 MNE/1000 erythrocytes in *Rhinella arenarum*. The frequency of MNE in *Odontophrynus cordobae* increased after 5 days of exposure to glyphosate (100 mg ai/L) and cyclophosphamide. After 2 and 5 days of exposure to glyphosate (200, 400 and 800 mg ai/L), the MNE frequency in *Rhinella arenarum* was higher than the basal frequency, as it occurred in the group exposed to cyclophosphamide. Regarding acute toxicity and genotoxicity, the results show that *Odontophrynus cordobae* is more sensitive to cyclophosphamide and glyphosate exposure than *Rhinella arenarum*. A correlation was detected between exposure concentration and MNE frequency in *Rhinella arenarum*.

Key words: Genotoxicity, micronucleus test, roundup®, glyphosate, amphibians.

INTRODUCTION

Many authors have reported that in the past 30 years there has been a significant decline in amphibian populations in diverse parts of the world (Hayes et al., 2010). Multiple causes have been suggested to explain this decline and among them, environmental pollution due to chemicals is gaining attention. Amphibians, like other organisms inhabiting agroecosystems, are highly exposed to agrochemicals and a correlation has been reported between the use of pesticides and the decline of amphibian populations (Beebee and Griffiths, 2005;

Jones et al., 2010). Pesticides, including insecticides and herbicides, are particularly detrimental to amphibians due to its aquatic habitat, sensitive skin and unprotected eggs (Govindarajulu, 2008; Bouhafs et al., 2009).

Glyphosate based herbicides are the most widely non-selective, broad-spectrum herbicides used in the world. In Argentina, glyphosate use on its wide variety of commercial formulations has increased dramatically. With the increased use of glyphosate-based herbicides containing the surfactant polyethoxylated tallowamine (POEA) increased concern about the potential impact that the formulations may have on amphibians' populations. Some studies indicate that the toxic effect of glyphosate herbicides containing surfactants is higher than the active ingredient (ai) per se. Thus, the toxicity of

*Corresponding author. E-mail: betinabosch@gmail.com.

some formulated glyphosate products to amphibians is greater than that caused by the active ingredient (Relyea, 2005a; Mann et al., 2009; Bernal et al., 2009; Modesto and Martinez, 2010). Furthermore, there is a global concern regarding the adverse genetic response that the formulated herbicide may have on non-target organisms (Clements et al., 1997; Holečková, 2006; Çavaş and Könen, 2007; Cavalcante et al., 2008; Andreikénaité et al., 2007; Poletta et al., 2007; Baršienė et al., 2008; Poletta et al., 2009; Vera et al., 2010a). Recent studies have shown that amphibians are one of the most sensitive vertebrate groups to the toxicological effects of this herbicide (Govindarajulu, 2008).

Several end-points have been used to assess genotoxicity in aquatic and terrestrial organisms. Among them, the micronucleus test (MN) has been widely used in various species to detect clastogenic or aneugenic effects, both under laboratory conditions or in situ studies (Udroiu, 2008). Micronuclei are chromosome fragments or whole chromosomes which were not incorporated into the main nucleus, in cells of any actively dividing tissue. Increasing frequency of MN is considered as a biomarker of genotoxic effect at subcellular level and an early response of chromosomal damage (Fenech, 2000; (Cuenca and Ramírez, 2004; Garaj-Vrhovac et al., 2008). This biomarker has been measured in amphibian species that inhabit environments where large quantities of pesticides and other chemical substances are periodically poured (Lajmanovich et al., 2005; Cabagna et al., 2006; Vera Candiotti et al., 2010b).

The micronucleus test has been frequently used to detect genotoxicity induced by clastogenic and aneugenic agents in pre-metamorphic anuran amphibians and urodeles (Jaylet et al., 1986; Marty et al., 1989; Cabagna et al., 2006; Mouchet et al., 2006; Vera et al., 2010b), but few used in post-metamorphic amphibians (Zhuleva and Dubinin, 1994; Matson et al., 2005). The interest of the study of post-metamorphic anuran amphibians is related to the temporal and spatial preponderance of this stage in agroecosystems, which added to the characteristics of this group and turned them into the organisms that are recommended for developing of biomarkers for environmental monitoring (Burggren et al., 2007).

The objectives of this study were:

i) to define the basal MN frequency in peripheral blood erythrocytes of adult Odontophrynus cordobae (Anura: Cyclorhamphidae) and Rhinella arenarum Hensel, 1867 (Anura: Bufonidae), amphibians widely distributed in the agroecosystems of the Province of Córdoba, Argentina.
ii) propose the variation of these species in MN frequency as a biomarker of genotoxic effect using cyclophosphamide as positive control.
iii) to test the MN frequency with four different concentrations of glyphosate herbicide in a commercial formulation.
iv) to compare the sensitivity of both amphibians to these compounds and propose the micronucleus test in these

species as a diagnostic tool and monitoring of environmental quality.

MATERIALS AND METHODS

Animals

Two species of anuran amphibians in post-metamorphic stage, O. cordobae and R. arenarum were selected as bioassay organisms. These anurans have an extensive Neotropical distribution, and inhabit natural areas, agricultural land and urban territories with availability of temporary ponds (Leynaud et al., 2006). Both species are easily collected by hand and acclimated to laboratory conditions.

Specimens of O. cordobae were manually collected in temporary ponds in a pristine zone (Villa de Las Rosas, Province of Córdoba, lat. S 31° 56'; leng. W 65° 4'; alt. 713 m.a.s.l.) during September and October, 2005. Specimens of R. arenarum were manually collected in temporary ponds in a suburban area, in the east of Río Cuarto, (Province of Córdoba, 33°06' lat. S, 64°25' leng. O; alt. 438, 62 m.a.s.l.) during November and December, 2008 (Figure 1).

The average total size (snout-vent) was 10.45±0.15 cm for R. arenarum and 6.20± 0.35 cm for O. cordobae. The animals collected were those considered young adults in reproductive age, according to their body size. The animals were kept under laboratory conditions at ambient temperature and natural photoperiod, according to the time of the year in which each test was performed, for acclimatization before starting the assays. They were placed in 10 L plastic containers that were lined with damp paper towels to keep the animals hydrated. During this time, the animals were fed on larvae of Tenebrio molitor (Coleoptera: Tenebrionidae) bred in the laboratory.

Chemicals

The following substances were used: lyophilized cyclophosphamide monohydrate (CAS Nº 6055-19-2), Microsules® and isopropilamine salt of N-phosphomethyl-glycine, glyphosate, (CAS Nº 38641-94-0) Roundup® (glyphosate 48%). All test solutions were freshly prepared before use and renewed every 48 h.

Treatments

Hayashi (2007) recommendations for the treatments were followed regarding the number of animals per treatment group, the route of exposure to the substances tested, the animal preparation and the test conditions. After acclimatization the animals were assigned to groups of equal size (n = 5) as follows:

(I) Negative control group (C).
(ii) Positive control group (PC).
(iii) Groups exposed to glyphosate (G).

Each experimental group consisted of three males and two females; sex was determined by the presence/absence of vocalization activity and position in amplexus, observed in the field.

The negative control group was kept in water. Cyclophosphamide, well known as a genotoxic substance (clastogenic effect), was used as a positive control. The positive control group was kept in 40 mg/L cyclophosphamide during 5 days. The groups exposed to the herbicide were kept in Roundup® solutions prepared in water, at the following concentrations: glyphosate 100 mg (G100), 200 mg (G200), 400 mg (G400) and 800 mg (G800) active ingredient (a.i.)/ L. Dermal exposure was achieved by keeping the animals in the corresponding solutions in a volume sufficient to cover half of the distal humerus.

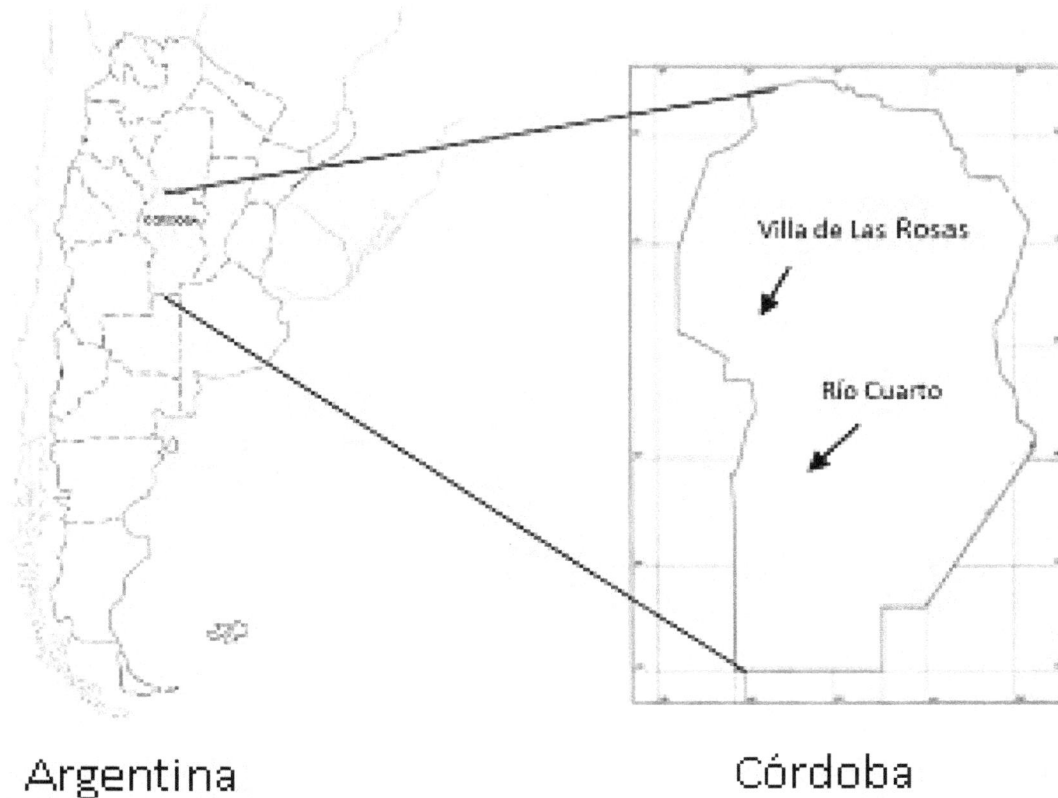

Figure 1. Study localities.

Blood samples were obtained from Days 2 and 5 of treatment, through a small incision in the angularis vein. A drop of blood was immediately smeared on a clean slide which was dried at room temperature for 24 h. After fixation with absolute methanol for 15 min, the slides were stained with May Grünwald-Giemsa, washed with distilled water and air dried.

Protocols described for anuran amphibian larvae (Campana et al., 2003; Feng et al., 2004) were used for the MN test and adapted to post-metamorphic stages.

The slides were coded and blind-scored by one researcher at 1000× magnification with an optic microscope (Nikon Labophot 2), and were photographed using a digital camera (HP Photosmart 935). The number of erythrocytes with MN was determined by analyzing 3000 erythrocytes per animal. Only non-overlapping cells with intact cellular and nuclear membrane were counted.

For the MN identification, the following criteria were used:

(a) The diameter of MN usually varies between 1/16th and 1/3rd of the mean diameter of the main nuclei.
(b) MN are non-refractive and, therefore, they can be readily distinguished from artifacts such as staining particles.
(c) MN are not linked or connected to the main nuclei.
(d) MN may touch but not overlap with the main nuclei, and the micronuclear boundary should be distinguished from the nuclear limit.
(e) MN usually has the same staining intensity as the main nuclei but, occasionally, it may be more intense (Fenech et al., 2003). The frequency of micronucleated erythrocytes was expressed as the number of micronucleated erythrocytes (MNE)/1000 analyzed erythrocytes.

Statistical analysis

The Kolmogorov–Smirnov test was performed to verify whether the results follow a normal distribution. Student´s t test was used to analyze the data from cyclophosphamide assays. The data from the Roundup® experiments were analyzed by one-way ANOVA with Dunnett's post-test. The statistical analysis was performed using GraphPad Prism version 5.00. The relationship between the Roundup® concentrations (mg a.i. /L) tested in *R. arenarum* and the observed frequency of MNE was determined by a correlation and regression analysis.

RESULTS, DISCUSSION AND CONCLUSION

The erythrocytes from both species showed a characteristic elliptical shape, with an approximate size of 22 ×15 µm, and an elliptical nucleus in a central position. We found one MN per cell which was located close to the main nucleus (Figures 2 and 3).

The basal frequency of MNE in *O. cordobae* was higher, 0.40 ± 0.18 MNE/1000 analyzed erythrocytes, than that found in *R. arenarum*, 0.30 ± 0.09 MNE/1000 analyzed erythrocytes (Figures 4 and 5). These parameters have not been previously reported and could

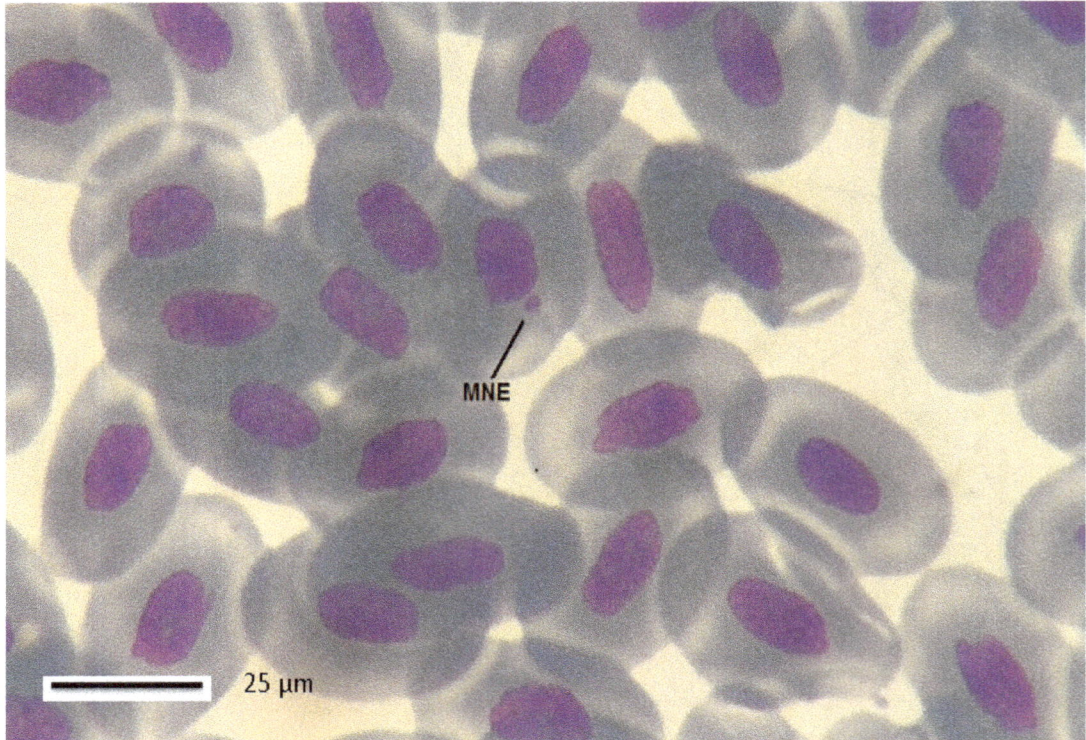

Figure 2. Micronucleated erythrocyte (MNE) in peripheral blood smear from *Odontophrynus* cordobae.

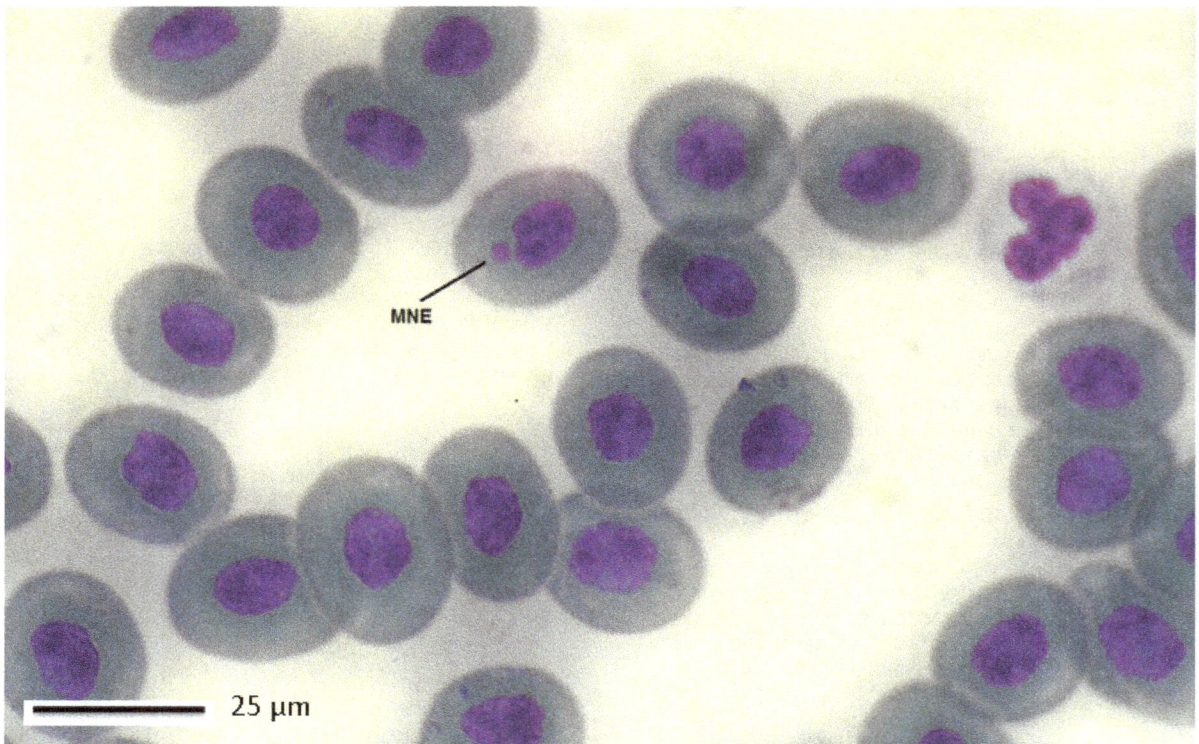

Figure 3. Micronucleated erythrocyte (MNE) in peripheral blood smear from *Rhinella arenarum.*

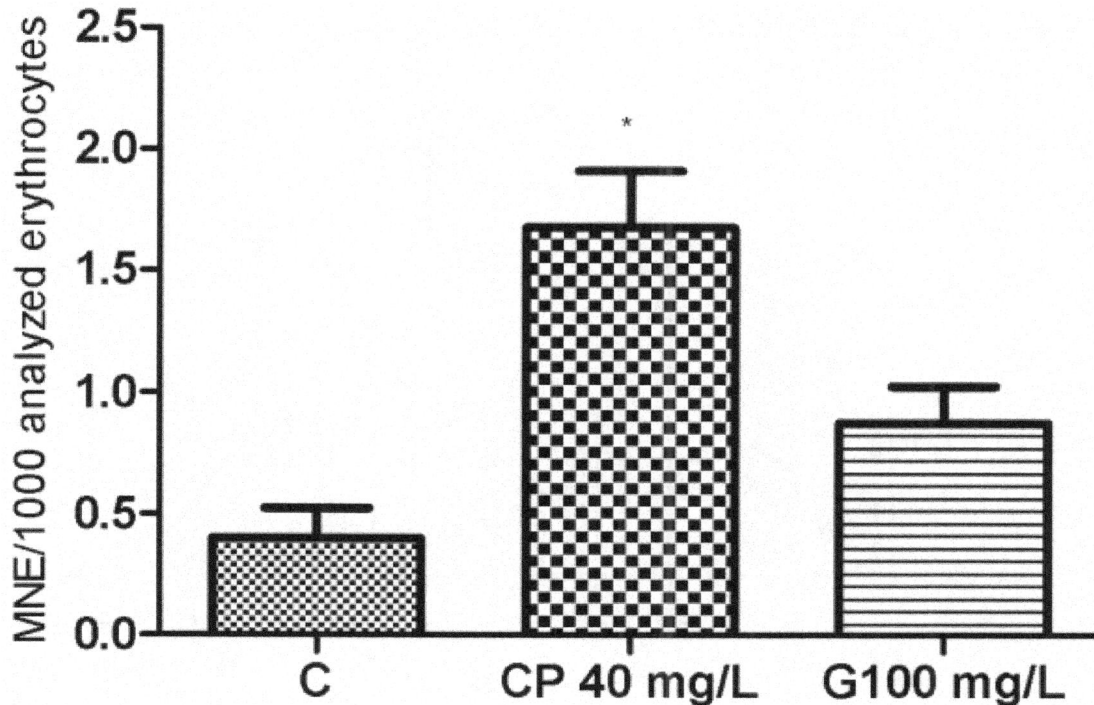

Figure 4. Mean frequencies of micronucleated erythrocytes per 1000 analyzed erythrocytes (MNE/1000 analyzed erythrocytes) in *O. cordobae.* for the negative control group (C), cyclophosphamide positive control group (CP) and groups exposed to Roundup® at 100 mg a.i./L (G100) at day 5 of treatment (*statistically significant difference $p<0.05$ as compared to the negative control), n= 5 animals per group.

be used as a reference for the sampling of both species in other agricultural areas of the province of Córdoba. In Argentina, other research groups have selected other species of the genus *Odontophrynus* (*O. americanus*) and *R. arenarum* in pro-metamorphic stage as bioassay organisms to perform the MN test (Cabagna et al., 2006; Vera et al., 2010a). Cabagna et al. (2006) reported lower basal MNE frequencies in tadpoles of the species *Odontophrynus americanus* than those observed in post-metamorphic stages of *O. cordobae* in this study. In contrast, the basal frequencies reported for larvae of the species *R. arenarum* (Vera et al., 2010a) are higher than those found in this study, suggesting a greater sensitivity of larvae in *R. arenarum*.

Concentrations of glyphosate G200, G400 and G800 were lethal to *O. cordobae*. The group exposed to cyclophosphamide for 5 days showed significant differences in MNE frequency when compared to the control group (1.68 ± 0.52 MNE/1000 analyzed erythrocytes, $p <0.05$; t= 4,824; Figure 4). In *R. arenarum*, the basal frequency of MNE was duplicated in the group of amphibians that received dermal exposure to cyclophosphamide, with a statistically significant difference ($p<0.05$) for 2 (t=2.33) and 5 (t= 3.77) days. After the application of the test with this known

clastogenic as a positive control, we can ensure that micronucleus test in post-metamorphic *O. cordobae* and *R. arenarum* can be used as a biomarker of genotoxic effect for environmental monitoring as it has been assayed for pro-metamorphic stages (Cabagna et al., 2006; Vera et al., 2010a).

Originally, blood samples were planned to be obtained at Days 5 and 9 for *O. cordobae* and *R. arenarum*. During the first day of treatment, the groups of *O. cordobae* exposed to glyphosate (G200, G400 and G800) showed signs associated with acute toxicity (massive blood shedding and changes in the color of the skin), and in the second day, the effects were lethal in the three groups. At Day 5 of the treatment, only samples from the negative control, positive control, and G100 group could be obtained (Figure 4). Govindarajulu (2008) reported a direct effect (lysis) on the skin of amphibians in post-metamorphic stages by glyphosate based formulations. As a consequence of these results, sampling for the tests in *R. arenarum* were brought forward to Days 2 and 5 of treatment (Figure 5). Our results, however, show that *O. cordobae* presented a higher acute toxicity than *R. arenarum*. Concentrations of exposure which were lethal for *O. cordobae*. (G200, G400 and G800 mg a.i./L) did not affect *R. arenarum*. The latter only showed mild signs

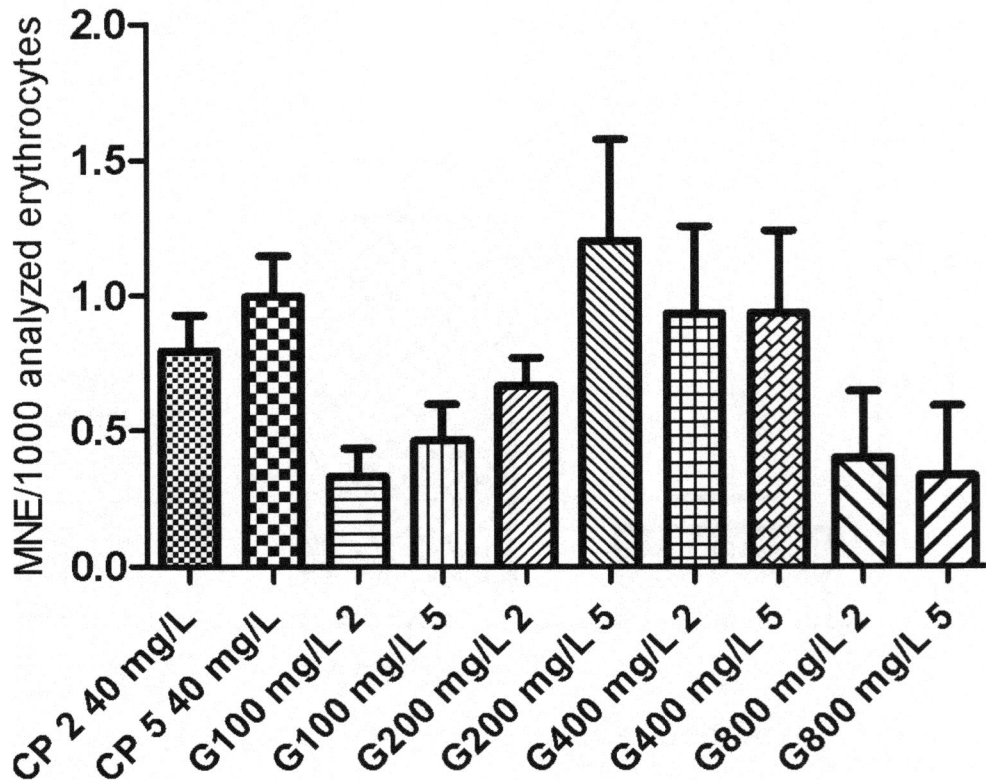

Figure 5. Mean frequencies of micronucleated erythrocytes (MNE/1000 analyzed erythrocytes) in the treatment groups of *R. arenarum* with cyclophosphamide at day 2 (CP 2) and at day 5 (CP 5); with Roundup® 100 mg a.i /L at day 2 (G100 2) and at day 5 (G100 5); 200 mg a.i./L at day 2 (G200 2) and at day 5 (G200 5); 400 mg a.i./L at day 2 (G400 2) and at day 5 (G400 5); and 800 mg a.i../L at day 2 (G800 2) and at day 5 (G800 5), n= 5 animals per group.

of toxicity.

Besides, at a concentration of 100 mg a.i /L (G100), *O. cordobae* had a higher frequency of MNE (0.88 ± 0.33 MNE / 1000 analyzed erythrocytes) than *R. arenarum* (0.46 ± 0.16 MNE/1000 analyzed erythrocytes). Some studies have examined whether it is the active ingredient (glyphosate) or other components in the commercial formulations of the herbicides the cause of toxicity in anurans and other amphibians (Govindarajulu, 2008). There are a number of formulations of glyphosate-based herbicides all of which have the same basic ingredients: isopropylamine salt of glyphosate, a surfactant often undisclosed, unspecified inert substances and water. Polyethoxylated tallow amine (POEA) is a surfactant derived from animal fat, mainly used in glyphosate-based products (Giesy et al., 2000, Modesto and Martinez, 2010). It is worth noting that laboratory studies have shown that POEA is a substance that causes high mortality in fish and amphibians, even though it is legally classified as an inert ingredient (Jones et al., 2010).

According to the results, we could assume that the commercial formulation of glyphosate used in the present

work has a clastogenic effect in the studied anurans, agreeing with the report of Clements et al. (1997), who applied the comet assay in tadpoles of *Lithobates catesbeianus* (bullfrog) exposed to a glyphosate-based formulation.

The concentrations used in this work were approximately 10 times higher than those utilized in previous studies with glyphosate in pro-metamorphic amphibians. We considered the observations of other authors who suggested that the post-metamorphic stages of the Australian Sign-bearing Froglet (*Crinia insignifera*) were 14 times less sensitive to glyphosate than the tadpoles of the same species (Mann and Bidwell, 1999). It is important to note that the concentrations recommended by the agrochemical industry, ranges from 1 to 2% (that is, 10 to 20 ml/L of the solution) which is equivalent to 4800 to 9600 mg a.i./L, a concentration 10 to 20 times higher than those tested in the present work.

Previous studies in amphibians using glyphosate report acute toxicity in tadpoles (Smith, 2001; Lajmanovich et al., 2003, Howe et al., 2004; Cauble and Wagner, 2005, Relyea, 2005a, 2005b). However, many amphibians

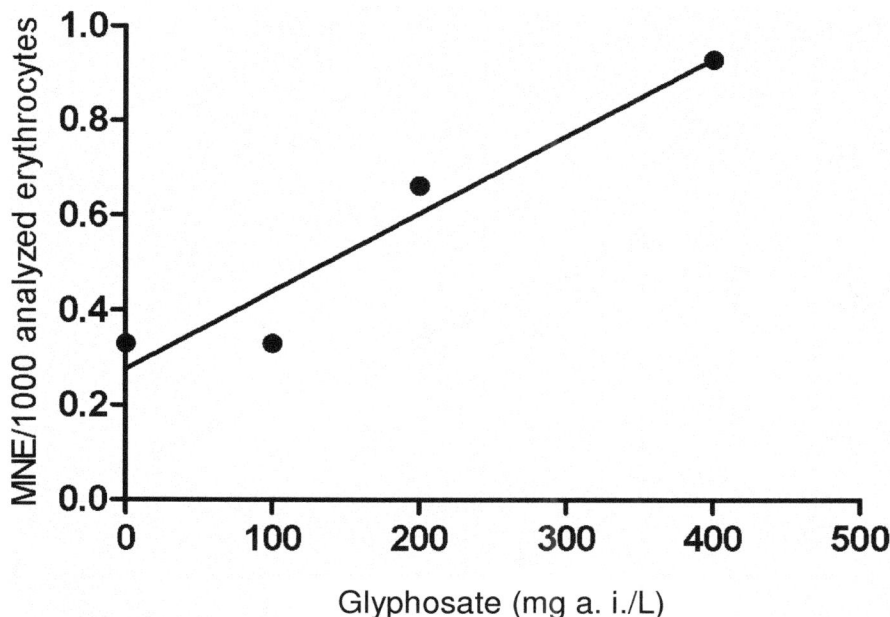

Figure 6. Regression analysis between Roundup® concentration (100, 200 and 400 mg a.i./L) and the frequency of micronucleated erythrocytes per 1000 analyzed erythrocytes (MNE/1000 analyzed erythrocytes) in *R. arenarum*, at day 2 of treatment (r^2= 0.96, p<0.05), n= 5 animals per group.

spend a large fraction of their life in the post-metamorphic stage. References on the effect assessment of glyphosate in amphibians in post-metamorphic stages, as in the current genotoxicity study, appears to be restricted to only two Australian species (Mann and Bidwell, 1999). In Relyea (2005a) study, the maximum dosage of domestic use Roundup® caused a decrease on survival in three species of anurans within 24 h of spraying. It is suggested that herbicide exposure and its impact on the amphibian's post-metamorphic stages may be non-trivial and should not be ignored, and the frequency and magnitude of these effects need further investigation (Relyea, 2005b). Recently, the impact of environmental exposure to glyphosate on frog populations in Colombia's coca fields has also been discussed (Solomon et al., 2007).

To our knowledge there is only one study that evaluated genotoxicity of glyphosate in amphibians using another test. In that study, the comet assay was performed in erythrocytes from *L. catesbeianus* tadpoles to asses DNA damage following exposure to the herbicide Roundup® (Clements et al., 1997). All tadpoles treated with 108 mg a.i./L Roundup ® died within 24 h, and animals treated with 1.7 to 27 mg a.i./L Roundup ® showed significant DNA damage when compared with unexposed control animals.

No statistically significant differences were found in the mean frequency of MNE among the experimental groups in *R. arenarum* (p>0.05; ANOVA). Despite this, it is

interesting to note that the herbicide seems to produce a variation in the frequency of MNE in a concentration response manner (Figures 6 and 7). At day 2 of treatment, increments in the average frequency of MNE were observed with increasing concentrations of the herbicide up to G400 included (r^2 = 0.96, α = 0.05) (Figure 6). At Day 5 of treatment, a decrease in the mean frequency of MNE was observed from G200 to G800 mg ai/L, (r^2 = 1, p <0.05) (Figure 7). This difference in the behavior of the average MNE frequencies observed at Days 2 and 5 may be due to a toxic effect of the herbicide on hematopoietic organs. Analysis of more replications and appropriated histological methods can reveal more information on such effects.

Given the biological and ecological characteristics of the studied species, it would be convenient to extend this evaluation to pure glyphosate to rule out the effect of other components present in the commercial formulation. In addition, future studies should look into the biological effects of AMPA, the main environmental breakdown product of glyphosate, often found as a pollutant. To this end, we have detected, by three different assays, the potential genotoxic effect of AMPA in mammalian cells *in vitro* and *in vivo* (Mañas et al., 2009).

The MN test is a valuable tool to monitor and to identify the genotoxic effect of environmental pollutants, as well to predict and, probably, to prevent the consequences of exposure to them. Since cytogenetic manifestations are considered an early genotoxic finding, the MN test allows

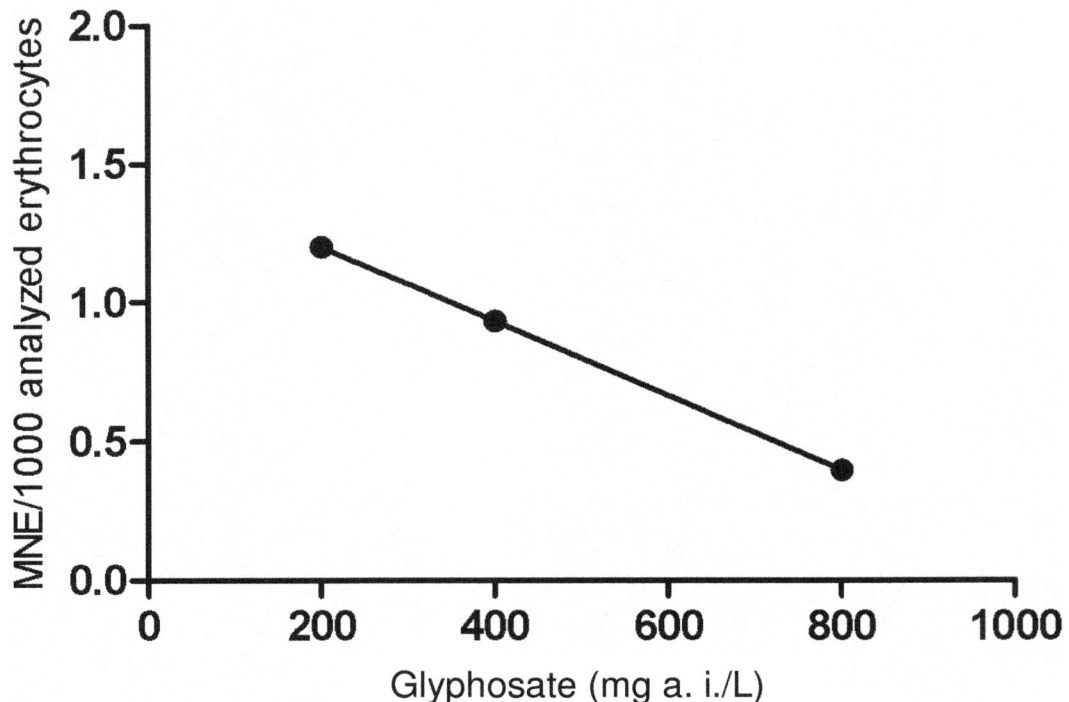

Figure 7. Regression analysis between the groups exposed to Roundup® (200, 400 and 800 mg a.i./L) and the frequency of micronucleated erythrocytes per 1000 analyzed erythrocytes (MNE/1000 analyzed erythrocytes) in *R. arenarum*, at day 5 of treatment ($r^2=1$, $p<0.05$), n= 5 animals per group.

detecting a level of genetic damage when it is still reversible (Cuenca and Ramírez, 2004; Zhu et al., 2005).

Most of the studies of genotoxicity in amphibian have focused on anuran larvae; therefore, little information is available on the impact of glyphosate on adult amphibians, as reported here. The major concern is related to the excessive amounts of glyphosate in commercial formulations which are incorporated to the soil and the ecosystems in Argentina, and to their possible negative effect on amphibian populations.

Our results show that both species in post-metamorphic stage have a basal spontaneous occurrence of MNE, which can also be induced by cyclophosphamide and a glyphosate-based formulation Therefore, the MN test offers a useful tool to evaluate the genotoxic effect of environmental polluents. However, as this is an exploratory work, it needs future investigation under other experimental conditions such as different exposure concentrations and duration of treatment. The development and validation of biomarkers *in vivo*, like the MN test in peripheral blood in *O. cordobae* and *R. arenarum*, provides a basis for the *in situ* assessment of the potential risk of environmental exposure to genotoxic agents, being this a present concern at ecological as well as public health level.

We concluded that both amphibians can be used for environmental monitoring of pesticide contamination.

REFERENCES

Andreikėnaitė L, Baršienė J, Vosylienė M (2007). Studies of micronuclei and other nuclear abnormalities in blood of rainbow trout (*Oncorhynchus mykiss*) treated with heavy metal mixture and road maintenance salts. Acta Zool. Lit., 17: 213-219.

Beebee T, Griffiths R (2005). The amphibian decline crisis: A watershed for conservation biology? Biol. Conserv., 125: 271-285.

Baršienė J, Rybakovas A, Förlin L, Šyvokienė J (2008). Environmental genotoxicity studies in mussels and fish from the Göteborg area of the North Sea. Acta Zool. Lit., 18: 240-247.

Bernal M, Solomon K, Carrasquilla G (2009). Toxicity of Formulated Glyphosate (Glyphos) and Cosmo-Flux to Larval Colombian Frogs. Lab. Acute Toxicity J. Toxicol. Environ. Health, 72: 961-965.

Burggren W, Warburton S (2007). Amphibians as animal models for laboratory research in physiology. ILAR J., 48: 260-269.

Bouhafs N, Berrebbah H, Devaux A, Rouabhi R, Djebar J (2009). Micronucleus Induction in Erythrocytes of Tadpole *Rana saharica* (Green Frog of North Africa) Exposed to Artea 330EC. Am. Eur. J. Toxicol. Sci., 1: 7-12.

Cabagna M, Lajmanovich R, Peltzer P, Attademo A, Ale A (2006). Induction of Micronucleus in Tadpoles of *Odontophrynus americanus* (Amphibia: Leptodactylidae) by the Pyrethroid Insecticide Cypermethrin. Toxicol. Environ. Chem., 88: 729-737.

Campana M, Panzeri A, Moreno V, Dulout F (2003). Micronuclei induction in *Rana catesbiana* tadpoles by the pyrethroid insecticide lambda-cyhalothrin. Genet. Mol. Biol., 26: 99-103.

Cauble K, Wagner R (2005). Sublethal effects of the herbicide Glyphosate on amphibian metamorphosis and development.

Micronucleus test in post metamorphic Odontophrynus cordobae and Rhinella arenarum (Amphibia: Anura)...

81

Bull. Environ. Contam. Toxicol., 75: 429-435.

Çavaş T, Könen S (2007). Detection of cytogenetic and DNA damage in peripheral erythrocytes of goldfish (Carassius auratus) exposed to a glyphosate formulation using the micronucleus test and the comet assay. Mutagenesis, 22: 263-268.

Cavalcante D, Martinez C, Sofia, S (2008). Genotoxic effects of Roundup® on the fish Prochilodus lineatus. Mutation Research/Genetic Toxicol. Environ. Mutagen., 655: 41-46.

Clements C, Ralph S, Petras M (1997). Genotoxicity of selected herbicide in Rana catesbeiana tadpoles using the alkaline single-cell gel DNA electrophoresis (comet) assay. Environ. Mol. Mutagen., 29: 277-288.

Cuenca P, Ramírez V (2004). Environmental mutagenesis and use of biomarkers to predict cancer risk. Rev. Biol. Trop., 52: 585-590.

Fenech M (2000). The in vitro micronucleus technique. Mutat. Res., 445: 81-95.

Fenech M, Chang W, Kirsch-Volders M, Hollandd N, Bonassi S, Zeigerf E (2003). HUMN project: Detailed description of the scoring criteria for the cytokinesis-block micronucleus assay using isolated human lymphocyte cultures. Mutat. Res., 534: 65-75.

Feng S, Kong Z, Wang X, Zhao L, Peng P (2004). Acute toxicity of two novel pesticides on amphibian, Rana N. Hallowel. Chemosphere, 56: 475-463.

Garaj-Vrhovac V, Gajski G, Ravlić S (2008). Efficacy of HUMN criteria for scoring the micronucleus assay in human lymphocytes exposed to a low concentration of p,p'-DDT. Braz. J. Med. Biol. Res., 41: 473-476.

Giesy J, Dobson S, Solomon K (2000). Ecotoxicological risk assessment for Roundup® herbicide. Environ. Contam. Toxicol., 167: 35–120.

Govindarajulu P (2008). Literature review of impacts of Glyphosate herbicide on amphibians: What risks can the silvicultural use of this herbicide pose for amphibians in B.C.? B.C. Ministry of Environment, Victoria, BC. Wildlife Report N° R-28.

Hayashi M (2007). Three Rs in mutation research- From in vivo to in silico evaluation - AATEX 14, 9-13. Proc. 6th World Congress on Alternatives and Animal Use in the Life Sciences.

Hayes TB, Falso P, Gallipeau S, Stice M (2010). The cause of global amphibian declines: a developmental endocrinologist's perspective. J. Exp. Biol., 213: 921-933.

Holečková B (2006). Evaluation of the in vitro effect of glyphosate-based herbicide on bovine lymphocytes using chromosome painting. Bull. Vet. Inst. Pulawy, 50: 533-536.

Howe CM, Berrill M, Pauli BD, Helbing CC, Werry, K, Veldhoen N (2004). Toxicity of Glyphosate-based pesticides to four North American frog species. Environ. Toxicol. Chem., 23: 1928-1938.

Jaylet A, Deparis P, Ferrier V, Grinfeld S, Siboulet R (1986). A new micronucleus test using peripheral blood erythrocytes of the newt Pleurodeles waltl to detect mutagens in fresh-water. Mutat. Res., 164: 245-257.

Jones D, Hammond J, Relyea R (2010). Roundup ® and amphibians: the importance of concentration, application time and stratification. Environ. Toxicol. Chem., 29: 2016-2025.

Lajmanovich R, Sandoval M, Peltzer P (2003). Induction of Mortality and Malformation in Scinax nasicus Tadpoles Exposed to Glyphosate Formulations. Bull. Environ. Contam. Toxicol., 70: 612–618.

Lajmanovich R, Cabagna M, Peltzer P, Gabriela A, Stringhini G, Attademo A (2005). Micronucleus induction in erythrocytes of the Hyla pulchella (Amphibia: Hylidae) tadpoles exposed to insecticide endosulfan. Mutat. Res., 587: 67-72.

Leynaud G, Pelegrin N, Lescano J (2006). Amphibians and Reptiles. In: Maeshlands of Río Dulce and lagoon Mar Chiquita (Córdoba, Argentina). (ed Bucher E.H.), National Academy of Sciences (Córdoba, Argentina), pp. 219-235.

Mañas F, Peralta L, Raviolo J, García Ovando H, Weyers A, Ugnia L, Gonzalez Cid M, Larripa I, Gorla N (2009). Genotoxicity of AMPA, environmental metabolite of glyphosate, assessed by the comet assay and cytogenetic tests, Ecotoxicol. Environ. Saf., 72: 834-883.

Mann RM, Bidwell JR (1999). The toxicity of Glyphosate and several Glyphosate formulations to four species of southwestern Australian frogs. Arch. Environ. Contam. Toxicol. 36: 193-199.

Mann RM, Hyne R, Choung C, Wilson S. (2009). Amphibians and agricultural chemicals: Review of the risks in a complex environment. Environ. Pollut., pp. 1–25.

Marty J, Lesca P, Jaylet A, Ardourel C, Rivière JL (1989). In vivo and in vitro metabolism of benzo(a)pyrene by the larva of the newt, Pleurodeles waltl Comp. Biochem. Physiol. C., 93: 213-219.

Matson C, Palatnikov G, Mc Donald T, Autenrieth R, Donnelly K, Anderson T, Canas J, Islamzadeh A, Bickham J (2005). Patterns of genotoxicity and contaminant exposure: Evidence of genomic instability in the marsh frogs (Rana ridibunda) of Sumgayit, Azerbaijan. Environ. Toxicol. Chem., 24: 2055-2064.

Modesto KA, Martinez C (2010). Effects of Roundup Transorb on fish: Hematology, antioxidant defenses and acetylcholinesterase activity. Chemosphere, pp. 781-787. doi:10.1016/j..2010.07.005.

Mouchet F, Gauthier L, Mailhes C, Jourdain M, Ferrier V, Triffault G, Devauxe A (2006). Biomonitoring of the genotoxic potential of aqueous extracts of soils and bottom ash resulting from municipal solid waste incineration, using the comet and micronucleus tests on amphibian (Xenopus laevis) larvae and bacterial assays (Mutatox® and Ames tests). Sci. Total Environ., 355: 232-246.

Poletta G, Larriera A, Kleinsorge E, Mudry M (2009). Genotoxicity of the herbicide formulation Roundup® (glyphosate) in broad-snouted caiman (Caiman latirostris) evidenced by the Comet assay and the Micronucleus test. Mutat. Res./Genet. Toxicol. Environ. Mutagen., 672: 5-102.

Poletta G, Larriera A, Kleinsorge E, Mudry M (2007). Caiman latirostris (broad-snouted caiman) as a sentinel organism for genotoxic monitoring: Basal values determination of micronucleus and comet assay. Mutat. Res./Genet. Toxicol. Environ. Mutagen., 650: 202-209.

Relyea R (2005a). The lethal impacts of Roundup and Predatory stress on Six Species of North American Tadpoles. Arch. Environ. Contam. Toxicol., 48. 351-357.

Relyea R (2005b). The lethal impact of Roundup on aquatic and terrestrial amphibians. Ecol. Appl., 15: 1118-1124.

Smith GR (2001). Effects of acute exposure to a commercial formulation of Glyphosate on the tadpoles of two species of anurans. Bull. Environ. Contam. Toxicol., 67: 483-488.

Solomon K, Anadon A, Carrasquilla G, Cerdeira AL, Marshall J, Sanin LH (2007). Coca and poppy eradication in Colombia: environmental and human health assessment of aerially applied Glyphosate. Rev. Environ. Contam. Toxicol., 190: 43-125.

Udroiu I (2008). The micronucleus test for aquatic toxicology. En: Aquatic Toxicology Research Focus. Elías Svensson (ed). Nova Science Publishers, Inc., pp. 145-160.

Vera Candioti JV, Soloneski S, Larramendy ML (2010a). Genotoxic and cytotoxic effects of the formulated insecticide Aficida® on Cnesterodon decemmaculatus (Jenyns, 1842) (Pisces: Poeciliidae). Mutat. Res./Genet.Toxicol. Environ. Mutagen., 703: 180-186.

Vera CJV, Natale GS, Soloneski S, Ronco AE, Larramendy ML (2010b). Sublethal and lethal effects on Rhinella arenarum (Anura, Bufonidae) tadpoles exerted by the pirimicarb-containing technical formulation insecticide Aficida®. Chemosphere, 78: 249-255.

Zhu L, Huang Y, Liu G (2005). Using DNA damage to monitor water environment. Chin. J. Ocean. Limn., 3: 340-348.

Zhuleva L, Dubinin N (1994). Use of the micronucleus test for assessing the ecological situation in regions of the Astrakhan district. Genetika, pp. 999-1004.

11

The effects of lihocin toxicity on protein metabolism of the fresh water edible fish, *Channa punctatus* (Bloch)

Abdul Naveed[1]*, C. Janaiah[2] and P. Venkateshwarlu[3]

[1]Department of Zoology, Panchsheel College of Education, Nirmal. A.P., India.
[2]Department of Zoology, Kakatiya University, Warangl 506 009, India.
[3]Department of Zoology, K. D. C. Warangal, Andhra Pradesh, India.

In vivo evaluations were made to assess the pesticide activity of lihocin against fresh water fish *Channa punctatus* and its ultimate mode of action on fish protein metabolism. Biochemical studies show that after exposing the fish to sub lethal dose of lihocin, total protein levels significantly decreased while FAA, glutamine, alkaline phosphatases, acid phosphatases, AlAT, AAT, GDH, AMP deaminase and adenosine deaminase were significantly enhanced in the liver, brain and kidney tissues of *C. punctatus*. The alterations in all the aforementioned biochemical parameters were significantly ($p < 0.05$) time and dose dependent. As such, the negative impact of lihocin was shown on the respiratory, as well as, energy production of the fish.

Keywords: *Channa punctatus*, enzyme activity, Lc50, lihocin (OC) pesticidal impact, protein metabolism.

INTRODUCTION

Channa punctatus is a common predatory fish which have low food value and due to its predatory nature, it engulfs the fingerlings of cultured carps at several stages of their rearing (Jhingran, 1975). Thus, it adversely affects the cultured carp production and set a great loss to the fish farmer. The indiscriminate and extensive use of insecticides to protect crops poses a serious threat to humans and the surrounding environment. The pesticides which are liberated into aquatic environment have a deleterious effect on fish and subsequently on man (Metelev et al., 1983).

The organochlorine insecticide 2-Chloro N,N,N, trimethyl ethano ammonium, commercially available as lihocin (OC), is used as a treatment against ectoparasite and as an insecticide for crops. Lihocin is poorly hydrolyze and as such, it biodegrades slowly in the environment. So, this compound persists for longer time in the food chain and cause severe effects at different levels of food chains. The toxicity of the aqueous latex extract of *N. Inidcum* on fresh water snails (*Lymaaea*

acuminate and *Indoplanoribs excustus*) and fish (*Channa punctatus* and *Colisa fasciatus*) has been established by Singh et al. (1993) and Singh and Singh (2000). A review of the toxicological literature reveals that the exposure to toxic chemicals can produce unexpected effects in non target animals (Veronica and Collins, 2003; Gonzalez et al., 2004; Lehtonen and Leimio, 2003; Salah, 1983). Most of the OC compounds and their derivatives adversely affect the nervous system.

The present study was undertaken to identify the effects of lihocin on liver, brain and gill tissues of the insecticide exposed fish *C. Punctatus*.

MATERIALS AND METHODS

Healthy fresh water fish, *Channa punctatus* (15 to 25 cm in length and 80 to 120 gr in body weight), were collected locally from the Nirmal of Adilabad district and used as the test animal. These fish were kept in cement tanks (6 x 3 x 3 feet), at least 3 weeks, for acclimatization under continuous water flow. The average temperature of water was 22 ± 1.0C. The tank water was aerated continuously and food was provided in the form of dried groundnut cake, powdered small prawn, goat liver, etc., and water was changed every 24 h. The dead animals, if any, were removed as soon as possible from the test container to prevent water fouling.

---

*Corresponding author. E-mail: naveed_nrml@yahoo.co.in.

The physico-chemical parameters of water were as follows: atmospheric temperature (29 to 35°C, pH 7.2), electrical conductivity (0.052 mnhos), calcium (5 mg/L), sodium (2.1 mg/L), bicarbonates (142 mg/L), total alkalinity (69 mg/L), sulphates (7.1 mg/L), nitrates (3.4 mg/L), iodine (0.01 ppm), chlorides (37 mg/L), dissolved oxygen (4.2 mg/L), BOD (1.6 mg/L), COD (0.008 mg/L) and fluoride (0.03 ppm).

Toxicity experiment was performed according to Bayne et al. (1977). The fishes were exposed for 24, 48, 72 and 96 h at four different concentrations of tap water. Six tubs were set up for each concentration and each tub contains six fishes in 6L- de-chlorinated tap water. Control animals were kept in similar condition without any treatment. Mortality was recorded at every 24 h up to 96 h exposure period. Fishes were considered dead if any failed to respond to stimulus provided with glass rod.

The LC_{50} of values were evaluated according to Finney (1971). The formula was used to assess the mortality of the fish and this was recommended by the WHO and FAO.

Channa punctatus were kept in a plastic containing 6 L of tap water. Each tub contains six experimental animals. Fishes were exposed for 24 h or 96 h exposure period to sub-lethal concentration of LC_{50} doses of lihocin (6.61 ppm). Control animals were kept in a similar condition without any treatment. After completion of treatment, fishes were removed from tubs, washed with water and killed by severe blow on the head. The liver, brain and gill tissues were quickly dissected out in ice tray and used for biochemical analysis. However, each experiment was repeated at least six times.

Protein levels were estimated according to Lowry et al. (1951) using bovine serum albumin as a standard. Homogenates (2 ml w/v) cold distilled water was prepared in 30% TCA. As such, values have been expressed as mg/100 mg wet.wt of tissue.

Free amino acids were estimated using the method of Moor and Stein (1954). Homogenates (5% w/v) were prepared in 10% (w/v) TCA and centrifuged at 300 rpm, whereas supernatant was used for amino acid estimation. FAA has been expressed as mg/100 mg wet.wt of the tissue.

Glutamine was estimated using the method of acid hydrolysis described by Colowick and Kaplan (1967). Homogenates (10% w/v) cold distilled water were prepared in 10% H_2SO_4, and so, their values were expressed as moles of glutamine/g wet.wt of the tissue.

AAT and AIAT were estimated using the method of Reitman and Frankel (1957). Homogenates (10% w/v) tissue were prepared in cold 0.25 M sucrose solution and centrifuged at 3000 rpm for 15 min to obtain a clear supernatant which was used as enzyme source. Values were expressed in micro moles of pyruvate formed/mg protein /h.

Glutamate dehydrogenase (GDH) activity levels were estimated following the method of Lee and Lardy (1965). Homogenates (10% w/v) were prepared in 0.25 M sucrose solution and centrifuged at 3000 rpm, while supernatant was used for the enzyme source. As such, values were expressed as micro moles of formazan formed / mg protein / hr of tissue.

AMP deaminase was estimated using the method of Weil-Malherbe and Green (1955) modified by Wagelin et al. (1978). Homogenates (10% w/v) were prepared in cold distilled water and centrifuged at 3000 rpm, while supernatant was used for enzyme source. AMP deaminase was expressed as micro moles of ammonia formed / mg protein / h.

For assaying adenosine deaminase activity, the method of Agarwal and Parks (1978) was used. Homogenates tissue (10% w/v) were prepared in ice cold distilled water and centrifuged at 3000 rpm for 15 min to obtain a clear supernatant which was used as enzyme source.

Alkaline and acid phosphatases were estimated using the method developed by Kind and King (1954). The enzyme assays were made after preliminary standardization regarding linearity with respect to time of incubation of enzyme concentration.

RESULTS AND DISCUSSION

Exposure of lihocin caused significant behavioral changes in the fish *C. punctatus*. On the introduction of lihocin, all the fish immediately settled down at the bottom of the tub. Within 10 to 15 min, the fish felt suffocation and they came to the water surface to gasp for air. As exposure period increased, the surfacing phenomenon of fish also increased. Also, the rates of operculum movement, mucous secretion from skin and respiration through gill also increased. After some time, the opercula movement of fish slowed down, although they tried to stay at the upper water surface, but the loss of body equilibrium was pronounced. Finally, all the body activity decreased and they settled down at the base of the aquaria and died. Moreover, control animals were free from such behavioral changes.

The protein content decreased in the liver, brain and kidney tissues during lihocin treatment (Tables 1, 2 and 3). According to Nelson and Cox (2005) and Sathyanarayana (2005), the physiological status of animal is usually indicated by the metabolic status of proteins. Jrueger et al. (1968) reported that the fish can get the energy through the catabolism of proteins. Proteins are mainly involved in the architecture of the cell, which is the chief source of nitrogenous metabolism. Thus, the depletion of protein fraction in liver, brain and kidney tissues may have been due to their degradation and possible utilization for metabolic purposes. Increases in free amino acid levels were the result of breakdown of protein for energy and impaired incorporation of amino acids in protein synthesis (Singh et al., 1996). The toxicants may have effect on hormonal balance, which could directly or indirectly affect the tissue protein levels (Murthy and Priyamvada, 1982; Khilare and Wagh, 1988).

The free amino acid (FAA) pool was increased in the tissues of the fish during exposure to lihocin, while the elevated FAA levels were utilized for energy production by supplying them as keto acids into TCA cycle through aminotransferases to contribute energy needs during toxic stress. Increases in free amino acid levels were the result of breakdown of protein for energy and impaired incorporation of amino acids in protein synthesis (Singh et al., 1996). It is also attributed to lesser use of amino acids (Seshagiri et al., 1987) and their involvement in the maintenance of an acid-base balance (Moorthy et al., 1984). Natarajan (1985) suggested that stress conditions induce elevation in the transanimation pathway. The increase in FAA levels of tissues indicates stepped up proteases activities and fixation of ammonia into keto acids (Srinivasa et al., 1986). Tripathi and Singh (2003) reported that the enhanced FAA may be due to depletion of reserved glycogen, so the fish can try to yield metabolic

Table 1. Effect of lihocin on the protein metabolism in the liver tissue of *Channa punctatus* (Bloch).

Parameter	Control	Lihocin			
		24 h	48 h	72 h	96 h
Total proteins	163.6 ± 2.11	156.16 ± 1.34* PC = -4.29	149.83 ± 2.79* PC = -8.16	132.81 ± 3.61 PC = -18.82	124.84 ± 2.63 PC = -23.69
Free amino acids	719 ± 8.96	742 ± 42.14 PC = 3.19	796 ± 30.29 PC = 10.70	891 ± 81.62 PC = 23.92	978 ± 94.36 PC = 36.02
Glutamine	84.60 ± 6.42	87.29 ± 5.22* PC = 3.17	93 ± 6.45* PC = 9.92	98.30 ± 4.85 PC = 16.19	117.38 ± 9.85 PC = 38.74
Alkaline phosphatases	9.33 ± 0.99	10.46 ± 0.76* PC = 12.11	12.63 ± 0.54 PC = 35.36	20.64 ± 0.94 PC = 121.22	21.55 ± 2.61 PC = 130.97
Acid phosphatases	8.36 ± 0.36	9.24 ± 0.11* PC = 10.52	13.08 ± 0.85 PC = 56.45	15.08 ± 0.28 PC = 80.38	17.75 ± 0.85 PC = 112.32
AlAT	23.5 ± 1.60	28.43 ± 1.62 PC = 20.97	32.33 ± 2.13 PC = 37.57	33.20 ± 1.85 PC = 41.27	40.16 ± 1.46 PC = 170.89
AAT	19.75 ± 0.75	22.78 ± 1.18 PC = 15.34	28.36 ± 1.36 PC = 43.59	31.46 ± 4.65 PC = 59.29	33.84 ± 1.71 PC = 71.34
GDH	0.71 ± 0.02	0.83 ± 0.01 PC = 16.90	1.06 ± 0.02 PC = 49.29	1.12 ± 0.12 PC = 57.74	1.28 ± 0.23 PC = 80.28
AMP deaminase	0.39 ± 0.02	0.42 ± 0.01* PC = 7.69	0.62 ± 0.01 PC = 58.97	0.68 ± 0.03 PC = 74.35	0.74 ± 0.042 PC = 89.74
Adenosine deaminase	0.132 ± 0.04	0.145 ± 0.016* PC = 9.84	0.196 ± 0.023 PC = 48.48	0.203 ± 0.001 PC = 53.78	0.283 ± 0.002 PC = 114.39

Each value is mean SD of 6 observations. All values are statistically significant from control at 5% level ($p < 0.05$). PC denotes percent change over control. * Not significant.

Table 2. Effect of lihocin on the protein metabolism in the brain tissue of *Channa punctatus* (Bloch).

Parameter	Control	Lihocin			
		24 h	48 h	72 h	96 h
Total proteins	103.83 ± 2.26	96.16 ± 1.34* PC = -7.39	92.33 ± 1.88* PC = -11.07	86.32 ± 2.80 PC = -16.86	69.83 ± 3.49 PC = -32.74
Free amino acids	562 ± 20.46	571 ± 13.22 PC = 1.60	598 ± 14.65 PC = 6.40	617 ± 8.44 PC = 9.78	645 ± 19.04 PC = 14.76
Glutamine	58.34 ± 5.99	62.80 ± 6.43* PC = 7.64	69.85 ± 5.65 PC = 19.72	71.823 ± 6.42 PC = 23.10	79.88 ± 8.09 PC = 36.92
Alkaline phosphatases	7.03 ± 0.93	8.20 ± 1.09 PC = 16.64	10.93 ± 0.72 PC = 55.47	12.11 ± 0.32 PC = 72.26	12.59 ± 0.86 PC = 79.08
Acid phosphatases	6.25 ± 0.73	7.24 ± 0.07 PC = 15.84	9.40 ± 1.20 PC = 50.40	12.80 ± 0.15 PC = 104.8	14.66 ± 1.11 PC = 134.56

Table 2. Contd.

AIAT	12.16 ± 1.21	14.35 ± 0.73 PC = 18.00	19.16 ± 1.34 PC = 57.56	22.80 ± 1.10 PC = 87.50	27.00 ± 1.64 PC = 122.03
AAT	8.48 ± 0.95	9.48 ± 0.99 PC = 11.79	12.55 ± 1.35 PC = 47.99	16.45 ± 1.89 PC = 93.98	19.53 ± 1.24 PC = 130.30
GDH	0.43 ± 0.02	0.62 ± 0.01 PC = 44.18	0.81 ± 0.01 PC = 88.37	0.921 ± 0.01 PC = 114.18	1.03 ± 0.02 PC = 139.53
AMP deaminase	0.29 ± 0.01	0.36 ± 0.02 PC = 24.14	0.37 ± 0.021 PC = 27.58	0.44 ± 0.03 PC = 51.72	0.64 ± 0.001 PC = 120.68
Adenosine deaminase	0.098 ± 0.001	0.108 ± 0.011* PC = 10.20	0.123 ± 0.023 PC = 25.51	0.143 ± 0.012 PC = 45.91	0.193 ± 0.001 PC = 96.93

Each value is mean SD of 6 observations. All values are statistically significant from control at 5% level ($p < 0.05$). PC denotes percent change over control. * Not significant.

Table 3. Effect of lihocin on the protein metabolism in the kidney tissue of *Channa punctatus* (Bloch).

Parameter	Control	Lihocin			
		24 h	48 h	72 h	96 h
Total proteins Micro grams/100 mg wet.wt	213.83 ± 2.60	204.5 ± 1.25* PC = -4.36	195.83 ± 1.77* PC = -8.41	182.24 ± 0.73 PC = -28.80	164.23 ± 0.65 3.191.29
Free amino acids Micro grams/100mg wet.wt	342 ± 12.59	362 ± 14.09 PC = 5.84	379 ± 12.49 PC = 10.81	399 ± 12.49 PC = 16.66	399 ± 15.09 PC = 21.92
Glutamine Micro grams/100mg wet.wt	36.85 ± 6.21	39.11 ± 3.84* PC = 16.13	43.29 ± 5.11 PC = 17.47	47.82 ± 44.10 PC = 29.76	53.86 ± 3.17 PC = 46.16
Alkaline phosphatases ip formed/mg protein/hr	9.64 ± 1.08	10.42 ± 0.14* PC = 8.09	11.36 ± 1.87 PC = 17.84	13.80 ± 0.16 PC = 43.15	15.03 ± 0.98 PC = 55.91
Acid phosphatases ip formed/mg protein/hr	5.50 ± 0.53	6.62 ± 0.12 PC = 20.36	8.07 ± 0.15 PC = 46.72	9.42 ± 0.28 PC = 71.27	10.24 ± 0.87 PC = 86.18
AIAT Micro grams/100mg wet.wt	17.33 ± 1.37	18.29 ± 0.24* PC = 5.53	20.83 ± 1.34 PC = 20.19	24.22 ± 0.18 PC = 39.75	26.5 ± 2.14 PC = 52.91
AAT Micromoles of pyruvate formed/mg protein/hr	14.29 ± 1.99	15.42 ± 0.86 PC = 7.90	18.07 ± 0.53 PC = 26.45	19.34 ± 0.46 PC = 35.33	21.91 ± 1.53 PC = 53.32
GDH Micromoles of formazon formed/mg protein/hr	0.09 ± 0.01	0.13 ± 0.01 PC = 44.44	0.22 ± 0.01 PC = 144.44	0.263 ± 0.02 PC = 192.22	0.312 ± 0.03 PC = 246.66
AMP deaminase Micromoles of ammonia formed/mg protein/hr	0.24 ± 0.001	0.29 ± 0.03 PC = 20.83	0.35 ± 0.017 PC = 45.83	0.38 ± 0.06 PC = 58.33	0.41 ± 0.025 PC = 70.83
Adenosine deaminase Micromoles of ammonia formed/mg protein/hr	0.045 ± 0.002	0.072 ± 0.012 PC = 60.00	0.096 ± 0.002 PC = 113.33	0.120 ± 0.01 PC = 166.66	0.136 ± 0.001 PC = 202.22

Each value is mean SD of 6 observations. All values are statistically significant from the control at 5% level ($p < 0.05$). PC denotes percent change over control. * Not significant.

energy by gluconeogenesis process. Similar findings were observed by Vijuen and Steyn (2003) in various animals during different toxic conditions.

The elevated levels of glutamine in liver, brain and kidney tissues reveal an enhancement of the biosynthesis of glutamine (Tables 1, 2 and 3). The elevated glutamine may be utilized in the formation of amino acids in protein synthesis. According to Narsaimha and Raman (1985), the organochlorine pesticides may initiate the synthesis of glutamine during toxic conditions. The present finding suggests that the fish has inherent tissue specific resistance on the potentiality to withstand ambient pesticide toxicity by suitably modulating its metabolic profiles.

The increase in AMP deaminsae and adenosine deaminase activities could contribute ammonia to the tissue through the purine nucleotide metabolism. According to Nelson and Cox (2005), these deaminases contribute little amount of nucleotides in the different tissues through purine nucleotide metabolism. Enhanced activity of deaminases may be due to tissue damage under xenobiotic action.

The lihocin pesticide caused increase in GDH activity in the tissues during initial periods of exposure. The important function of GDH is that the amino group of most amino acids is transferred to α-ketoglutarate to produce glutamate. According to Nelson and Cox (2005) and Sathyanarayana (2005), increased GDH activity may indicate an increased rapid utilization of amino acids. The oxidation of glutamate in Kreb's cycle leads to increased energy, though it is small (Narasimha and Rama, 1985).

The activities of AAT and AlAT during the toxic exposure of lihocin pesticide were enhanced (Tables 1, 2 and 3). The elevated activities of AAT and AlAT were observed by Narasimha et al. (1986) in *Anabas testudineus* during treatment of organophosphates pesticides in *T. mossambica* under linden toxicity and by Sajal et al. (1988) in gastropoda, *Thiara lineata*, during methyl parathion toxicity.

The alkaline and acid phosphatases were enhanced during the toxic exposure period and under stress conditions. The elevated levels of phosphatases may indicate an increase in the rate of phosphorylation and transport of molecules across the cell membrane. As such, the enhanced phosphatases activity revealed an increase in the transportation of metabolites through the cellular cellular membrane (Venkateshwarlu et al., 1990) and it was reported that the pesticides cause significant increase in the cellular damage which enhanced the activity of phosphatases activity. Carla et al. (2005) reported that phosphatases activity was altered in kupffer-melanomacro phagic cells of *Rana esculenta* during environmental pollution. As such, the fish can utilize stored proteins to overcome toxic stress. During toxic stress, the levels of key enzymes involved in protein metabolism are changed.

The present study concludes that lihocin, in sub lethal concentration, affects tissue protein metabolism in *Channa panctatus*. As a consequence of lihocin toxicity, the fish shifts to alternate methods of metabolism to overcome the toxic stress and maintain its survival in the polluted environment.

ACKNOWLEDGEMENT

One of the authors (Abdul Naveed) gratefully acknowledges Sri, Abdul Shukur, founder of Panchsheel College of Education, Nirmal, for providing laboratory facilities and needful help.

REFERENCES

Agarwal RP, Parks RE (1978). Adenosine deaminase from human erythrocytes in: Hoffee PA, Jones ME (eds) methods in enzymology Vol 2 Academic Press New York, pp. 502-507.

Bayne BL, Widdow L, Worral C (1977). Physiological responses of marinebiota to Pollutanants. J. Academic Press New York, p. 379

Fenoglio   C,, Boncompagni E, Fasola M, Gandini C, Comizzoli S, Milaness G, Barni S(2005) Effect of environmental Pollution on the liver parenchyma cells and kupffer melanomacrophagic cell of the frog *Rana esculenta*. J. Eco. Toxic Environ. Saf., 60(3): 259-268.

Colowick SP, Kaplan NO (1967). In: methods in enzymology (ed. Hirs, C.H.W). Determination of amide residues by chemical methods, 11: 63.

Finney DJ (1971). In: probit Analysis, 3rd Ed., Cambridge University Press. Cambridge, p. 333.

Gonzalez R, Martienez L, Tenrron O (2004). Effect of themtidne and phedeartital on methyl parathion metabolism. *I. Hyallella azteca*. Bull. Environ. Contam. Toxicol., 72(B): 1247-1252.

Jhingran VG (1975). Fish and Fisheries of India. Hindustan Publishing Corporation (India), New Delhi, pp. 954.

Jrueger HW, Saddler JB, Chapman GA, Tinsely IJ, Lowry RR (1968). Bioenergetics, exercise and fatty acids of fish. J. Am. zool., 8: 119.

Khilare YK, Wagh SB (1988). Long term effects of pesticides Endosulfan, Malathion and sevin on the fish *Puntius stiqma*. J. Environ. Ecol., 6(3): 589-593.

Kind PRN, King EJ (1954). *In vitro* determination of serum alkaline phosphatase. J. Clin. Pathol., 7: 322.

Lee YL, Lardy H (1965) Influence of thyroid hormones on L. lglysaro phosphate dehydrogenase in various organ of the rat. J. Biol. Chem. 240: 1427.

Lehtonen KK, Leimio, S (2003). Effect of copper and malathion on metallothionin in the levels of acetyl cholinesterase activity of the mussel *Mytilus radulis* and clam *Hacomg batthica* from the north Baltic sea. J. Bull. Environ. Contam. Toxicol., 71(3): 489-496.

Lowry OH, Rosebrough NJ, Farr WL, Randall RJ (1951) Protein measurement with the folin phenol regent. J. Biol. Chem., 182: 265-275.

Metelev VV, Kanaev AI,   Dzaskhova NG (1983) Water toxicology Amerind pub co pvt Ltd New Delhi, p. 3.

Moor S, Stein WH (1954) A modified ninhydrin reagent for the photometric determination of amino acids and related compounds. J. Biol. Chem., 211: 907-913.

Murthy AS, Priyamvada DA (1982). The effect of endosulfan and its isomers on tissue protein glycogen and lipids in the fish *Channa punctatus*. J. Pestic. Biochem. Physiol., 17: 280-286.

Moorthy KS, Kashi RB, Swamy KS, Chetty CS (1984) Changes in respiration and ionic content in the tissues of fresh water mussel exposed to methyl-parathion toxicity. J Toxicol. Lett., 21: 287-291.

Narasimha MB, Sathya Prasad K, Madhu C, Ramana RKV (1986) Toxicity of Lindane to fresh water fish *Tilapia mossambica*. J. Environ. Ecol., 4: 1.

Narasimha MB, Ramana RKV (1985) Alteration in the nitrogen metabolism in liver of fish *T. mossambica* under lindane toxicity. J. Pids 6th Nat. Sym. Life Sci., pp. 17-19.

Natarajan GM (1985) Inhibition of branchial enzymes in Snake head fish (*Channa striatus*) by oxy demetom-methyl. Pesti. Biochem. Physiol., 23: 41-46.

Reitman S, Frankel S (1957). Calorimetric determination of serum glutamic oxaloacetic acid and glutamate pyruvate transaminase. J. Ann. Clin. Pathol., 28: 560.

Sajal R, Chakrabarty T, Bose A, Robin S (1988) Effect of sublethal concentration of methyl parathion of enzymes. J. Environ. Ecol. 6(3): 563-567.

Salah AH (1983). Comparative studies on the neurotoxicity of O.P compounds in different animal species. J. Neuro. Toxicol. 40: 107-116.

Sathyanarayana U (2005). Biochemistry book and allied (P) Ltd. 8/1 Chintamani Das Lane Kolkata 700009 India, pp. 349

Seshagiri RK, Srinivas M, Kashi Reddy B, Swamy KS, Chetty CS (1987) Effect of benthiocarb on protein metabolism of teleost, *Sarotherodon mossambica*. Ind. J. Environ. Health, 29: 440-450.

Singh DK, Singh A, Agarwa RA (1993). Neriumindicum as a potent moulluscicide of plant origin. J. Med. App. Malacol., 5: 93-95.

Singh A, Singh DK, Mishra TN, Agarwal RA (1996) Molluscicides of plant origin. J. Biological Agric. Hortic., 13: 205-252.

Singh D, Singh A (2000). The acute toxicity of plant origin pesticides into the fresh water fish Channa punctatus. J. acta hydrochim. Hydrobiol., 28(2): 92-94.

Srinivasa MK, Kasi Reddy, Swamy KS, Sri Ramulu C (1986) Dichlorovos induced metabolism change in tissues of fresh water murrel *L. marginalis*. J. Environ. Ecol., 3: 278-279.

Tripathi PK, Singh A (2003) Toxic effects of dimethoate and carbaryl pesticides on reproduction and related enzymes of the fresh water Snail *Lymnaea acuminate*. J. Bull. Environ. Contam. Toxicol., 71: 0535-0542.

Venkateshwarlu P, Shobha RVJ, Janaiah C, Prasad MSK (1990) Effect of Endosulfan and Kelthane on haematology and serum biochemical parameters of the teleost, *Clarias batuachus (Linn)*. Indian J. Comp. Anim. Physiol., 8(1): 8-13.

Veronica W, Collins PA (2003). Effect of cypermethrin on the fresh water Crab, *Trichodactylus*. J. Bull. Environ. Contam. Toxicol., 71(1): 106-113.

Vijuen A, Steyn GJ (2003) Zinc effects on the embryos and larvae of the sharp tooth catfish, *Clarias gaiepinus*. J. Burchell, p.1022.

Wagclin I, Manzolin FA, Pane G (1978). Enzymes involved in AMP metabolism in chick embryo muscles, J. Comp. Biochem. Physiol., 59B: 55-57.

Weil-Malharbe H, Green RH (1955) Ammonia formation in brain adenylic deaminase. J. Biochem., 61: 218-224.

12

Ribosomal protein S3 gene expression of *Chironomus riparius* under cadmium, copper and lead stress

Kiyun Park and Inn-Sil Kwak*

Department of Fisheries and Ocean Science, Chonnam National University, San 96-1, Dundeok-dong, Yeosu, Jeonnam 550-749, Republic of Korea.

Heavy metals are of interest because they are often present at significant levels in the environment and can have severe effects on the aquatic environment. To examine the effects of oxidative stress induced by heavy metals on chironomids, the full-length cDNA of ribosomal protein S3 (RpS3) from *Chironomus riparius* was determined using molecular cloning. The basal expression of the RpS3 gene was not affected by changes in environmental rearing conditions such as temperature, media, and sediment type. However, RpS3 gene expression in *C. riparius* increased significantly in response to exposure to cadmium, copper, and lead, regardless of the exposure level. These results suggest that expression of the RpS3 gene, which is associated with the DNA repair process, may be used as an indicator of oxidative stress induced by heavy metals during environmental risk assessment.

Key words: *Chironomus riparius*, ribosomal protein S3, cadmium, copper; lead, environmental risk assessment.

INTRODUCTION

Environmental pollutants such as heavy metals pose serious risks to many aquatic organisms via changing neurophysiologic, biochemical and behavioral parameters (Scott and Sloman, 2004). Heavy metals from natural and anthropogenic sources accumulate in aquatic sediments in the form of trace elements, where they pose a threat to sediment biotic communities (Muntau and Baudo, 1992; Cheng 2003; Besser et al., 2008). Additionally, heavy metals such as cadmium, copper, lead, chromium, nickel, arsenic and mercury are frequently detected as groundwater contaminants (Clements and Kiffney, 1994). Aquatic species take up and accumulate both essential and nonessential trace elements from sediments. The heavy metals not assimilated into the aquatic organisms or not easily degraded or excreted are transferred to higher trophic level organisms (Reynoldson, 1987; Tessier and Campbell, 1987; Landrum and Robbins, 1990; Eimers et al., 2002). Therefore, understanding the responses of aquatic organisms to heavy metal toxicity is

important to monitoring water quality.

For the purpose of this study, cadmium was chosen as the contaminant of interest because it is widely recognized as an environmental pollutant (Aoki et al., 1984) and is highly toxic, affecting a wide range of physiological processes such as plasma membrane transport and the transcription of genes (Maroni et al., 1986). Exposure to cadmium via the air and food can also lead to renal tubular dysfunction (Korenekova et al., 2002) or reproductive complications (Massanyi et al., 1996; Lukac et al., 2003; Henson and Chedrese, 2004).

Copper pollution appears in the aquatic environment as a result of mine washing or agricultural leaching. Although copper is an essential trace element involved in biological functions such as iron absorption, hemopoiesis, and fermentation (Skalicka et al., 2005), it is also one of the most toxic heavy metals (To´th et al., 1996). Organs of aquatic animals may accumulate copper (Rojik et al., 1983; B´alint et al., 1997), which can lead to redox reactions generating free radicals that ultimately cause morphological alterations and change certain physiological processes.

Due to its high abundance and physical characteristics, such as ductility and high density, lead poses a major threat to genomic processes in vertebrates (Mortada et

*Corresponding author. E-mail: iskwak@chonnam.ac.kr, inkwak@hotmail.com.

al., 2004). In aquatic habitats, lead can be highly toxic for aquatic species since it can disrupt metabolic pathways and cause ionoregulatory damage (Rogers et al., 2005). Despite significant reductions in its use in paint production and as a fuel additive, lead continues to enter the environment through anthropogenic means, thereby retaining its status as a priority pollutant (USEPA, 2006). Lead acts as a Ca^{2+} antagonist (Busselberg et al., 1991; Rogers and Wood, 2004). At the physiological level, lead accumulates in cellular organelles (Qian and Tiffany-Castiglioni, 2003), impairing the properties of some calcium-dependent proteins such as heat shock proteins (HSPs) with lethal consequences on reproductive behaviors (Feder and Hofmann, 1999). The principal effects of chronic lead exposure on aquatic organisms are presumably hematological (Hodson et al., 1978), neurological (Davies et al., 1976) and renal (Patel et al., 2006) impairment.

Ribosomal protein genes are essential for cellular development (Chen and Ioannou, 1999). The four rRNAs (28S, 18S, 5.8S and 5S) are used in the ribosome machinery for protein biosynthesis. Because ribosomes are protein-rRNA complexes, this also implies that there is a high degree of amino acid sequence conservation between equivalent ribosomal proteins in different species (Draper and Reynaldo, 1999). Ribosomal proteins have the complex task of coordinating ribosome structure and protein biosynthesis to maintain cell homeostasis and survival. Some ribosomal proteins have secondary functions independent of their involvement in protein biosynthesis. A number of these proteins function as cell proliferation regulators, and in some instances, as inducers of cell death (Chen and Ioannou, 1999). Among the ribosomal proteins, ribosomal protein S3 (RpS3) is unusual in that it has multiple functions, including an apurinic/apyrimidinic (AP) lyase activity that participates in the repair of DNA damage. RpS3 shuttles between the cytoplasm and the nucleus to function in both compartments. As RpS3 has a nuclear localization signal in the N-terminal region, it is believed that its translation function operates in the cytosol, while its repair function operates in the nucleus (Lee et al., 2002). These activities repair the DNA damage caused by oxidizing agents and ionizing radiation. In addition to their role in ribosomal functions, many ribosomal proteins have secondary functions in replication, transcription, RNA processing, DNA repair, and malignant transformation (Wool, 1996). More than 80 different types of ribosomal proteins have been identified in eukaryotes, but only a few genes have been sequenced in Chironomus (Govinda et al., 2000; Martínez-Guitarte et al., 2007).

Toxins and other stressors can cause changes in gene expression, which have proven useful as biomarkers. For example, the metallothionein gene in Drosophila species was induced by a number of heavy metals, including zinc, cadmium, copper, silver, and mercury (Maroni et al., 1986). Metallothionein, heat shock proteins, and glutathione-S-transferase are involved in regulating the interactive effects of metal/metalloid mixtures at low dose levels (Wang and Fowler, 2008). However, few environmental studies of the Chironomus family have been conducted at the molecular level (Martínez-Guitartea et al., 2007; Park and Kwak, 2010), even though many studies have been conducted to evaluate their general biological responses to heavy metals (Martinez et al., 2001, 2003; Nowak et al., 2007). This is probably because there is little sequence information available regarding environmentally responsive genes. In C. riparius, characterized ribosomal proteins were only four genes of RpL8 (Govinda et al., 2000), RpL11 and RpL13 (Martínez-Guitartea et al., 2007), and RpL15 (Nair and Choi, 2011).

Chironomids are an ecologically diverse family of dipterans that are probably the most widespread aquatic macroinvertebrates. This is due to their physiological tolerance of various environmental conditions, such as extreme salinity or temperature, extreme pH levels, and reduced levels of dissolved oxygen (Anderson, 1977). Chironomids are increasingly used in toxicity experiments because of their widespread distribution, short life-cycle, ability to be reared in the laboratory and their easily identifiable life-cycle stages (Anderson, 1977). Given that they are benthic macroinvertebrates, chronomids can also be used for evaluation of sediment and water toxicity (Ibrahim et al., 1998). Indeed, morphological abnormalities have been observed in Chironomus larvae exposed to heavy metals and endocrine disrupting chemicals (Martinez et al., 2001, 2003; Kwak and Lee, 2005; Park and Kwak, 2008; Park et al., 2010). Thus, Chironomids are a good aquatic model for assessment of the toxicity of freshwater that has been contaminated with heavy metals.

In the present study, the RpS3 gene from C. riparius was characterized to determine the effects of oxidative stress induced heavy metals on chironomids. Comparative phylogenetic and molecular studies were conducted to analyze the homologies within insects. RpS3 expression was analyzed by means of real-time RT-PCR during different stages of life-cycle development and under various environmental conditions. Additionally, changes in RpS3 expression in response to exposure to various concentrations of cadmium, copper and lead were evaluated.

MATERIALS AND METHODS

Organisms

C. riparius were reared using the methodologies outlined by Streloke and Köpp (1995). We obtained C. riparius larvae from adults reared in the laboratory. An original strain was provided by the Korea Institute of Toxicology (Daejeon, Korea). The larvae were reared in an environmental chamber under long-day conditions with a light:dark cycle of 16:8 h and a light intensity of about 500 lx. The water in the incubator chamber was maintained at a constant temperature of 20±1°C (Sanyo, Osaka, Japan). After the larvae emerged from eggs, they were kept in Duran crystallizing dishes

(Schott, Mainz, Germany) with approximately 500 ml of M4 culture medium (Elendt, 1990) and a sediment layer of composed of 1 cm of fine sand (< 63 µm particle size) under continuous aeration. All dishes received 5 mg of food that had been ground in a blender daily (0.5 mg Larva^{-1}; Tetra-Werke, Melle, Germany), which resulted in food not being provided in a limited condition (Pery et al., 2002).

Exposure conditions

All experimental larvae were acquired by the eleventh day after hatching from the same control egg masses. The larvae were then exposed to water enriched with cadmium (CdCl$_2$), copper (CuSO$_4$) and lead (Pb(NO$_3$)$_2$)(Sigma-Aldrich Co., St Louis, USA). The nominal metal concentrations were based on available data regarding the toxicity values for *C. riparius* and actual concentrations in the Anam River (Janssens et al., 1998, 2001; Milani et al., 2003; Igwilo et al., 2006). The nominal concentrations for cadmium were 3.27 and 100 µgL^{-1}, while they were 1.10 and 100 µgL^{-1} for copper and 5. 50 and 100 µgL^{-1} for lead. All treatments were prepared from stock solutions of 0.1 gL^{-1} Cu^{2+}, 0.1 gL^{-1} Cd^{2+} and 0.5 gL^{-1} Pb^{2+}. Additionally, we characterize RpS3 expression variability under several environmental conditions during development. To accomplish this, the embryos were reared at 18°C (Martínez-Guitarte et al., 2007), using dechlorinated tap water as culture media (Nair and Choi, 2011) or using an artificial sediment (150 to 300 µm particle size) (Dias et al., 2008).

Thirteen fourth-instar *C. riparius* larvae were transferred into 300 mL Duran crystallizing dishes (Schott, Mainz, Germany) filled with 200 ml of M4 media and then treated them with one of the three aforementioned concentrations of copper, cadmium or lead. All organisms were exposed to the treatment for 24 h and all experiments were conducted in triplicate using independent samples (for example, three boxes each containing 3 µg L^{-1} of cadmium for 24 h). Each group of thirteen larvae was then utilized for subsequent analyses. Untreated larvae that were used as a control were also evaluated in triplicate. Exposure was conducted under a constant temperature (20±1°C) and a photoperiod of 16:8 h light:dark for all experiments.

RpS3 gene characterization

To amplify sequences of the RpS3 gene from *C. riparius*, polymerase chain reaction (PCR) was conducted using primers specific for higher Diptera (*Aedes aegypti* and *Anopheles gambiae* in Figure 1) consensus sequences. Multiple sequence alignments were then conducted using ClustalW (Thompson et al., 1994). The specific primers used were 5'TTGAARGGYCGCGTGTTTGAAGTT3' and 5' GAATTGCAYGGAGATGG WGGC3' for RpS3. 'R' represents mixture of A and G, 'W' represents a mixture of A and T, and 'Y' represents a mixture of T and C. The PCR mixture, which had a total volume of 50 µl, contained 1× Taq DNA polymerase buffer (SolGent, Daejeon, Korea), 200 µM dNTP, 2 units of Taq DNA polymerase (SolGent, Daejeon, Korea), and primers at a concentration of 20 µM. The template was composed of *C. riparius* DNA or cDNA. DNA was extracted from fourth-instar larvae using an AccuPrep$^®$ Genomic DNA extraction kit (Bioneer, Daejeon, Korea). The RNA extraction and cDNA synthesis were conducted as described in section 2.5. PCR was conducted by subjecting the samples to the following conditions: 5 min at 94°C, 35 cycles of 1 min at 94°C, 1 min at 55°C and 2 min at 72°C, and 7 min at 72°C using a MyCycler^{TM} thermal cycler (Bio-Rad, Hercules, USA). The reaction products, which consisted of 551bp of DNA or 480 bp of cDNA, were then cloned using a TOPO TA cloning kit (Invitrogen, Carlsbad, USA), after which they were sequenced using an ABI

3700 Genetic analyzer. To acquire the full-length RpS3 cDNA, we used a GeneRacer kit (Invitrogen, Carlsbad, USA) according to the manufacturers' instructions.

Phylogenetic analysis

Translation of the RpS3 cDNA sequence was conducted using a web tool available online (nucleic acid to amino acid translation) (http://www.biochem.ucl.ac.uk/cgi-bin/ mcdonald/cgina2aa.pl). The amino acid sequences were aligned with those of other organisms using Clustal X version 1.8 and then displayed using the GeneDoc Program (ver 2.6.001). A phylogenetic tree was constructed by neighbor-joining analyses using software available online (TreeTop) (Brodsky et al., 1993). Bootstrap values were calculated based on 1000 replicates.

Gene expression analysis

Total RNA was isolated from *C. riparius* fourth instar larvae (ten animals) using TRIZOL$^®$ reagent (Invitrogen, Scotland, UK) according to the manufacturers' instructions. Single-strand cDNA was synthesized from 4 µg of total RNA using random hexamer primer for reverse transcription in a 20 µl reaction mix using the SuperScriptTM III RT kit (Invitrogen, Scotland, UK). The cDNAs obtained were used as templates for PCR reactions with gene-specific primers for RpS3. In addition, PCR was conducted using primers specific for glyceraldehyde-3-phosphate dehydrogenase (GAPDH) as an internal control. The sequences of the oligonucleotide primers were: RpS3 forward 5'-GATGTCAAGACAACTGATGGA-3'; RpS3 reverse 5'-CACCATGGCATGCCTTCT CGA-3'; GAPDH forward 5'-GGTATTTCATTGAATG ATCACTTTG-3'; GAPDH reverse 5'-TAATCCTTGGATTGCATGTACTTG -3' (GenBank accession no. EU999991). The relative expression level of genes was measured using real-time RT-PCR, which was conducted on an iCyclerIQ thermocycler (Bio-Rad, Hercules, USA) with SYBR Green (Bio-Rad, Hercules, USA). The PCR program consisted of 1 cycle of 94°C for 3 min followed by 32 cycles of 94°C for 30 s, 55°C for 30 s, and 72°C for 30 s. The data were then analyzed using the delta-delta C$_t$ method in microsoft excel. Each test consisted of at least three replicates and the values were normalized using GAPDH as an internal control. The relative amount of RpS3 under several environmental conditions was calculated in compared to RpS3 expression in the normal rearing condition as a control.

Data analysis

All results are expressed as the mean ± SD, unless otherwise stated. The level of RpS3 mRNA in each sample was normalized against its own level of GAPDH based on standard curves. The levels of the RpS3 transcripts within each metal-treated group relative to the non-exposed control were estimated using normalized values. Differences in the RpS3 mRNA levels among groups were assessed by ANOVA followed by Tukey's multiple range test using SPSS 12.0KO (SPSS Inc., Chicago, IL, USA). Differences were considered significant at $p < 0.05$.

RESULTS

Identification of RpS3 gene and phylogenetic analysis

Partial sequences of the *C. riparius* RpS3 gene were

A

Species	Protein (%)	Accession number
C. riparius	100	EU683898
C. sonorensis	91	AAU06483
A. aegypti	89	XP_001651594
A. gambiae	88	XP_313275
P. americana	87	AAW57773
T. balearica	87	CAJ17219
B. mandarina	86	ABQ42714
A. mellifera	85	XP_396741
D. melanogaster	85	NP_524618

B

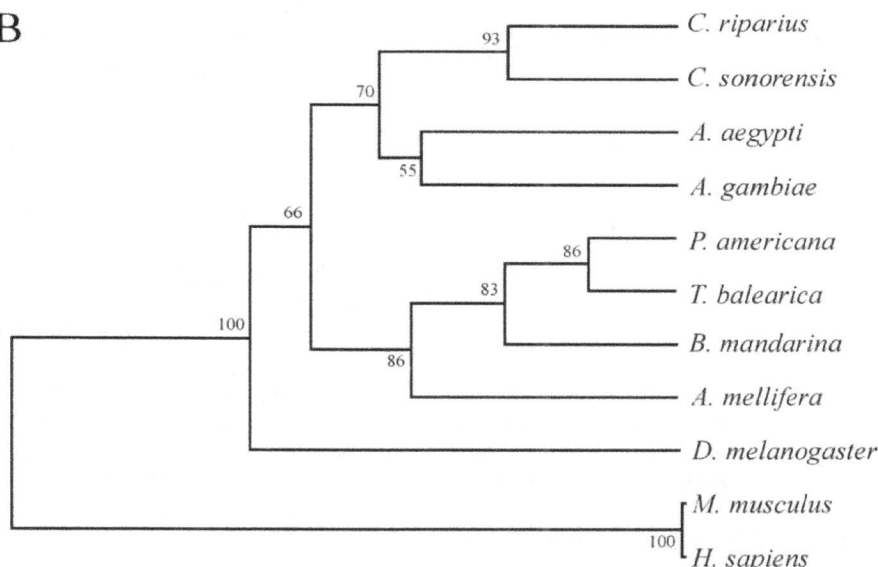

Figure 1. Phylogenetic trees of RpS3 genes constructed by neighbor-joining analysis (bootstrap value 1000). (A) Percentage of sequence similarity among RpS3 genes. (B) The numbers at the nodes are the percentage bootstrap values. Amino acid sequences were aligned using Clustal X (ver 1.8).

amplified by PCR using primers designed from higher diptera consensus sequences. The full-length of cDNA from *C. riparius* was then determined using molecular cloning and rapid amplification of cDNA ends (RACE). The complete cDNA sequence of *C. riparius* RpS3 gene was 813 bp and encoded a deduced amino acid sequence of 271. The complete coding sequences of *C. riparius* RpS3 were deposited in GenBank under accession No. EU683898. The genomic structures for RpS3 constitute two exons and one intron of 71 bp between the first 341 bp and second 472 bp exons. The sequence of the entire nucleotide region of RpS3 identified in this study was found to be 78% homologous with that of the mosquito, *Armigeres subalbatus*

(EU205451), 74% homologous with that of the mosquito, *Anopheles gambia* (BX058714), and 74% with that of the roundworm, *Brugia malayi* (XM_001896856). Although the homology of the DNA sequence was low, the deduced amino acid sequence was approximately 88% identical to that of other insect RpS3 proteins (Figure 1).

To evaluate the relationship of *C. riparius* RpS3 with homologues from other insects, we conducted a phylogenetic analysis (Figure 1). *C. riparius* RpS3 clustered with mosquitoes, such as *Culicoides sonorensis*, *Aedes aegypti* and *A. gambiae*, while RpS3 from a honey bee (*Apis mellifera*) formed another cluster with that of an American cockroach (*Periplaneta Americana*), a beetle (*Timarcha balearica*), and the wild silkmoth (*Bombyx mandarina*) (Figure 1). Overall, these results indicated that the RpS3 sequence from *Chironomus* is most closely related to those of mosquitoes, such as *C. sonorensis* (Figure 1).

RpS3 gene expression under several environmental conditions

To characterize RpS3 expression variability, its expression was analyzed under several environmental conditions during development (Figure 2). Specifically, the effects of temperature, media, and sediment type on RpS3 expression were evaluated. No significant differences were observed in the expression of the RpS3 gene during different developmental stages in response to the altered rearing conditions ($p > 0.05$) (Figure 2). Additionally, the RpS3 gene was stably expressed during different developmental stages under various environmental conditions.

RpS3 expression in response to heavy metals exposure

To examine the possible environmental regulation of RpS3 expression, real time RT-PCR analysis was conducted to evaluate the expression levels of transcripts under different conditions of cellular stress. Specifically, we analyzed the response of the RpS3 gene of *C. riparius* exposed to cadmium, copper and lead at three different concentrations (Figure 3). RpS3 gene expression increased significantly after cadmium exposure, regardless of the treatment dose. The response of RpS3 was greatest in *C. riparius* exposed to 100 μgL^{-1} cadmium for 24 h. After copper exposure, RpS3 gene expression significantly increased across all copper concentrations in a dose dependent fashion. Expression of the RpS3 gene also increased more than two-fold in *C. riparius* that were exposed to lead (Figure 3). Indeed, expression of the RpS3 gene was significantly higher in *C. riparius* that were exposed to lead than in those exposed to cadmium or copper ($p < 0.01$). There were no significant differences in expression observed among non-treated groups ($p > 0.05$).

DISCUSSION

Chironomids, benthic macro-invertebrates, are used extensively to assess the acute and sublethal toxicity of contaminated sediments and water (Kahl et al., 1997; Matthew and David, 1998; Matthew et al., 2001; Bettinetti et al., 2002; Choi et al., 2002). However, few studies have been conducted to evaluate the response of chironomids to chemical toxicity at the molecular level

(Park et al., 2009; Park and Kwak 2010; Nair and Choi, 2011). In this study, RpS3 cDNA from *C. riparius* was characterized, and transcript level of expression under several environmental conditions and different heavy metal conditions was analysed.

The *C. riparius* RpS3 cDNA encodes 271 aa with a theoretical pI of 5.14. The molecular phylogenetic relationships between *C. riparius* RpS3 and other insects showed that the RpS3 of *C. riparius* was most closely related to the RpS3 genes of mosquitoes, especially *Culicoides* species (Figure 1). This is reasonable because *C. riparius* is closer, in evolutionary terms, to *Culicoides* than to *Drosophila*, and both *Chironomus* and *Culicoides* belong to the suborder, Nematocera. A molecular phylogenetic study revealed that the *C. riparius* L11 and L13 genes were more closely related to those of *Anopheles*, belonging to the suborder Nematocera (Martínez-Guitarte et al., 2007). Recently, a study of expressed sequence tags from *C. tentans* (belonging to the subgenus *Camptochironomus*) also found that the greatest similarity (59%) was to that of sequence tags from *A. gambiae*, while only 24% similarity with the tags from *Drosophila melanogaster* was observed (Arvestad et al., 2005).

The RpS3 protein has been shown to be remarkably versatile in its ability to influence both ribosomal function and DNA repair transactions. Recent studies have shown that RpS3 is an integral part of the organization of the pre-40S subunit in yeast (Schafer et al., 2006). Beyond its role in ribosomal maturation, numerous studies have shown that RpS3 is involved in the repair of DNA damage. The RpS3 protein of *D. melanogaster* possesses various DNA repair activities, including the capacity to incise at apurinic/apyrimidinic (AP) sites and 8-oxo-7,8-dihydroguanine (8-oxoG) residues (Cappelli et al., 2003). Unlike *Drosophila* RpS3, the human RpS3 protein lacks the ability to liberate 8-oxoG from damaged DNA substrates; nevertheless, it possesses a remarkably high binding affinity for 8-oxoG (Hegde et al., 2004). In addition to its role in DNA repair, RpS3 is also involved in apoptosis (Jang et al., 2004). These functions of RpS3 indicate that RpS3 may be useful as an indicator of mutagenic agents or pollutants that cause oxidative stress.

Heavy metal ions, lead and cadmium are well known carcinogens with different natural origins (Rojas et al., 1999; Huff et al., 2007). Cadmium might contribute to increased risk for tumor formation via its inhibition of excision and mismatch DNA repair processes (Giaginis et al., 2006). Metal ion-DNA interactions are important in nature because they can alter the structure and function of genetic material. Pb^{2+} and Cd^{2+} bind to dsDNA, which results in different modifications of the dsDNA structure. Pb^{2+} interacts with dsDNA preferentially at adenine-containing segments, leading to oxidative damage and the formation of 2, 8-dihydroxyadenine, which is the oxidation product of adenine residues and a biomarker of

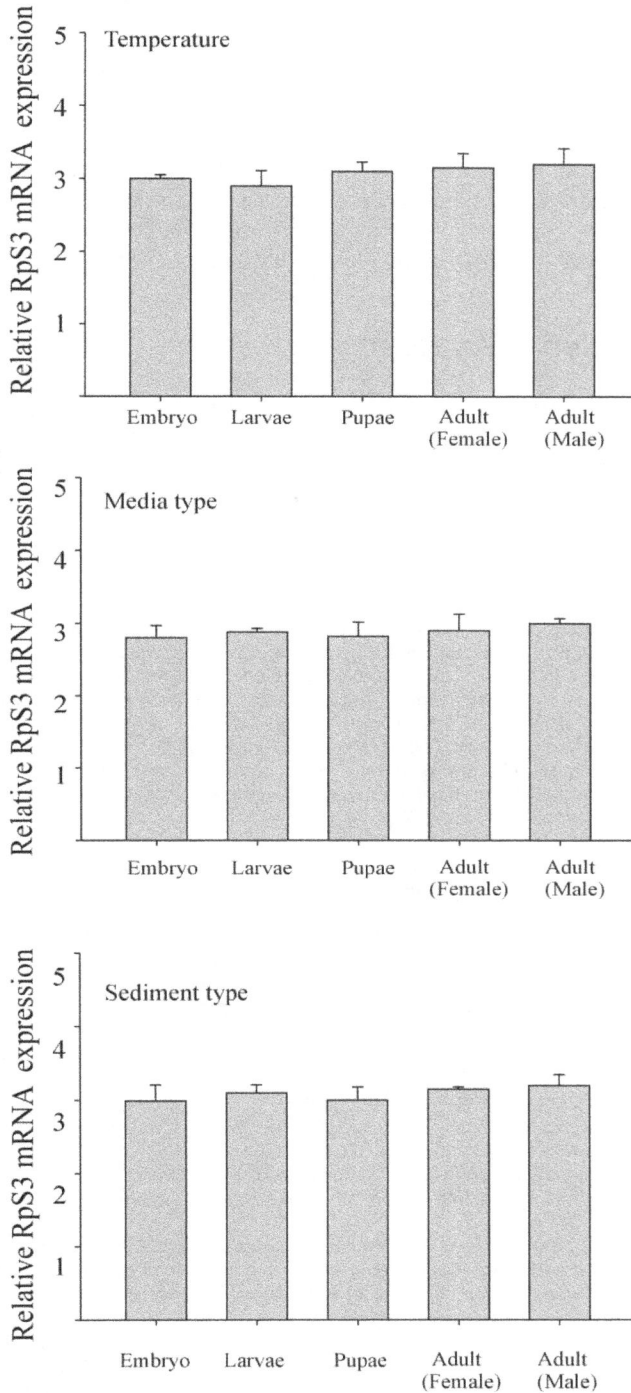

Figure 2. Analysis of RpS3 gene expression under several environmental conditions. RpS3 gene expression during life-history development at 18°C (upper), on media with dechlorinated tap water (middle) or on artificial sediment (150 to 300 μm particle sizes) (lower). The raw values were normalized against GAPDH, and the data were then used to calculate the relative expression levels. The relative amount of RpS3 under several environmental conditions was calculated in compared to RpS3 expression in the normal rearing condition as a control. The experiment was performed in triplicate and the values shown represent the mean ± standard error of the mean.

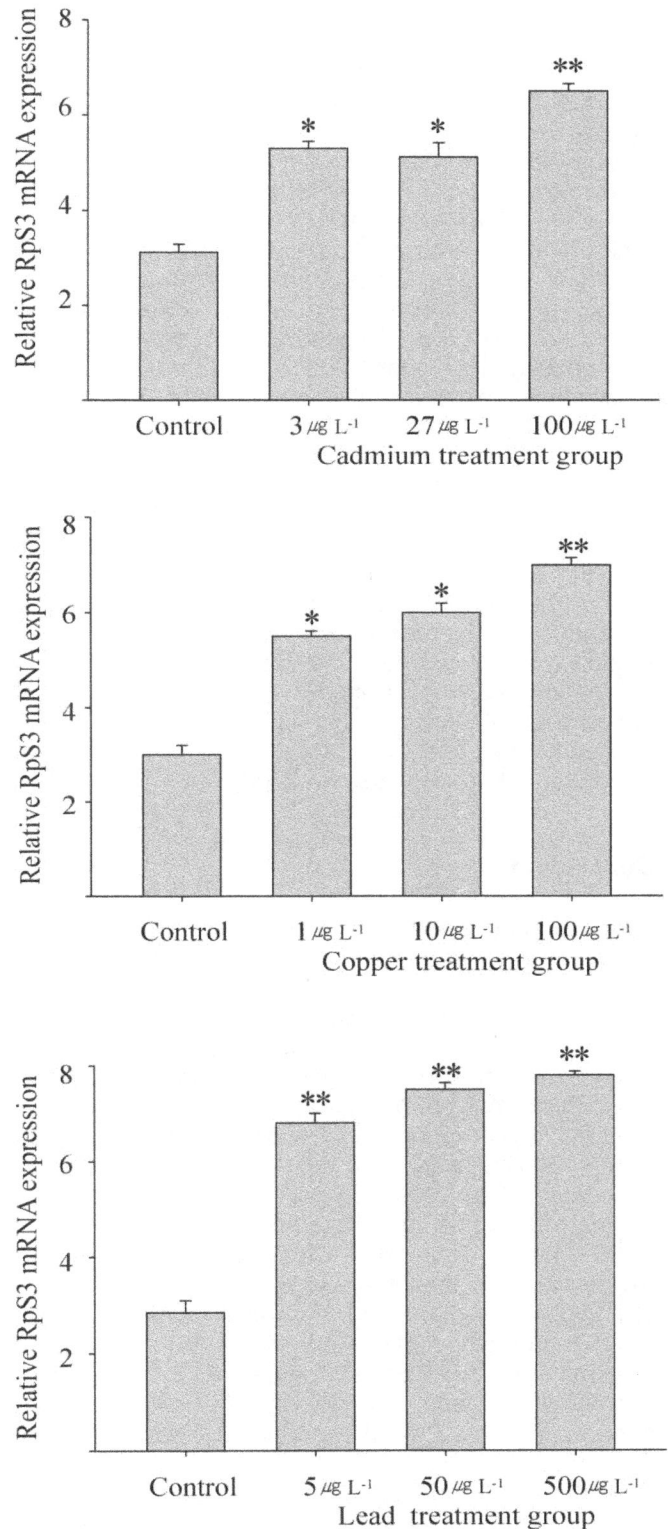

Figure 3. Expression of the RpS3 gene in the fourth-instar larvae of *Chironomus riparius* exposed to cadmium, copper and lead. mRNA expression is shown as relative to GAPDH expression after normalization. The experiment was performed in triplicate and the values shown represent the mean ± standard error of the mean (*indicates statistical significance at $p < 0.05$; ** indicates $p < 0.01$*).

DNA oxidative damage. The interaction with Cd^{2+} only causes conformational changes, which leads to destabilization of the double helix and can enable the action of other oxidative agents on DNA (Oliveira et al., 2008). Copper, which is a redox-active trace metal ion, induces an increase in oxidative stress that results in DNA damage and activation of p53-dependent cell death (Du et al., 2008). Additionally, copper is required as a co-factor for many enzymes that catalyze oxidation/reduction reactions, including those involved in electron transport (cytochrome c oxidase) (Prá et al., 2008), antioxidant enzymes (Cu/Zn superoxide dismutase and ceruloplasmin) (Linder, 2001), melanin and collagen biosynthetic pathways (tyrosinase and lysyl oxidase, respectively) (Lutsenko et al., 2007) and hormones (dopamine-monooxygenase and a-amidating monooxygenase) (Prá et al., 2008). The exposure of *C. riparius* to copper has been found to induce a delay in larval growth in both sexes and a reduction in the lifespan of males (Servia et al., 2006). Additionally, a significantly higher frequency of functional alterations, specifically decondensed centromeres and telomeres, and a reduction in the activity of Balbiani rings, was observed in *C. riparius* that were treated with copper. Finally, a dose dependent relationship was observed between copper concentration and the frequency of chromosomal aberrations (Michailova et al., 2006).

Conclusion

Previous studies have also shown that the expressions of ribosomal protein genes could be altered by heavy metal stress. The RPL23aB and RPL2 transcript level were decreased rapidly in soybean by heavy metal stress of copper (Ludwig and Tenhaken, 2001; McIntosh and Bonham-Smith, 2005). In contrast, it has been shown that other ribosomal protein genes, such as that of S9, was up-regulated in response to cadmium and repressed by zinc in molluscs (Achard et al., 2006). Recent report described that reduction of ribosomal protein L15 transcript occurred in *C. riparius* as a result of exposure to cadmium (Nair and Choi, 2011). However, the expressions of ribosomal protein genes L8 (Govinda et al., 2000), L11 and L13 (Martínez-Guitarte al., 2007) were unaffected after exposure to cadmium. To determine the possible environmental regulation of RpS3 expression, we analyzed responses to cadmium, copper and lead exposure. After cadmium and copper exposure, *C. riparius* RpS3 gene expression increased significantly, regardless of the exposure conditions (Figure 3). Additionally, expression of the RpS3 gene was significantly higher in *C. riparius* that were exposed to lead than in those exposed to cadmium or copper of relative low concentrations (Figure 3). The level of RpS3 expression was induced by heavy metal exposures, regardless of different heavy metal treatments. The

response to Pb^{2+} occurred as a result of its interaction with segments of dsDNA, which led to oxidative damage. These results suggest that the expression of RpS3 in *C. riparius* could be employed as an indicator of mutagenic agents, such as heavy metals that cause oxidative stress. Furthermore, to determine if RpS3 is useful as a biomarker, its expression should be investigated under field-conditions through characterization of RpS3 sequence information in wild species of *Chironomus*.

ACKNOWLEDGMENT

This study was supported by the National Research Foundation of Korea Grant funded by the Korean Government [NRF-2011-0007657] and [NRF-2011-0005878].

REFERENCES

Achard-Joris M, Gonzalez P, Marie V, Baudrimont M, Bourdineaud P (2006). cDNA cloning and gene expression of ribosomal S9 protein gene in the mollusk *Corbicula fluminea*: A new potential biomarker of metal contamination up-regulated by cadmium and repressed by zinc. Environ. Toxicol. Chem., 25: 527-533.

Anderson RL (1977). Chironomidae toxicity tests – biological background and procedures. In: Buikema, A.L. and Cairns, J., (Eds.). Aquatic Invertebrate Bioassays, American Society for Testing and Materials (ASTM), p. 209.

Aoki Y, Suzuki KT, Kubota K (1984). Accumulation of cadmium and induction of its binding protein in the digestive tract of fleshfly (*Sarcophaga peregrina*) larvae. Comp. Biochem. Physiol., 77: 279-282.

Arvestad L, Visa N, Lundeberg J, Wieslander L, Savolainen P (2005). Expressed sequence tags from the midgut and an epithelial cell line of *Chironomus tentans*: Annotation, bioinformatic classification of unknown transcripts and analysis of expression levels. Insect Mol. Biol., 14: 689-695.

B´alint T, Ferenczy J, Kátai F, Kiss I, Kráczer L, Kufcsák O, Láng G, Polyhos C, Szabó I, Szegletes T, Nemcsók J (1997). Similarities and differences between the massive eel (*Anguilla anguilla* L.) devastations that occured in lake Balaton in 1991 and 1995. Ecotoxicol. Environ. Saf., 37: 17-23.

Brodsky LI, Drachev AL, Leontovich AM, Feranchuk SI (1993). A novel method of multiple alignment of biopolymer sequences. Biosys., 30: 65-79.

Busselberg D, Evans ML, Rahmann H, Carpenter DO (1991). Effects of inorganic and triethyl lead and inorganic mercury on the voltage activated calcium channel of Aplysia neurons. Neurotoxicol., 12: 733-744.

Cappelli E, D'Osualdo A, Bogliolo M, Kelley MR, Frosina G (2003). *Drosophila* S3 ribosomal protein accelerates repair of 8-oxoguanine performed by human and mouse cell extracts. Environ. Mol. Mutagenetics 42: 50-58.

Chen FW, Ioannou YA (1999). Ribosomal proteins in cell proliferation and apoptosis. Int. Rev. Immunol., 18: 429-448.

Cheng S (2003). Heavy metal pollution in China: Origin, pattern and control. Environ. Sci. Pollut. Res. Int., 3: 192-198.

Clements WH, Kiffney PM (1994). Integrated laboratory and field approach for assessing impacts of heavy metals at the Arkansas River, Colorado. Environ. Toxicol. Chem., 13: 397-404.

Davies PH, Goettl JP, Sinley JR, Smith NF (1976). Acute and chronic toxicity of lead to rainbow trout *Salmo gairdneri*, in hard and soft water. Water Res. 10: 199-206.

Dias V, Vasseur C, Bonzom JM (2008). Exposure of *Chironomus riparius* larvae to uranium: Effects on survival, development time, growth, and mouthpart deformities. Chemosphere, 71: 574-581.

Draper DE, Reynaldo LP (1999). RNA binding strategies of ribosomal proteins. Nucleic Acid Res., 27: 381-387.

Du T, Ciccotosto GD, Cranston GA, Kocak G, Masters CL, Crouch PJ, Cappai R, White AR (2008). Neurotoxicity from glutathione depletion is mediated by Cu-dependent p53 activation. Free Radic. Biol. Med., 44: 44-55.

Feder M, Hofmann GE (1999). Heat shock proteins, molecular chaperones and the stress response: Evolutionary and ecological physiology. Ann. Rev. Physiol., 61: 243-282.

Giaginis C, Gatzidou E, Theocharis S (2006). DNA repair systems as targets of cadmium toxicity. Toxicol. Appl. Pharmacol., 213: 282-290.

Govinda S, Kutlow T, Bentivegna CS (2000). Identification of a putative ribosomal protein mRNA in Chironomus riparius and its response to cadmium, heat shock, and actinomycin D. J. Biochem. Mol. Toxicol., 14: 195-203.

Hegde V, Wang M, Deutsch WA (2004). Characterization of human ribosomal protein S3 binding to 7,8-dihydro-8-oxoguanine and abasic sites by surface Plasmon resonance. DNA Repair (Amst.), 3: 121-126.

Henson MC, Chedrese PJ (2004). Endocrine disruption by cadmium, a common environmental toxicant with paradoxical effects on reproduction. Exp. Biol. Med., 229: 383-392.

Hodson PV, Blunt BR, Spry DJ (1978). Chronic toxicity of water-borne and dietary lead to rainbow trout (Salmo gairdneri) in Lake Ontario water. Water Res., 12: 869-878.

Huff J, Lunn RM, Waalkes MP, Tomatis L, Infante PF (2007). Cadmium-induced cancers in animals and in humans. Int. J. Occup. Environ. Health, 13: 202-212.

Ibrahim H, Kheir R, Helmi S, Lewis J, Crane M (1998). Effects of organophosphorus, carbamate, pyrethroid and organochlorine pesticides, and a heavy metal on survival and cholinesterase activity of Chironomus riparius Meigen. Bull. Environ. Contam. Toxicol., 60: 448-455.

Jang CY, Lee JY, Kim J (2004). RpS3, a DNA repair endonuclease and ribosomal protein, is involved in apoptosis. FEBS Lett., 560: 81-85.

Korenekova B, Skalicka M, Nad P (2002). Cadmium exposure of cattle after long-term emission from polluted area. Trace Elem. Electrolytes 19: 97-99.

Kwak IS, Lee W (2005). Mouthpart deformity and developmental retardation exposure of Chironomus plumosus (Diptera: Chironomidae) to tebufenozide. Bull. Environ. Contam. Toxicol., 75: 859-865.

Lee CH, Kim SH, Choi JI, Choi JY, Lee CE, Kim J (2002). Electron paramagnetic resonance study reveals a putative iron–sulfur cluster in human rpS3 protein. Mol. Cell 13: 154-156.

Linder MC (2001) Copper and genomic stability in mammals. Mutat. Res., 475: 141–152.

Lukac N, Massanyi P, Toman R, Trandzik J (2003). Effect of cadmium on spermatozoa motility. Sovremena poljoprivreda, 3-4: 215-217.

Lutsenko S, Barnes NL, Bartee MY, Dmitriev OY (2007). Function and regulation of human copper-transporting ATPases. Physiol. Rev., 87: 1011-1046.

Maroni G, Lastowski-Perry D, Otto E, Watson D (1986). Effects of heavy metals on Drosophila larvae and a metallothionein cDNA. Environ. Health Perspect. 65: 108-116.

Martinez EA, Moore BC, Schaumloffel J, Dasgupta N (2001). Induction of morphological deformities in Chironomus tentans exposed to zinc- and lead-spiked sediments. Environ. Toxicol. Chem., 20: 2475-2481.

Martinez EA, Moore BC, Schaumloffel J, Dasgupta N (2003). Morphological abnormalities in Chironomus tentans exposed to cadmium-and copper-spiked sediments. Ecotoxicol. Environ. Saf., 55: 204-212.

Martínez-Guitarte JL, Planelló R, Morcillo G (2007). Characterization and expression during development and under environmental stress of the genes encoding ribosomal proteins L11 and L13 in Chironomus riparius. Comp. Biochem. Physiol. B Biochem. Mol. Biol., 147: 590-596.

Massanyi P, Lukac N, Trandzik J (1996). In vitro inhibition of the motility of bovine spermatozoa by cadmium chloride. J. Environ. Sci. Health Part A 31: 52-55.

Michailova P, Petrova N, Ilkova J, Bovero S, Brunetti S, White K, Sella G (2006). Genotoxic effect of copper on salivary gland polytene chromosomes of Chironomus riparius Meigen 1804 (Diptera, Chironomidae). Environ. Pollut., 144: 647-654.

Mortada WI, Sobh MA, El-Defrawy MM (2004). The exposure to cadmium, lead and mercury from smoking and its impact on renal integrity. Med. Sci. Monit., 10: 112-116.

Nair PM, Choi J (2011). Characterization of a ribosomal protein L15 cDNA from Chironomus riparius (Diptera; Chironomidae): transcriptional regulation by cadmium and silver nanoparticles. Comp. Biochem. Physiol. B Biochem. Mol. Biol., 159: 157-162.

Nowak C, Jost D, Vogt C, Oetken M, Schwenk K, Oehlmann J (2007). Consequences of inbreeding and reduced genetic variation on tolerance to cadmium stress in the midge Chironomus riparius. Aquat. Toxicol., 85: 278-284.

Oliveira SC, Corduneanu O, Oliveira-Brett AM (2008). In situ evaluation of heavy metal-DNA interactions using an electrochemical DNA biosensor. Bioelectrochem., 72: 53-58.

Park K, Bang HW, Park J, Kwak IS (2009). Ecotoxicological multilevel-evaluation of the effects of fenbendazole exposure to Chironomus riparius larvae. Chemosphere, 77: 359-367.

Park K, Kwak IS (2008). Characterization of heat shock protein 40 and 90 in Chironomus riparius larvae: effects of di(2-ethylhexyl) phthalate exposure on gene expressions and mouthpart deformities. Chemosphere, 74: 89-95.

Park K, Kwak IS (2009). Alcohol dehydrogenase gene expression Chironomus riparius exposed to di(2-ethylhexyl) phthalate. Comp. Biochem. Physiol. C Toxicol. Pharmacol., 150: 361-367.

Park K, Kwak IS (2010). Molecular effects of endocrine-disrupting chemicals on the Chironomus riparius estrogen-related receptor gene. Chemosphere, 79: 934-941.

Park K, Park J, Kim J, Kwak IS (2010). Biological and molecular responses of Chironomus riparius (Diptera, Chironomidae) to herbicide 2,4-D (2,4-Dichlorophenoxyacetic acid). Comp. Biochem. Physiol. C Toxicol. Pharmacol., 151: 439-446.

Patel M, Rogers JT, Pane EF, Wood CM (2006). Renal responses to acute lwad waterborne exposure in the freshwater rainbow trout (Oncorhynchus mykiss). Aquat. Toxicol., 80: 362-371.

Prá D, Franke SI, Giulian R, Yoneama ML, Dias JF, Erdtmann B, Henriques JA (2008). Genotoxicity and mutagenicity of iron and copper in mice. Biometals 21: 289-297.

Qian Y, Tiffany-Castiglioni E (2003). Lead-induced endoplasmic reticulum (ER) stress responses in the nervous system. Neurochem. Res., 28: 153-162.

Rogers JT, Patel M, Gilmour KM, Wood CM (2005). Mechanisms behind Pb-induced disruption of Na$^+$ and Cl$^-$ balance in rainbow trout (Oncorhynchus mykiss). Am. J. Physiol. Regul. Integr. Comp. Physiol., 289: 463-472.

Rogers JT, Wood CM (2004). Characterization of branchial lead-calcium interaction in the freshwater rainbow trout Oncorhynchus mykiss. J. Exp. Biol., 207: 813-825.

Rojas E, Herrera LA, Poirier LA, Ostrosky-Wegman P (1999). Are metals dietary carcinogens? Mutat. Res., 443: 157-181.

Rojik I, Nemcs'ok J, Boross L (1983). Morphological and biochemical studies on liver, kidney and gill of fishes affected by pesticides. Acta Biol. Hun., 34: 81-92.

Schafer T, Maco B, Petfalski E, Tollervey D, Bottcher B, Aebi U, Hurt E (2006). Hrr25-dependent phosphorylation state regulates organization of the pre-40S subunit. Nature, 441: 651-655.

Scott GR, Sloman KA (2004). The effects of environmental pollutants on complex fish behaviour: integrating behavioural and physiological indicators of toxicity. Aquat. Toxicol., 68: 369-392.

Servia MJ, Péry AR, Heydorff M, Garric J, Lagadic L (2006). Effects of copper on energy metabolism and larval development in the midge Chironomus riparius. Ecotoxicol., 15: 229-240.

Skalicka M, Korenekova B, Nad P (2005). Copper in livestock from polluted area. Bull. Environ. Contam. Toxicol., 74: 740-744.

Streloke M., Köpp H (1995). Long-term toxicity test with Chironomus riparius development and validation of a new test system. Blackwell Wissenschaftsverlag, Berlin/Vienna, P. 315 (Mitt A D Biol Bundesanst).

Thompson JD, Higgins DG, Gibson TJ (1994). CLUSTAL W: Improving the sensitivity of progressive multiple sequence alignment through sequence weighting, positions-specific gap penalties and weight matrix choice. Nucl. Acids Res., 22: 4673-4680.

To'th L, Juha'sz M, Varga T, Csikkel-Szolnoki A, Nemcso'k J (1996). Some effect of CuSO₄ on carp. J. Environ. Sci. Health B 31: 627-635.

USEPA (2006). National Recommended Water Quality Criteria. Office of Water, Washington, DC.

Wang G, Fowler BA (2008). Roles of biomarkers in evaluating interactions among mixtures of lead, cadmium and arsenic. Toxicol. Appl. Pharmacol., 233: 92-99.

Wool IG (1996). Extraribosomal functions of ribosomal proteins. Trends Biochem. Sci., 21: 164-165.

13

Metal quantification in cattle: A case of cattle at slaughter at Ota Abattoir, Nigeria

D. O. Nwude, J. O. Babayemi* and I. O. Abhulimen

Department of Chemical Sciences, Bells University of Technology, Ota, Ogun State, Nigeria.

In this study, cattle were assessed for the levels of metals in some organs. Six metals, Arsenic (As), Cadmium (Cd), Cobalt (Co), Chromium (Cr), Nickel (Ni) and Lead (Pb) were determined in twenty samples comprising four different parts (liver, kidney, muscle tissue and blood) from five cows at slaughter in a major abattoir in Ota, using flame atomic absorption spectrophotometer (S-series 712354v1.27), following digestion. The heavy metal concentrations ranged from 12.864 to 18.475 mg/kg for As; 0.522 to 2.131 mg/kg, Cr; ND-1.227mg/kg, Pb; 0.463 to 0.844 mg/kg, Cd; ND-0.112 mg/kg, Co; and ND-1.075 mg/kg, Ni. Since these observed levels were higher than the WHO standards, gross contamination of the cattle could be inferred; and the levels of metals in cattle parts could be used as biomarkers of metal pollution, though highly influenced by several factors.

Key words: Cattle, biomarker, heavy metal, metal pollution.

INTRODUCTION

Pollution is currently a global problem that exists in various dimensions (Adekola et al., 2002; Asonye et al., 2007). The type and degree of pollution varies with time and space. The sources of heavy metal pollution include natural sources (Miranda et al., 2009; Hobbelen et al., 2006), ore mining or metal smelting (Uhlig et al., 2001; Haimi and Mätäsniemi, 2002), municipal waste, industrial effluents, application of sewage sludge and animal manure on agricultural land (del Val, 1999; Blanco-Penedo et al., 2006; Salehi and Tabari, 2008), and aerial deposition of particulates from vehicular emission (Ward and Savage, 1994). The problem becomes worse when it becomes difficult to determine when, how and to what extent is the level of pollution in a particular environment. Several researchers far back in history have used various objects as markers of environmental pollution. Such markers could be changes in availability, levels and characteristics of the objects under consideration. Recently, biological indicators have become very common and have been found very useful. This is because living organisms have high sensitivity to

changes in environmental components and conditions that constitute their living. However, different organisms respond at different degree to changes in some environmental parameters. Therefore, the biodiversity needs to be well understood when considering biomarkers of environmental pollution.

Several researchers have investigated levels of pollutants in microbes, invertebrates, mammals, apes, plants and even human tissues as biomarkers of pollution. The use of moss (Acar, 2006), earthworm (Wang et al., 2009a; Elaigwu et al., 2007), fishes (Azmat et al., 2008), great tits (Geens et al., 2010), some gastropod and bivalve species (Liang et al., 2004), horse (Maia et al., 2006), microbial biomass carbon (Zhang et al., 2010; Wang et al., 2009a; Kizilkaya et al., 2004), human scalp hair (Wang et al 2009b), human nails (Were et al., 2008), blood and wool or hair from sheep, horses and alpacas (Ward and Savage, 1994), buffalo (Abou-Arab, 2001), and forest reindeer (Medvedev, 1995) have been reported.

The effectiveness and reliability of using the levels of metals in living organisms as biomarker of pollution are functions of several factors such as species, age of the organism and sex (Parker and Hamr, 2001), animal parts and tissues considered (Miranda et al., 2009; Nwude et al., 2010a), feeding habit, animal management system,

---

*Corresponding author.E-mail: babayemola@yahoo.co.uk.

Table 1. Variation of metal concentrations (mg/kg) in the Liver of the cows.

Cow	As	Cd	Co	Cr	Pb	Ni
A	12.864	0.698	0.03	0.522	0.417	0.446
B	16.564	0.581	ND	1.036	0.994	0.093
C	16.007	0.762	ND	1.515	0.882	0.061
D	16.635	0.776	ND	1.306	0.318	0.196
E	18.441	0.844	ND	1.305	0.276	ND

ND = Not detectable range.

Table 2. Variation of metal concentrations (mg/kg) in the kidney of the cows.

Cow	As	Cd	Co	Cr	Pb	Ni
A	17.39	0.739	0.021	1.256	0.276	0.573
B	15.334	0.869	ND	1.71	0.419	0.561
C	17.346	0.888	ND	1.311	1.227	0.394
D	18.475	0.996	ND	1.378	0.378	0.306
E	16.767	1.342	ND	1.151	ND	0.316

ND = Not detectable range.

Table 3. Variation of metal concentrations (mg/kg) in the muscle tissue of the cows.

COW	As	Cd	Co	Cr	Pb	Ni
A	15.931	0.463	ND	1.242	0.994	0.272
B	17.803	0.764	ND	2.131	0.981	1.075
C	17.221	0.715	ND	1.261	ND	0.216
D	17.034	0.733	ND	1.34	0.349	0.24
E	17.062	0.831	ND	0.955	0.18	0.378

ND = Not detectable range.

animal physiology, season, level of pollution, type and chemistry of the metal, metal uptake potential of the dominant plant species the animals feed on (Guala et al., 2010a; 2010b; Weis and Weis, 2004) in the environment under consideration, nature of the soil at the polluted site, leaching potentials of the metals (Chen et al., 2006), influence of the presence and /or level of the a metal on the other in the living organism (Blanco-Penedo et al., 2006) and in the concerned environmental media. Assessing the levels of metals in some organs of cattle as biomarker should put into consideration which cattle organ is to be analyzed. Examining investigations by Miranda et al. (2005), kidney had the highest accumulation of some metals, Cd, Pb and As, in the industrialized area; and Cd as well in the rural area. The trends were followed by the levels in the liver for all the metals, except in the rural area where Pb and As were

highest in the liver. Cattle, in addition to being the most common source of meat in South-West Nigeria, are grazing animals which are constantly in contact with the polluted environment; hence the need to assess the levels of pollutants, especially heavy metals in cow parts. This study therefore aimed at assessing the levels of metals which could be toxic at higher concentrations in cattle parts, since the most common source of meat in Nigeria, especially in the South-West, is cow; and to see if the levels in cattle could reflect pollution.

MATERIALS AND METHODS

Four different parts (Liver, kidney, muscle tissue and blood) from five different cows (A, B, C, D and E) were collected randomly in polythene bags immediately as the cattle were slaughtered in an abattoir located at Toll Gate, Ota, Ogun state. Since the cows were meant for consumption as meat (beef), and there was no law against that in the country, approval from animal ethical committee then became unnecessary. The samples were transferred to the laboratory and were stored in the refrigerator for about two weeks pending the time of analysis. 5 g of each sample (Liver, kidney, muscle tissue and blood) were introduced into the digestion flask. 5 mL of concentrated phosphoric acid were added and heated in the digestion chamber at temperature of 250 to 300°C for 2½ h. 20 ml deionized water were added, thoroughly shaken and filtered into a 50 mL standard flask. It was then made up to mark with deionized water (Nwude et al., 2010a; 2010b). As, Cd, Co, Cr, Pb and Ni in the filtrates were determined using Flame Atomic Absorption Spectrometer S- series 712354v1.27 at wavelengths 193.7, 228.8, 240.7, 357.9, 217.0, and 232.0 nm, respectively.

RESULTS AND DISCUSSION

Table 1 shows the metal concentrations in the liver ranging from 12.864 to 18.441 mg/kg for As; 0.581 to 0.844 mg/kg, Cd; ND- 0.03 mg/kg, Co; 0.522-1.515 mg/kg, Cr; 0.276-0.994 mg/kg, Pb; and 0.093 to 0.446 mg/kg, Ni. Levels in the kidney (Table 2) ranged from 15.334 to 18.475 mg/kg, As; 0.739 to 1.342 mg/kg, Cd; ND to 0.021 mg/kg, Co; 1.151 to 1.71 mg/kg, Cr; ND to 1.227 mg/kg, Pb; 0.306 to 0.573 mg/kg, Ni. From Table 3, muscle tissue gave 15.931 to 17.803 mg/kg, As; 0.463 to 0.831mg/kg, Cd; ND, Co; 0.955 to 2.131mg/kg, Cr; ND to 0.994 mg/kg, Pb; and 0.216 to 1.075 mg/kg, Ni. Blood concentration ranged from: 16.733 to 17.85 mg/kg, As; 0.523 to 0.843 mg/kg, Cd; ND to 0.112 mg/kg, Co; 0.942 to 1.357 mg/kg, Cr; ND to 0.742 mg/kg, Pb; and ND to 0.414 mg/kg, Ni (Table 4). Figure 1 compares the mean levels in the cow parts.

DISCUSSION

The concentrations of As were observed to be the highest in all parts of the 5 different cows. Concentration in the kidney was higher in cow D and lowest in cow B. The muscle tissue concentration had the highest value in

Table 4. Variation of metal concentrations (mg/kg) in the blood of the cows.

COW	As	Cd	Co	Cr	Pb	Ni
A	17.031	0.523	ND	1.285	0.742	0.414
B	17.521	0.72	ND	0.999	ND	0.166
C	17.85	0.638	ND	0.942	0.222	0.19
D	16.747	0.798	ND	1.023	ND	ND
E	16.733	0.843	ND	1.357	0.129	0.195

ND = Not detectable range.

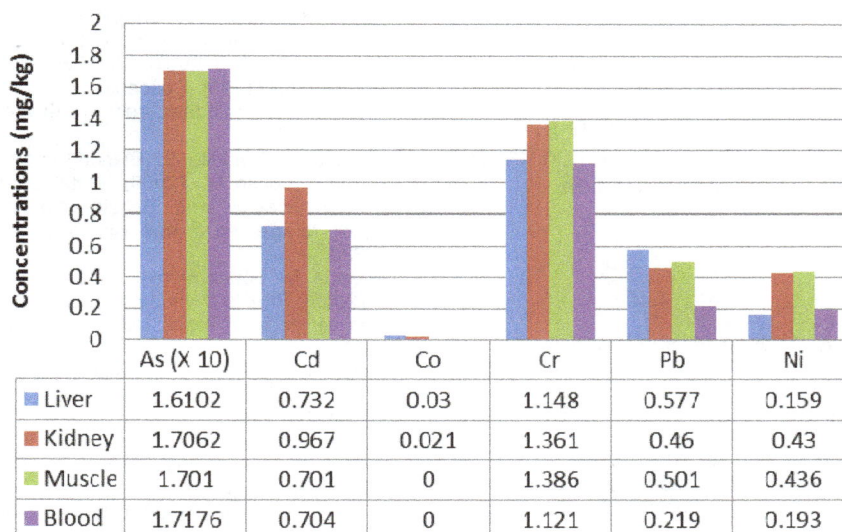

	As (X 10)	Cd	Co	Cr	Pb	Ni
Liver	1.6102	0.732	0.03	1.148	0.577	0.159
Kidney	1.7062	0.967	0.021	1.361	0.46	0.43
Muscle	1.701	0.701	0	1.386	0.501	0.436
Blood	1.7176	0.704	0	1.121	0.219	0.193

Figure 1. Mean concentrations of metals in cow parts.

cow B and lowest in cow A; while in the blood; maximum concentration was observed in cow D. As is a toxic metal which exert its toxic effect through an impairment of cellular respiration by inhibition of various mitochondrial enzymes (Sesha et al., 2007). The level observed in this study exceeds the permissible limit of 0.01 mg/kg WHO standard (WHO, 2002). Also, considering the concentration in various cows from each column, from cow A, it was observed that the levels were higher in the kidney and the blood; from cow B, the levels in the muscle tissue and the blood were closer. In cow C, accumulation was greatest in the blood. Cow D had the highest concentration in the kidney; and Cow E in the liver. It could be inferred that the target organs for As includes the blood and kidney. The tolerable limit for Cd as set by (WHO, 2001) is 0.005 mg/kg; thus from the analyzed results, Cd is not only at a higher concentration level in all parts, but could also be said to have considerably accumulated in the liver and kidney perhaps due to long term exposure (Lenntech., 2009). In cow A, lowest Cd concentration was in the muscle tissue; highest concentration was observed in the kidney for Cow D. In cow C, maximum concentration was as well in

the kidney.

The same were observed for Cows D and E. The varying levels of Cd concentrations in the parts of the different cows may however be related to the age factor. Cd is toxic in virtually every system in the animal and almost absent in the humans at birth, but accumulates with age (Akan et al., 2010). Aside the age factor, the nature of their feed (forage) and drinking from stagnant water and streams could contribute to their concentration levels, as the levels are significantly higher in all the five (5) cows. Co was detected only in the liver and kidney of cow A. Concentrations in other parts of the cows were below detection. The highest concentration of Cr in the liver was observed in cow C and lowest in cow A. In the kidney, the highest concentration was observed in cow B and lowest in cow E. Concentration in the muscle tissue was highest in cow B and lowest in cow E. Similar studies on the distribution of metals by Akan et al. (2010) revealed that the lowest levels of Pb were observed in the liver and the muscle tissue. Given the permissible limit of Pb concentration as 0.01 mg/kg (WHO, 2002), it could be said that the concentration levels in the various parts of the cows are harmful, except for those which were below detection limit. The source of the metal contamination could be traced to the nature of their feed, water, and location such as presence of refineries. The concentration of Ni in the liver was highest in cow A but below detection limit in cow E. The muscle tissue had the highest concentration in all the 5 cows. Figure 1 summarises the mean levels of the metals in the various cow parts. As had the highest concentrations in all the parts while Co was almost not detected.

Conclusion

Since the observed levels of metals in this study were higher than the WHO standards, pollution could be inferred; and the quantification of these metals in organs of cattle could be used as biomarkers of pollution, though several factors have to be put into consideration.

REFERENCES

Abou-Arab AAK (2001). 'Heavy metal contents in Egyptian meat and the role of detergent washing on their levels.' Food Chem. Toxicol., 39: 593-599.

Acar O (2006). 'Biomonitoring and annual variability of heavy metal concentration changes using moss (Hypnum cupressiforme L. ex. Hedw.) in Canakkale Province.' J. Biol. Sci., 6(1): 38-44.

Adekola FA, Eletta OA, Atanda SA (2002). 'Determination of the levels of some heavy metals in urban run-off sediments in Ilorin and Lagos, Nigeria.' J. Appl. Sci. Environ. Manage., 6(2): 23-26.

Akan JC, Abdulrahman FI, Sodipo OA, Chiroma YA (2010). Distribution of Heavy Metals in the Liver, Kidney and Meat of Beef, Mutton, Caprine and Chicken from Kasuwan Shanu Market in Maiduguri Metropolis, Borno State, Nigeria. Res. J. Appl. Sci. Eng. Technol., 2(8): 743-748, ISSN: 2040-7467.

Asonye CC, Okolie NP, Okenwa EE, Iwuanyanwu UG (2007). Some physico-chemical characteristics and heavy metal profiles of Nigerian rivers, streams and waterways. Afr. J. Biotechnol., 6(5): 617-624.

Azmat R, Aziz F, Yousfi M (2008). Monitoring the effect of water pollution on four bio-indicators of aquatic resources of Sindh Pakistan. Res. J. Environ. Sci., 2(6): 465-473.

Blanco-Penedo I, Cruz JM, López-Alonso M, Miranda M, Castillo C, Hernández J, Benedito JL (2006). Influence of copper status on the accumulation of toxic and essential metals in cattle. Environ. Int., 32: 901-906.

Chen GC, He ZL, Stoffella PJ, Yang XE, Yu S, Yang JY, Calvert DV (2006). Leaching potential of heavy metals (Cd, Ni, Pb, Cu and Zn) from acidic sandy soil amended with dolomite phosphate rock (DPR) fertilizers. J. Trace Elem. Med. Biol., 20: 127-133.

Del Val C, Barea JM, Azcón-Aguilar C (1999). Assessing the tolerance to heavy metals of arbuscular mycorrhizal fungi isolated from sewage sludge-contaminated soils. Appl. Soil Ecol., 11: 261-269.

Elaigwu SE, Ajibola VO, Folaranmi FM (2007). Earthworms (Eudrilus Eugenia Kingberg) as bio-indicator of the heavy metal pollution in two municipal dumpsites of two cities in northern Nigeria. Res. J. Environ. Sci., 1(5): 244-250.

Geens A, Dauwe T, Bervoets L, Blust R, Eens M (2010). Haematological status of wintering great tits (Parus major) along a metal pollution gradient. Sci. Total Environ., 408: 1174-1179.

Guala SD, Vega FA, Covelo EF (2010a). The dynamics of heavy metals in plant-soil interactions. Ecol. Model., 221: 1148-1152.

Guala SD, Vega FA, Covelo EF (2010b). Heavy metal concentrations in plants and different harvestable parts: A soil-plant equilibrium model. Environ. Pollut., 158: 2659-2663.

Haimi J, Mätäsniemi L (2002). Soil decomposeranimal community in heavy-metal contaminated coniferous forest with and without liming. Eur. J. Soil Biol., 38: 131-136.

Hobbelen PHF, van den Brink PJ, Hobbelen JF, van Gestel CAM. (2006). Effects of heavy metals on the structure and functioning of detritivore communities in a contaminated flood plain area. Soil Biol. Biochem., 38: 1596-1607.

Kizilkaya R, Aşkin T, Bayrakli B, Sağlam M (2004). Microbiological characteristics of soils contaminated with heavy metals. Eur. J. Soil Biol., 40: 95-102.

Lenntech (2009). http://www.lenntech.com/periodic/elements/pb.htm; accesed on the 3rd of February, 2011.

Liang LN, He B, Jiang GB, Chen DY, Yao ZW (2004). Evaluation of mollusks as biomonitors to investigate heavy metal contaminations along the Chinese Bohai Sea. Sci. Total Environ., 324: 105-113.

Maia L, de Souza MV, Fernandes RBA, Fontes MPF, Vianna MWS, Luz WV (2006). Heavy metals in horse blood, serum, and feed in Minas Gerais, Brazil. J. Equine Vet. Sci., 26(12): 578-583.

Medvedev N (1995). Concentrations of Cd, Pb and sulphur in tissues of wild, forest reindeer from North-West Russia. Environ. Pollution, 90(1): 1-5.

Miranda M, Benedito JL, Blanco-Penedo I, López-Lamas C, Merino A, López-Alonso M (2009). Metal accumulation in cattle raised in a serpentine-soil area: Relationship between metal concentrations in soil, forage and animal tissues. J. Trace Elem. Med. Biol., 23: 231-238.

Miranda M, López-Alonso M, Castillo C, Hernández J, Benedito JL (2005). Effects of moderate pollution on toxic and trace metal levels in calves from a polluted area of northern Spain. Environ. Int., 31: 543-548.

Nwude DO, Okoye PAC, Babayemi JO (2010b). Blood heavy metal levels in cows at slaughter at Awka abattoir, Nigeria. Int. J. Dairy Sci., 5(4): 264-270.

Nwude DO, Okoye PAC, Babayemi JO (2010a). Heavy metal levels in animal muscle tissue: A case study of Nigerian raised cattle. Res. J. Appl. Sci., 5(2): 146-150.

Parker GH, Hamr J (2001). Metal levels in body tissues, forage and fecal pellets of elk (Cervus elaphus) living near the ore smelters at Sudbury, Ontario. Environ. Pollut., 113: 347-355.

Salehi A, Tabari M (2008). Accumulation of Zn, Cu, Ni and Pb in soil and leaf of Pinus elderica Medw. Following irrigation with municipal effluent. Res. J. Environ. Sci., 2(4): 291-297.

Sesha SV, Arun Prabhath N, Raghavender M, Anjaneyulu Y (2007). Effect of As and Cr on the serum amino-transferases activity in Indian major camp, Labeo Rohita. Int. J. Environ. Res. Pub. Health, 4(3): 224-227.

Uhlig C, Salemaa M, Vanha-Majamaa I, Derome J (2001). Element distribution in Empetrum nigrum microsites at heavy metal contaminated sites in Harjavalta, western Finland. Environ. Pollut., 112: 435-442.

Wang Q, Zhou D, Cang L, Li L, Zhu H (2009a). Indication of soil heavy metal pollution with earth worms and soil microbial biomass carbon in the vicinity of an abandoned copper mine in Eastern Nanjing, China. Eur. J. Soil Biol., 45: 229-234.

Wang T, Fu J, Wang Y, Liao C, Tao Y, Jiang G (2009b). Use of scalp hair as indicator of human exposure to heavy metals in an electronic waste recycling area. Environ. Pollut., 157: 2445-2451.

Ward NI, Savage JM (1994). Elemental status of grazing animals located adjacent to the London Orbital (M25) motorway. Sci. Total Environ., 146-147: 185-189.

Weis JS, Weis P (2004). Metal uptake, transport and release by wetland plants: Implications for phytoremediation and restoration. Environ. Int., 30: 685-700.

Were FH, Njue W, Murungi J, Wanjua R (2008). Use of human nails as bio-indicators of heavy metals environmental exposure among school age children in Kenya. Sci. Total Environ., 393: 376-384.

Zhang F, Li C, Tong L, Yue L, Li P, Ciren Y, Cao C (2010). Response of microbial characteristics to heavy metal pollution of mining soils in central Tibet, China. Appl. Soil Ecol., 45: 144-151.

14

Iron and nitric oxide play key roles in the development of cardiovascular disorder

Adeyemi O. S.[1]* and Akanji M. A.[2]

[1]Department of Chemical Sciences, Redeemer's University, P. M. B 3005, Redemption City, Mowe – 121001, Nigeria.
[2]Department of Biochemistry, University of Ilorin, P. M. B 1515, Ilorin, Nigeria.

Iron (Fe) is an essential but potentially harmful nutrient. On the other hand, nitric oxide (NO) is an inorganic free-radical gaseous molecule which has been implicated to play an unprecedented variety of roles in biological systems. Although complex relationships between Fe and NO have been demonstrated, there are still controversies as to what are the influences of Fe on NO balance in the development of cardiovascular disorder. Both Fe and NO have dual but unique roles in the prevention and/or development of cardiovascular complications such as atherosclerosis and/or myocardial infarction. Sustained increase in the concentration of both Fe and NO has been associated with generation of free radical species, promoting oxidation of low density lipoproteins (LDL) which strongly correlated with cell oxidative damage. Moreover, the oxidation of LDL has been implicated as a risk factor in the genesis of cardiovascular disorders. In this light, the mechanistic interactions between Fe and NO in the development of and/or predisposition to cardiovascular disorder are discussed.

Key words: Iron, nitric oxide, oxidative damage, cardiovascular disorder.

INTRODUCTION

Myocardial infarction (MI) or acute myocardial infarction (AMI), commonly known as a heart attack is the interruption of blood supply to part of the heart, causing some heart cells to die. This is most commonly due to occlusion (blockage) of a coronary artery, following the rupture of a vulnerable atherosclerotic plaque, which is an unstable collection of lipids (fatty acids) and white blood cells (especially macrophages) in the wall of an artery. The resulting ischemia (restriction in blood supply) and oxygen shortage, if left untreated for a sufficient period of time, can cause damage or death (*infarction*) of heart muscle tissue (*myocardium*). Moreover, several investigators have studied the relationship between reactive oxygen species and various diseases, and it has been revealed that many of the reactions that involve reactive oxygen species often require the presence of a transition metal such as iron (Sengoelge et al., 2005; Orimadegun et al., 2007; Tavora et al., 2009).

Perhaps it is noteworthy to know that both iron and nitric oxide play important but opposite roles in the development of cardiovascular disorder. While an increased iron status may promote free hydroxyl radical generation in cellular systems and thus potentiate cellular damage and atherosclerosis, production of nitric oxide. However, a continuous and sustained production of nitric oxide could contribute to oxidative damage through the formation of peroxynitrite, a very reactive free radical which could promote lipid peroxidation. Because of the crucial roles that iron and nitric oxide play in cellular systems and the strong link that has been demonstrated between the duo (Richardson and Ponka, 1996; Richardson and Lok, 2008), it will not be out of place to emphasis that a delicate homeostatic balance exist between iron and nitric oxide in biological systems such that any disruption or perturbation to this strictly regulated balance could be catastrophic to the cellular system. Inspite of this, investigations may be required to elucidate and bring to light the outcome of iron and nitric oxide mechanistic interaction as relate to development of cardiovascular disorders. This will put to rest curiosity and questions relating to the likely consequences of a

---

*Corresponding author. E-mail: yomibowa@yahoo.com.

sustained high iron status and nitric oxide levels against the background that increased iron and nitric oxide could trigger generation of free radical species which may spell doom for the biological system if exposure is long enough. This article is an attempt aimed at discussing the probable roles of iron and/or nitric oxide toward development of cardiovascular disorder.

IRON (FE) AND NITRIC OXIDE (NO) IN LIVING CELLS

Iron is an essential but potentially harmful nutrient. It contributes to many important physiologic functions in the body and as well may increase biological markers of oxidative stress, cytotoxicity, and lipid peroxidation in biological systems. Iron not only affects the functions of leukocytes, endothelial cells and cytokine production but also causes oxidative stress and support microbial growth. More than 2 decades ago, it was proposed that iron depletion protects against ischemic heart disease and that this effect may explain the remarkably low incidence of cardiovascular disorders in menstruating women (Sullivan, 1981, 1989, 2003). Studies have been able to show that increase in cellular iron plays a major role in the formation of hydroxyl radicals which potentially contribute to cell damage and atherosclerosis (Sengoelge et al., 2005; Tavora et al., 2009). Free iron has been implicated in lipid peroxidation and ischemic myocardial damage, and it was reported that iron is an independent risk factor for myocardial infarction (Klipstein-Grobusch et al., 1999; Orimedegun et al., 2007). Though it was suggested that the availability of catalytic iron may be more important than overall body iron stores (Lauffer, 1990; Sengoelge et al., 2005) nevertheless an independent relationship between coronary heart disease and the level of serum iron status as a risk factor has been demonstrated (Salonen et al., 1992). Thus the question being asked is "how safe is stored iron?" (Sullivan, 2004). In a separate study, Ascherio et al. (1994) reported a link between an increased risk of fatal coronary heart disease and heme iron intake. This is even as indirect evidence that iron is involved in reperfusion injury after an ischemic event has been provided from animal studies (Klipstein-Grobusch et al., 1999).

This study revealed that free radicals were generated after blood flow was restored to ischemic myocardium, thus contributing to the subsequent myocardial injury (Klipstein-Grobusch et al., 1999). Events such as blood donation, which depletes iron stores in the donors, has been associated with reduced risk of myocardial infarction (Klipstein-Grobusch et al., 1999; Zheng et al., 2005) and cardiovascular disease (Orimadegun et al., 2007). Meanwhile it has been observed that maximum protection exists among iron deficient subjects since iron status affects the modification of LDL cholesterol and myocardial reperfusion injury (Sullivan, 1992; Facchini

and Saylor, 2002). More recent studies have also shown that men with moderately elevated ferritin level have a significantly worse coronary artery disease risk profile than men with lower level (Ramakrishnan et al., 2002).

The endothelium secretes a nitric oxide (NO) of vascular-relaxing substances as well as several vaso-constricting agents. However, one of the most potent endogenous vasodilators is endothelial-derived NO. NO is a critical modulator of blood flow and blood pressure (Vallance et al., 1989) and opposes the vaso-constricting effects of endothelin, angiotensin II, serotonin, and nor-epinephrine (Luscher et al., 1993). NO also suppresses the proliferation of vascular smooth muscle (Annuk et al., 2003). Deficiency of NO contributes not only to increased vascular resistance but to blood vessel medial thickening and/or myointimal hyperplasia, which may alter the structure of the vascular bed. Oxidative stress is one process that has been shown to decrease the expression of endothelial nitric oxide synthase (eNOS) (Luscher et al., 2003). eNOS limits monocyte/macrophage-endothelial cell interaction, therefore its loss or decreased expression may lead to the formation of a macrophage-rich atheroma. This results in a soft plaque that increases the risk of unstable angina, thrombosis, and acute myocardial infarction (Annuk et al., 2003). It has been demonstrated that excess number react with the superoxide anion to produce peroxynitrite, a very aggressive free radical species that contribute to lipid peroxidation and oxidative stress (Torreilles et al., 1999). Moreso a decreased production of number, or reduced sensitivity to the action of number, has consistently been shown to impair endothelial-dependent vasodilation, contributing to the pathogenesis of atherogenesis (Luscher et al., 1993).

POSSIBLE MECHANISMS BY WHICH IRON AND/OR NITRIC OXIDE MAY PREDISPOSE TO DEVELOPMENT OF CARDIOVASCULAR DISORDER

(1) Increased iron status may lead to the generation of free hydroxyl radicals which promotes oxidation of low-density lipoprotein (LDL). The importance of oxidized LDL in atherosclerosis (Orimadegun et al., 2007), and the potential for iron to act as a catalyst in processes leading to cellular oxidative damage (Reif, 1992; Silva and Aust, 1993; Satoh and Tokunaga, 2002), has been reported. This supports a potential role of iron in coronary heart disease. Previous studies have demonstrated that an association exist between iron and increased oxidation of LDL cholesterol with the former catalysing production of tissue-damaging free hydroxyl radicals (Klipstein-Grobusch et al., 1999). Moreso, that it has been established that LDL oxidation plays an important role in the pathogenesis of atherosclerosis and cardiovascular disease (Sengoelge et al., 2005). Oxidized LDL causes lipids to accumulate in macrophages and foam cells

(Ramakrishnan et al., 2002), and this process has been shown to be cytotoxic to many cell types and chemotactic for monocyte macrophages.

(2) High iron status may result in endothelial cell dysfunction. Experimental evidence have revealed that increase in cellular iron could down-regulate the expression of nitric oxide synthase (NOS) (Kim and Ponka, 2002; Richardson and Lok, 2008) through a strictly controlled mechanism. The enzyme NOS is responsible for the formation of NO starting from L-arginine. NO plays a critical role in the maintenance of vasculature in a state of vasodilation. The overall effect of inadequate amount of NO would be impaired endothelium-dependent vasorelaxation, blood pressure regulation and acceleration of atherogenesis with an onset of acute atherothrombotic events. Arterial smooth muscle proliferation has also been reported to be a complication of decreased activity of NOS secondary to increased iron status (Sengoelge et al., 2005). So whatever is capable of causing increases in iron and nitric oxide levels may abate the homeostatic balance between the iron and nitric oxide. This could have devastating effects on the cellular system as both iron and nitric oxide are capable of promoting the generation of free radical species which may lead to cell oxidative stress.

(3) Another possibility is that, if a factor raises the iron status, this may positively affects the activity of myeloperoxidase (MPO) and phagocyte associated NADPH oxidoreductase in the neutrophil and hence potentiate the formation, instability and rupture of plaque (Ikura et al., 2002; Tavora et al., 2009). Report by Vita et al. (2004) revealed that high cellular iron correlated with increased circulating MPO which plays a vital in neutrophilic inflammation. MPO which has emerged as a potential promoter of atherosclerosis (Hazen, 2004; Nichols and Hazen, 2005) is stored and secreted from activated neutrophils and monocytes.This as an important component in degranulation material of leukocytes, a critical process in human innate host defenses (Sugiyama et al., 2001; Naruko et al., 2002). The role of MPO in the initiation and propagation of atherosclerosis may not be unconnected with its potential to initiate lipid peroxidation as well as promote post-translational modification of target proteins (Podrez et al., 2000; Zhang et al., 2002). Recent studies have found that MPO is capable of promoting oxidation of lipoproteins which could lead to increase in cholesterol deposit and formation of foam cells in fatty streaks (Podrez et al., 2000; Roman et al., 2008). MPO was observed to be increased in human atheromas (Daugherty et al., 1994) and it has also been demonstrated in a variety of inflammatory conditions (Zhang et al., 2002) which could potentiate the development of cardiovascular disease. The emergence of myeloperoxidase as an important coronary risk factor underscores the links among iron, oxidative stress, and inflammation in cardiovascular

disorders. Myeloperoxidase contains iron, and its level could be increased in high iron status (Vita et al., 2004). In a separate report, Sullivan, (1989) proposed that decreased activity of potentially injurious iron-dependent enzymes could be one of the mechanisms by which iron depletion protects against ischemic heart disease. Moreso, diminished inflammatory injury in association with reduction in myeloperoxidase activity by induction of iron deficiency had been demonstrated by Baldus et al. (2003). Myeloperoxidase may thus be a potentially modifiable coronary risk factor whose level may be decreased by iron removal and vice versa (Figure 1).

HYPOTHESIS

Our hypothesis is that exposure capable of sustaining increases in iron levels and subsequently nitric oxide could predispose to or potentiate the development of cardiovascular disorders.This is based on the catalytic role of iron in the generation of free radical species which, if aided by sustained production of nitric oxide, may lead to cell oxidative damage or stress.

Testing the hypothesis

Our approach to testing the hypothesis would be experimental with subjects randomised to receive different concentrations of iron salts, nitric oxide donor or placebo. Increased tissue or serum iron and nitric oxide concentrations that correlate with sustained oxidative stress and cardiac damage biomarkers would be confirmatory.

CONCLUSION

We are of the view that a disruption of the delicate homeostatic balance existing between iron and nitric oxide could lead to potentiation of heart disease. Moreso all the evidences outlined further put together appear to support a potential association between iron and development of cardiovascular disorders viz-a-viz sustained increases in the generation of free radicals. Thus it is plausible to hypothesise that chronic exposure leading to sustained increases in the labile iron pool as well as nitric oxide could predispose to myocardial infarction and/or affects the fatality of existing heart disease. It may be necessary to consider the significance and potential effects on public health owing to the fact that coronary heart disease (CHD) which is one of the leading causes of death in most industrialized countries of the world is now also considered a prominent health problem in developing nations (WHO, 1981).

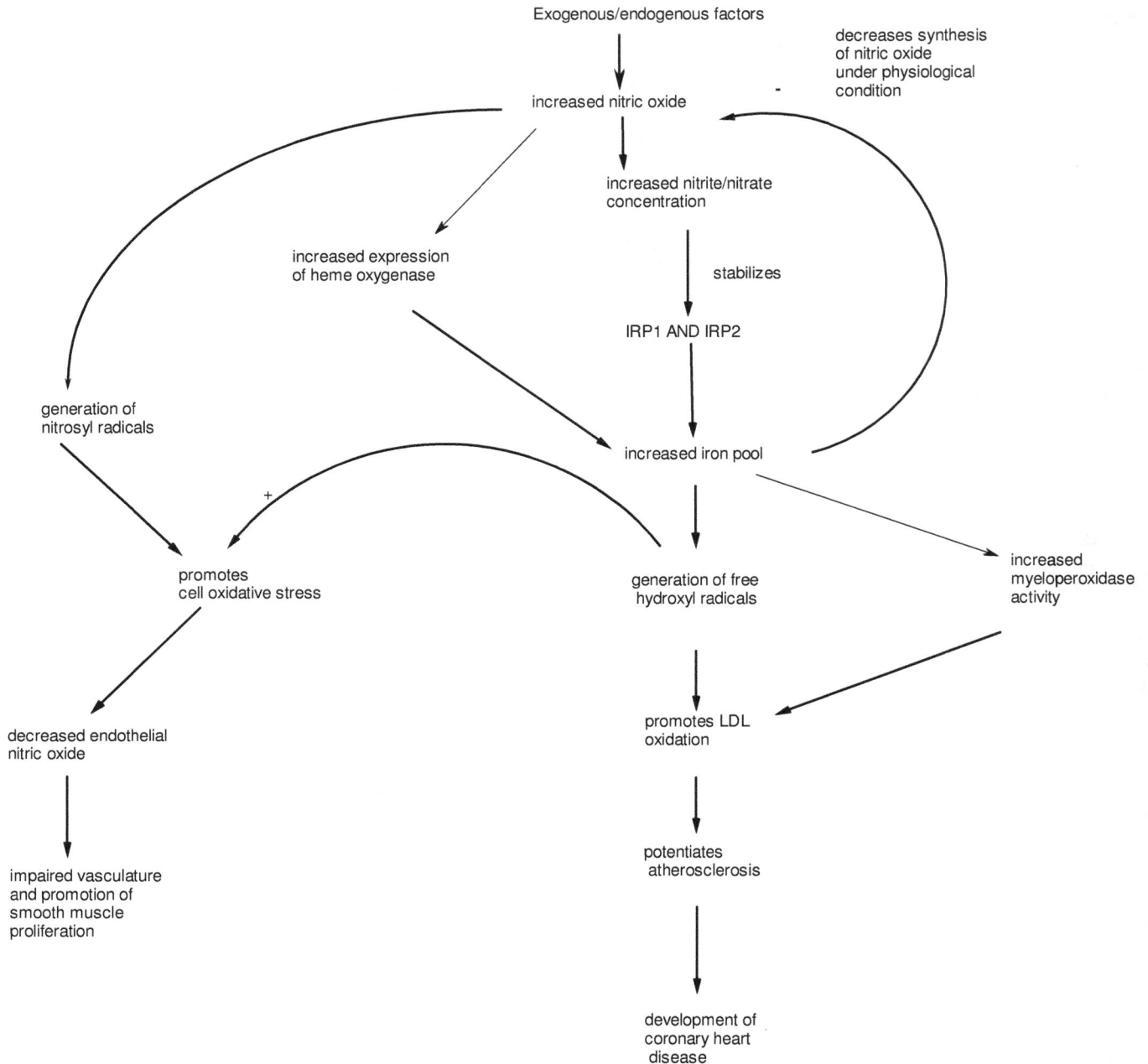

Figure 1. Proposed mechanistic interactions between iron and nitric oxide in living cells (IRP1 and IRP2: Iron regulatory protein 1 and 2).

REFERENCES

Adeyemi OS, Akanji MA, Johnson TO, Ekanem JT (2011). Iron and nitric oxide balance in African Trypanosomosis: Is there really a link? Asian J. Biochem., 6(1): 15-28.

Annuk M, Zilmer M, Fellstrom B (2003). Endothelium dependent vasodilation and oxidative stress in chronic renal failure: Impact on cardiovascular disease. Kidney Int. Suppl., 84: S50-S53.

Ascherio A, Willett WC, Rimm EB (1994). Dietary iron intake and risk of coronary disease among men. Circulation, 89: 969-974.

Baldus S, Heeschen C, Meinertz T, Zeiher AM, Eiserich JP, Munzel T,

Simoons ML, Hamm CW (2003). Myeloperoxidase serum levels predict risk in patients with acute coronary syndromes. Circulation, 108: 1440-1445.

Daugherty A, Dunn JL, Rateri DL, Heinecke JW (1994). Myeloperoxidase, a catalyst for lipoprotein oxidation, is expressed in human atherosclerotic lesions. J. Clin. Invest., 94(1): 437-444.

Facchini FS, Saylor KL (2002). Effect of iron depletion on cardiovascular risk factors: Studies in carbohydrate intolerant patients. Ann. N.Y. Acad. Sci., 967: 342-351.

Hazen SL (2004). Myeloperoxidase and plaque vulnerability. Arterioscler. Thromb. Vasc. Biol., 24(7): 1143-1146.

Ikura Y, Ogami M, Shimada Y (2002). Neutrophil infiltration of culprit lesions in acute coronary syndromes. Circulation, 106(23): 2894-2900.

Kim S, Ponka P (2002). Nitric Oxide-Mediated Modulation of Iron Regulatory Proteins: Implication for Cellular Iron Homeostasis. Blood Cells, Mol. Dis., 29(3): 400–410.

Klipstein-Grobusch K, Koster JF, Grobbee DE, Lindemans J, Boeing H, Hofman A, Witteman JCM (1999). Dietary Iron and Risk of Myocardial Infarction in the Rotterdam Study. Am. J. Epidemiol., 149: 421-428.

Lauffer RB (1990). Iron stores and the international variation in mortality from coronary artery disease. Med. Hypotheses, 35: 96-102.

Luscher TF, Tanner FC, Tschudi MR, Noll G (1993). Endothelial dysfunction in coronary artery disease. Annu. Rev. Med., 44: 395-418.

Naruko T, Ueda M, Haze K, van der Wal AC, van der Loos CM, Itoh A, Komatsu R, Nicholls SJ, Hazen SL (2005). Myeloperoxidase and cardiovascular disease. Arterioscler. Thromb. Vasc. Biol., 25(6): 1102-1111.

Orimadegun BE, Anetor JI, Adedapo DA, Taylor GO, Onuegbu JA, Olisekodiaka JM (2007). Increased Serum Iron Associated With Coronary Heart Disease Among Nigerian Adults. Pak J. Med. Sci., 23(4): 518-522.

Podrez EA, Abu-Soud HM, Hazen SL (2000). Myeloperoxidase-generated oxidants and atherosclerosis. Free Radic. Biol. Med., 28(12): 1717-1725.

Ramakrishnan U, Kuklina E, Stein AD (2002). Iron stores and cardiovascular disease risk factors in women of reproductive age in the United States. Am. J. Clin. Nutr., 76: 1256-1260.

Reif DW (1992). Ferritin as a source of iron for oxidative damage. Free Radic. Biol. Med., 12: 417-427.

Richardson DR, Ponka P (1996). Effects of nitrogen monoxide on cellular iron metabolism. In: M.D. Maines (Ed.), Nitric Oxide Synthase: Characterization Funct. Anal., 31: 329–345.

Richardson R, Lok HC (2008). The nitric oxide–iron interplay in mammalian cells: Transport and storage of dinitrosyl iron complexes. Biochim. Biophys. Acta, 1780: 638–651.

Roman RM, Wendland AE, Polanczyk CA (2008). Myeloperoxidase and coronary arterial disease: From research to clinical practice. Arq. Bras. Cardiol., 91(1): e11-19.

Salonen JT, Nyyssonen K, Korpela H, Tuomilehto J, Seppanen R, Salonen R (1992). High stored iron levels are associated with excess risk of myocardial infarction in Eastern Finish Men. Circulation, 86: 803-811.

Satoh T, Tokunaga O (2002). Intracellular oxidative modification of low density lipoprotein by endothelial cells. Virchows Arch., 440: 410-417.

Sengoelge G, Sunder-Plassmann G, Horl WH (2005). Potential risk for infection and atherosclerosis due to iron therapy. J. Renal Nutr., 15(1): 105–110.

Silva DM, Aust SD (1993). Ferritin and ceruloplasmin in oxidative damage: Review and recent findings. Can. J. Physiol. Pharmacol., 71: 715-720.

Sugiyama S, Okada Y, Sukhova GK, Virmani R, Heinecke JW, Libby P (2001). Macrophage myeloperoxidase regulation by granulocyte macrophage colony-stimulating factor in human atherosclerosis and implications in acute coronary syndromes. Am. J. Pathol., 158(3): 879-891.

Sullivan J (2004). Is stored iron safe? J. Lab. Clin. Med., 144: 280-284.

Sullivan JL (1981). Iron-and-the-sex-difference-in-heart-disease-risk. Lancet, 1: 1293-1294.

Sullivan JL (1989). The-iron-paradigm of ischemic heart disease. Am. Heart J., 117: 1177-1188.

Sullivan JL (1992). Stored iron and ischaemic heart disease empirical support for a new paradigm. Circulation, 86: 1035-1037.

Sullivan JL (2003). Are menstruating women protected from heart disease because of, or in spite of, estrogen? Relevance to the iron hypothesis. Am. Heart J., 145: 190-194.

Tavora FR, Ripple M, Li L, Burke AP (2009). Monocytes and neutrophils expressing myeloperoxidase occur in fibrous caps and thrombi in unstable coronary plaques. BMC Cardiovasc. Disord., 9: 27.

Torreilles F, Salman-Tabcheh S, Guerin MC, Torreilles J. (1999). Neurodegenerative disorders: The role of peroxynitite. Brain Rev., 30: 153-163.

Vallance P, Collier J, Moncada S (1989). Effects of endothelium-derived nitric oxide on peripheral arteriolar tone in man. Lancet, 2: 997-1000.

Vita JA, Brennan ML, Gokce N, Mann SA, Goormastic M, Shishehbor MH, Penn MS, Keaney JF Jr, Hazen SL (2004). Serum myeloperoxidase levels independently predict endothelial dysfunction in humans. Circulation, 110: 1134-1139.

WHO (1981). Prevention of Coronary Heart Disease. Report of a WHO Expert Committee on Prevention of CHD, Geneva, pp. 50-53.

Zhang R, Brennan ML, Shen Z, MacPherson JC, Schmitt D, Molenda CE, Hazen SL (2002). Myeloperoxidase functions as a major enzymatic catalyst for initiation of lipid peroxidation at sites of inflammation. J. Biol. Chem., 277(48): 46116-46122.

Zheng H, Cable R, Spencer B, Votto N, Katz SD (2005). Iron stores and vascular function in voluntary-blood-donors. Arterioscler. Thromb. Vasc. Biol., 25: 1577-1583.

15

Assessment of sorghum beer for alcohol and metal ions content and genotoxicity in mice bone marrow

Asita A. O.[1]*, Tanor E. B.[2], Magama S.[1] and Khoabane N. M.[2]

[1]Department of Biology, National University of Lesotho, P. O. Roma 180 Maseru, Lesotho, Southern Africa.
[2]Department of Chemistry, National University of Lesotho, P. O. Roma 180 Maseru, Southern Africa.

National well-being, life expectancy (LE), water quality, alcohol consumption etc, have been shown to be interrelated. In Lesotho, with LE below world average, one widely consumed beverage is a locally brewed sorghum beer, *sesotho*. Therefore, beers from four different sources were analyzed for metals, alcohols (mg/L) and mice bone marrow genotoxicity. The beers contained different ($p<0.05$) concentrations of As (0.012 to 0.059), Co (0.127 to 0.160), Cr (0.052 to 0.069), Cu (0.004 to 0.057), Fe (0.070 to 0.600), Ni (0.004 to 0.031), Pb (0.019 to 0.029), Se (0.645 to 0.942) and Zn (0.317 to 6.337). Cd (0.032 to 0.035), and Hg (0.131 to 0.150) concentrations were not different ($p>0.05$). The concentrations, in the beers, of Co, Cr, Cu, Fe, Ni and Pb were lower ($p<0.05$), while those of Se, As, Cd, Hg, Se As, Cd, Se and Zn were higher ($p<0.05$), than in the water used for brewing. Alcohol concentrations (mg/L) in the beers differed significantly ($p<0.05$) and were, total alcohol (53,000 to 74,200), methanol (800 to 2000) and ethanol (53,000 to 74,000). Beer from one source only was assessed for genotoxicity. It was nonclastogenic but toxic at 50% concentration only.

Key words: Sorghum beer, metals, alcohol, genotoxicity.

INTRODUCTION

The use of the gross domestic product (GDP) as a standard gauge of national economic health is being reexamined, mostly by development economists, most of whom now favour alternative yardsticks that combine indicators of human well-being, such as health, population, and wealth, with those of environmental sustainability (water quality, species diversity, and energy use) to generate a more integrated picture of the general wellbeing of nations and the world (Bourgeault-Tassé, 2011). Thus, interrelationships among some of the key indicators of well-being, such as life expectancy, human and environmental health and economic growth have been suggested (Aísa and Pueyo, 2004). Life expectancy can fall due to problems like famine, war, disease and poor health. The higher the life expectancy, the better socio economic conditions a country is in. For instance, the 2009 figures for life expectancy at birth for males and females in Lesotho were put at 46 and 50 years,

respectively whereas similar figures for males and females for the following developed countries were given as; USA (76 and 81), Germany (78 and 83), Norway (79 and 83) and Japan (80 and 86), respectively (WHO, 2011). Life expectancy was shown to correlate negatively with the behavioral factor of alcohol consumption, with larger amounts of alcohol consumption being related to slightly lower life expectancy (Organisation for Economic Co-operation and Development – OECD, 2006). Using information for 161 different countries, the lower life expectancy of males correlated with the higher per capita, alcohol consumption and cirrhosis of the liver deaths among males (Templer et al., 1993).

Water quality, as one aspect of the wider theme of environmental sustainability, is also related to the general well-being of individuals and nations and metal ions in water and beverages are of concern to many regulatory authorities. The widespread roles of metal ions in health and disease range from the requirement for intake of essential trace elements to toxicity associated with metal overload. Many metal ions are associated with enhanced oxidative stress, inflammation and cancer (Valko et al., 2006; Halliwell and Gutteridge, 2007).

Despite regulatory controls, numerous sources of metal

*Corresponding author. E-mail: ao.asita@nul.ls, aoasita@yahoo.co.uk.

ingestion have been recently reported including contaminated drinking water, seafood, breast milk, herbal medicines, smoking, together with plants and animals used in the diet (Wang et al., 2005; Ang et al., 2005; Sharma and Agrawal, 2005; Navarro and Alvarez, 2005; Zheng et al., 2007).

In Lesotho, indigenous fermented beers from sorghum (*Sorghum bicolor*) are the traditional drinks and are widely consumed. A survey in 1983/1984 found that between 40 to 60% of rural households in some districts in Lesotho brew and/or sell sorghum beer, which has traditionally lubricated much of the social and economic fabric of *Basotho* life such as facilitating *matsema* or work parties or by providing refreshments for rites and feasts connected with the ancestors (Senaoana et al., 1984). The occupation of beer brewing was connected with women migrants who brewed and sold beer illegally, thereby commercializing the traditional beer (Bonner, 1990). As migration rates increased, Basotho women settled in towns in Lesotho to brew and sell beer to migrants traveling to and from South Africa and local labourers (Larsson et al., 1998). The beer, which is called *sesotho* in Lesotho is prepared by boiling a mixture of germinated sorghum grains, warm water and maize meal in a metal drum, followed by cooling in the open air to produce a concoction called *sesotho*. To this concoction is added the inoculums of either yeast or a mixture of previously used sorghum which contains some wild yeast which ferments the sugars. The whole slurry mixture is kept in a warm place so that efficient fermentation is achieved (Larsson et al., 1998).

A survey that was carried out in 2000 of households in 29 villages in Lesotho showed that 26% of adults consumed alcohol and of the consumers, 65% reported consuming traditional home-brewed beer exclusively, with another 20% drinking both western-type and traditional beer (Siegfried, 2001). The daily volumes of traditional beer that were consumed on a typical drinking day ranged from 340 to 10 000 ml with a mean (SD) of (2634 ± 1737 ml) (Siegfried, 2001). In 2004, alcohol use disorders (15 + years) contributed to 1.35% of health problems among men and 0.15% among women in Lesotho (WHO, 2011).

The aim of this study therefore, was to assess the locally brewed sorghum beer and the water used for brewing from four different sources for the content of metal ions, alcohol, and *in vivo* mouse bone marrow erythropoietic cell toxicity and genotoxicity.

MATERIALS AND METHODS

Beer sample

The beer and water samples were collected from four brewing locations within the Maseru District of Lesotho, namely, Mafefooane, Mahlanyeng, Sea Point and Mafikeng. Sea Point is asuburb in Maseru city, while the other three locations are settlements in the periphery of the city within a 40 km radius.

Preparation of the beer samples

The beer samples were individually filtered through 8 layers of cheese cloth to remove large particulate matter, stored in a freezer and used throughout the duration of the experiment. Each filtered beer sample was further filtered using a vacuum pump, Rocker 400 by INSTRUVAC® of South Africa.

For the determination of metal ions, the temperature of the filtered beer sample was adjusted to 20 to 25°C and decarbonated by transferring the sample into a 250 ml Erlenmeyer flask with shaking, first gently and then vigorously, until all gas had been released. The sample was then filtered through Whatman No. 1 filter paper. A volume of 100 ml of the filtrate was transferred into a 250 ml beaker, 1.0 ml concentrated HNO_3 and 0.5 ml concentrated HCl were added and heated gently on a hot plate without boiling until the volume was reduced to 20 ml. The solution was refluxed for 30 min and cooled, transferred into a 50 ml graduated flask and diluted to the mark with deionized water.

Quantification of heavy metals

Preparation of solutions of heavy metals

All the reagents and chemicals used were high analytical grade and supplied by Merck, Johannesburg, South Africa. The stock solutions for the metals analysis were prepared according to the Varian AAS Handbook, 1998. The stock solutions contained 1000 mg/L of metal ion.

Determination of metal ions concentration

The concentrations of As, Cd, Co, Cr, Cu, Fe, Hg, Ni and Se in the beer and water samples were determined by induced coupled plasma atomic emission spectrophotometer (ICP-AES Varian Liberty AX).

The concentrations of Pb and Zn were determined by flame atomic absorption spectrophotometer (AAS, Varian spectra AA 220FS) as it gives netter detection for those metals (VFAASAM, 1989). Five standard solutions of each of the metal ions were prepared and calibration graphs constructed for determining the concentration of the metal ions in the samples.

Determination of percentage of total alcohol concentration (in v/v)

The total alcohol concentration was determined according to the back titration method 28.1.07 of the association of official analytical chemists, (AOAC, 1995).

The distillate of the beer sample was titrated with a standard ferrous ammonium sulphate. Blank titration was done by titrating 25 ml of the dichromate solution with ferrous ammonium solutions. The percentage of total alcohol concentration (in v/v) was calculated using the following equation:

$$\% \quad Alcohol \quad = 25.00 - \left(25.00 \times \frac{V_1}{V_2} \right)$$

Where, 25.00 was the amount of the dichromate solution used to collect the distillate, V_1 was the amount of the Fe^{2+} used to titrate the distillate and V_2 the amount of Fe^{2+} used in the blank titration.

Analysis of beer for different alcohols by nuclear magnetic resonance (NMR)

A 20% alcohol solution was prepared using deuterium oxide, D_2O

(99.9% atom D, containing 0.75% weight trimethylsilyl propionate [TSP]; Sigma Aldrich), in a 5 mm NMR tube. Spectra were acquired on a Varian Mercury 400 spectrophotometer (Varian Associates Inc, Palo Alto). The spectrophotometer consisted of a 54 mm bore size unshielded Oxford Magnet (Oxford Instruments Ltd, Oxfordshire) operating at a proton frequency of 400 MHz connected to a Varian mercury VX console with a performa 1 pulsed field gradient amplifier (20 gauss per centimeter) and run on Varian 6.1B software. A standard ^{1}H NMR spectrum was recorded at room temperature using a quantitative ^{1}H NMR pulse sequence. Standard 1 D processing with additional zero-filling to 64 k was applied and integrals measured. The integrals from the methyl signals of ethanol and methanol were used for comparison and quantification (Berger and Braun, 2004).

Toxic and genotoxic effects of beer to mouse bone marrow cells using the micronucleus assay

About 7 to 8 week-old male inbred NIH mice were purchased from the University of Free State, Proefdiereenhied animal unit (Bloemfontein, Republic of South Africa) and used in the experiment after 1 week of acclimatization. All the animals were allowed free access to pelleted horse feeds (Voernet (PTY) LTD Republic of South Africa) and tap water *ad libitum* through out the 1-week acclimatization period. Three groups of five animals per group were given filtered sorghum beer (100, 50 and 25%, respectively), diluted with tap water to the desired concentration as the only liquid source for 35 days. Both negative and positive control groups received water *ad libitum* during the 35 days period. On the 34th day, the positive control group animals were given the indirect-acting mutagen, cyclophosphamide monohydrate (CP) (Fluka Biochemika, Germany) intraperitoneally in a single dosing regimen at the volume of 10 ml/kg of body weight at the concentration of 40 mg/kg (Salvadori et al., 1992; Gimmler et al., 1999) dissolved in purified water BP (Medicolab, Republic of South Africa).

Slide preparations of the bone marrow smears were according to the method of Asita et al. (2008). Two smears were prepared for each animal. Slides were coded and scored blind using the OLYMPUS CXS21 light microscope.

To evaluate bone marrow toxicity, the ratio PCE/ (PCE + NCE) was calculated by counting a total of 1000 erythrocytes per animal using these slides. All together 2000 polychromatic erythrocytes per animal were examined for the presence of micronuclei for each individual animal, which means 10,000 PCEs scored per dose group, to determine the occurrence of MNPCEs. Frequency of MNPCEs per 1000 = (MNPCEs/ total number of PCEs counted) ×1000.

Statistical analysis

Data of the concentration of metal ions in the beer and water samples were expressed as mean ± S.D. The data were subjected to the statistical analysis by ANOVA (one-way), to compare for difference in the metal ions concentrations in the beer samples from one source collected on three different sampling days; to compare for difference in the concentrations of each metal ion in the beer and water samples from the four different sources; to compare for difference in the alcohol content in the beers from the four different sources. $P < 0.05$ was considered as the level of significance.

Student's t-test was used to compare the concentration of each metal ion in the beer from each source and its concentration in the water that was used for brewing. $P < 0.05$ was considered as the level of significance.

To evaluate bone marrow toxicity only the beer sample from the Sea Point site was used because of its location in the city.

Statistical differences between each concentration group and the concurrent negative control in the incidences of MNPCE/1000 PCEs and the PCE/ (PCE + NCE) ratio were determined by the Mann-Whitney U test for 2 independent samples. $P < 0.05$ was considered as the level of significance. The statistical analyses were performed using the SPSS 10.0 statistical program.

RESULTS

Metal ions concentrations

Metal ions in the beer and the corresponding water samples

The concentrations (mg/L) of the metal ions in the beer and the corresponding water samples used to brew the beers were as follows: As (0.012 − 0.059 and 0.033 − 0.383 respectively), Cd (0.032 − 0.035 and 0.023 − 0.260 respectively), Co (0.127 − 0.160 and 0.354 − 0.384 respectively, Cr (0.052 − 0.069 and 0.033 − 0.383 respectively), Cu (0.004 − 0.057 and 0.349 − 0.400 respectively), Fe (0.070 − 0.600 and 0.342 − 4.054 respectively), Hg (0.131 − 0.150 and 0.021 − 0.381 respectively), Ni (0.004 − 0.031 and 0.135 − 0.384 respectively), Pb (0.019 − 0.029 and 0.332 − 0.400 respectively), Se (0.645 − 0.942 and 0.371 − 0.420 respectively) and Zn (0.317 − 6.337 and 3.743 − 4.019 respectively) (Table 1).

1) The mean concentrations of the metal ions in the beer samples of the three sampling days did not differ significantly ($p > 0.05$) for the beers from Mafefooane, Mahlanyeng and Sea Point.
2) Only the mean concentrations of Cu ions in the beer samples of the three sampling days from Mafikeng differed significantly ($p < 0.05$).
3) The mean concentrations of As, Cr, Fe, Ni Co, Cu, Pb, Zn and Se in the beer samples from the four sources were significantly different ($p < 0.05$).
4) The mean concentrations of Cd and Hg in the beer samples from the four sources were not significantly different ($p > 0.05$).

Metal ions in the water samples

A Comparison of the concentrations of individual metal ion in the water samples from the four sources is presented in Table 2.
1) The concentrations of As, Cd, Cr, Fe, Hg, Ni and Se in the water samples from the four sources were significantly different ($p < 0.05$).
2) The concentrations of Co, Cu, Pb and Zn in the water samples from the four different sources were not significantly different ($p > 0.05$).
3) The mean concentrations of As, Co, Cr, Cu, Fe, Hg, Ni, and Zn in the water samples from Mahlanyeng were higher than their mean concentrations in the water

Table 1. Concentration of metals in the beers and waters (mg/L).

Metal	Beer/water	Mafefooane metal concentration (mg/L)		Mahlanyeng metal concentration (mg/L)		Sea point metal concentration (mg/L)		Mafikeng metal concentration (Mg/L)	
		Mean	S.D	Mean	S.D	Mean	S.D	Mean	S.D
As*	Beer	0.038	0.001	0.013	0.001	0.059‡	0	0.054‡	0.003
	Water	0.377	0.011	0.383	0.049	0.037	0.01	0.033	0.01
Cd	Beer	0.036	0.001	0.036	0.001	0.033‡	0.001	0.034‡	0.002
	Water	0.26	0.004	0.211	0.035	0.024	0.017	0.023	0.005
Co*	Beer	0.161	0.002	0.132	0.001	0.145	0.016	0.142	0.003
	Water	0.378	0.008	0.384	0.018	0.355	0.049	0.364	0.087
Cr*	Beer	0.075	0.007	0.053	0.001	0.066	0.001	0.061	0.003
	Water	0.385	0.015	0.391	0.02	0.364	0.001	0.351	0.004
Cu*	Beer	0.007	0.005	0.057	0	0.011	0.002	0.025†	0.003
	Water	0.395	0.001	0.4	0.036	0.392	0.005	0.349	0.1
Fe*	Beer	0.605	0.005	0.381	0.001	0.591	0.002	0.071	0.001
	Water	3.929	0.008	4.054	0.072	3.59	0.2	0.342	0.1
Hg	Beer	0.147	0.003	0.151	0.001	0.149‡	0.001	0.141	0.01
	Water	0.21	0.013	0.381	0.001	0.022	0.003	0.201	0.011
Ni*	Beer	0.004	0.001	0.01	0.001	0.032	0.001	0.028	0.002
	Water	0.381	0.065	0.384	0.01	0.351	0.02	0.135	0.042
Pb*	Beer	0.031	0.002	0.02	0.001	0.022	0.001	0.027	0.002
	Water	0.373	0.003	0.399	0.009	0.4	0.011	0.332	0.1
Se*	Beer	0.646‡	0.001	0.94‡	0.001	0.766‡	0.002	0.776‡	0.016
	Water	0.42	0.01	0.409	0.008	0.371	0	0.407	0.003
Zn*	Beer	0.318	0.001	0.773	0.019	0.775	0.002	6.345‡	0.009
	Water	3.746	0.1	4.019	0.176	3.916	0.224	3.743	0.764

n = sample size (3); S.D = Standard deviation * = statistically significant difference in metal ions concentrations in the beers from the different sources; † = statistically significant difference in metal ions concentrations in the beers of the three sampling days from one source; ‡ = statistically higher metal ion concentration in beer sample than in the water used for brewing.

samples from Mafefooane, Sea Point and Mafekeng. The mean mean concentration of Pb in the water sample from Mahlanyeng was the same as that from Sea Point 4). The concentrations of Cd and Se in the water samples from Mafooane were higher than their concentrations

Table 2. Comparison of the concentrations of individual metal ions in the water from the four different sources.

Water source	Sample	Metal ions concentration in beer samples (mg/L)										
		As	Cd	Co	Cr	Cu	Fe	Hg	Ni	Pb	Se	Zn
Mafefooane (bore hole)	a	0.387	0.264	0.381	0.382	0.395	3.936	0.195	0.309	0.376	0.430	3.746
	b	0.365	0.256	0.369	0.372	0.395	3.929	0.218	0.399	0.370	0.420	3.646
	c	0.380	0.260	0.383	0.402	0.396	3.921	0.216	0.436	0.373	0.410	3.846
Mahlanyeng (tap water)	a	0.328	0.250	0.384	0.411	0.364	4.054	0.381	0.394	0.390	0.410	4.149
	b	0.422	0.181	0.367	0.372	0.436	3.982	0.382	0.384	0.408	0.416	4.089
	c	0.398	0.202	0.402	0.391	0.400	4.126	0.380	0.374	0.400	0.401	3.819
Sea Point (tap water)	a	0.047	0.044	0.376	0.363	0.398	3.790	0.021	0.351	0.404	0.371	3.688
	b	0.037	0.014	0.389	0.364	0.390	3.590	0.019	0.331	0.408	0.371	4.135
	c	0.027	0.014	0.299	0.365	0.389	3.390	0.025	0.371	0.388	0.371	3.926
Mafikeng (bore hole)	a	0.033	0.023	0.465	0.351	0.349	0.342	0.206	0.112	0.232	0.410	3.076
	b	0.023	0.028	0.311	0.347	0.449	0.242	0.189	0.110	0.332	0.407	4.576
	c	0.043	0.018	0.317	0.355	0.249	0.442	0.209	0.184	0.432	0.404	3.576
One-way ANOVA result	Test stat	175.607*	115.942*	0.206	6.68*	0.593	679.546*	894.288*	26.114*	1.202	32.741*	0.327

* = Statistically significant difference in metal ion concentrations in the waters from the four different sources.

in the water samples from Mahlanyeng, Sea Point or Mafekeng. And the water samples from the Mafefooane site contained higher concentrations of As, Co, Cr, Cu, Fe, Hg and Ni than the water samples from Sea Point and Mafekeng.
5) The water samples from the Sea point site contained higher concentrations of As, Cd, Cr, Cu, Fe, Hg, Ni, Pb and Zn than the water samples from the Mafekeng site. However, water from the Mafekeng site contained higher concentrations of Co and Se than the water samples from the Sea Point site.

According to the number of metal ions with the highest concentration, the water samples from the four sites were ranked as follows: Mahlanyeng > Mafefooane > Sea Point > Mafekeng.

Comparison of metal ions concentrations in beer and the water used for brewing

The results are presented in Table 3 and summarized as follows:

1) The concentrations of Co, Cr, Cu, Fe, Ni and Pb in the beer samples were significantly lower (p< 0.05) than their concentrations in the waters from the 4 water sources. The observation suggested that the brewing process resulted in a lowering of the concentrations of these metal ions in the beers.
2) The concentration of Se ions in the beers from Mafefooane and Mahlanyeng, As, Cd, Hg and Se ions in the beers from Sea Point and As, Cd, Se and Zn ions in the beers from Mafekeng, were

significantly higher (p< 0.05), than their concentrations in the waters used for brewing. The observation suggested that these metal ions were introduced into the beers from extraneous sources during the brewing process.

Comparison of metal ions in the beer and water samples with World Health Organization (WHO, 2008) guideline values for drinking water

The concentrations of Hg, Pb and Se in the water and beer samples were higher than the WHO values. The concentrations of Cu in all the water and beer samples were lower than the WHO value. With the exception of beer samples from

Table 3. Comparison of the concentration of each metal ion in the beer with their concentration in the water for each source (t-values).

Metal	Beer and water source			
	Mafefooane	**Mahlanyeng**	**Sea point**	**Mafikeng**
As	52.274 ↓	13.107 ↓	- 3.811 ↑	- 3.516↑
Cd	96.143 ↓	8.585 ↓	-0.931	- 3.755 ↑
Co	48.509 ↓	24.884 ↓	7.116↓	4.418↓
Cr	31.726 ↓	30.009 ↓	446.5↓	99.681↓
Cu	135.312 ↓	16.503 ↓	127.791↓	5.61↓
Fe	648.992 ↓	88.364 ↓	25.968↓	4.7↓
Hg	8.292 ↓	345.5 ↓	-67.352 ↑	7.309↓
Ni	9.997 ↓	64.729↓	27.644↓	4.406↓
Pb	162.383 ↓	72.2 ↓	61.827↓	5.282↓
Se	-39.022 ↑	-122.075 ↑	-342.08 ↑	- 38.195 ↑
Zn	59.379 ↓	31.805 ↓	24.329 ↓	- 5.901↑

↓ = Metal ion in beer sample was significantly lower than the concentration in the water used for brewing). ↑ = Metal ion in beer sample was statistically significantly higher than the concentration in the water used for brewing.

Mahlanyeng and water samples from Sea Point, the concentrations of As, Cd and Cr in the water and beer samples were higher than the WHO values. The concentrations of Ni in the water samples were higher than, and their concentrations in the beer samples were lower than, the WHO guideline values for drinking water. No guideline values are available for Co, Fe and Zn (Table 4).

Alcohol concentrations in one day samples (n = 5) of beer from the four sources

Only the total alcohol (determined by Back titration method), methanol and ethanol (determined by NMR) concentrations (% alcohol by volume) could be detected in the methods used.

1) The mean total alcohol, methanol and ethanol concentrations in the beers from the four sources were as follows: Mafefooane (6.36, 0.2, and 6.4 respectively); Mahlanyeng (5.30, 0.18, and 5.30 respectively); Sea Point (7.42, 0.08 and 7.4 respectively) and Mafikeng (7.42, 0.2 and 6.4 respectively). The concentration of ethanol was higher than that of methanol in all the beer samples.
2) The mean total alcohol, methanol and ethanol concentrations in the beer samples from the four sources were statistically different (p< 0.05) (Table 5).

Mice bone marrow erythropoietic cell toxicity and clastogenicity (micronucleus assay)

The results obtained in the micronucleus test of the effect of indigenous sorghum beer on the incidence of MNPCE,

MNNCE and frequency of PCEs in total erythrocytes are summarized in Table 6, for the beer sample from the Sea point source only. No increases in the frequency of MNPCE were observed in any of the groups of mice that received sorghum beer for 35 days. However, only in the group that received 50% beer was any statistically significant decrease in the PCE/NCE ratio observed (Table 6).

DISCUSSION

As presented in Table 1, locally brewed sorghum beer from Mafefooane, Mahlanyeng, Sea Point and Mafikeng sources in Lesotho contained significantly different (p< 0.05) concentrations of As, Cr, Fe, Ni Co, Cu, Pb, Zn and Se in the ranges of 0.004 mg/L for Cu and Ni to 0.942 and 6.337 mg/L for Se and Zn respectively. The concentrations of Cd and Hg in the beer samples however, were not significantly different (p> 0.05). No significant differences (p > 0.05) in the mean metal ions concentrations in the beer samples of the three sampling days for the samples from Mafefooane, Mahlanyeng and Sea Point. Wide variations in the levels of metals were thus detected in the beers. In a study of the concentrations of metal ions in wine samples from the province of Mendoza in Argentina, the concentrations of aluminium, cadmium, calcium, copper, iron, lead, zinc, and chromium were between 0.017-0.018 µg/L, 0.001-0.0047 µg/L, 0.010 - 0.015 µg/L, 0.023-0.028 µg/L, 0.480-0.790 µg/L, 0.050-0.090 µg/L, 0.024-0.130 µg/L, and <0.0002-0.00625 µg/L, respectively (Lara et al., 2005). In another study, different concentrations and diversity of metals were found in beverages; highest for red wine samples (30 metals totaling 5620.54 ± 123.86 ppb, that is, µg/L) followed by apple juice (15 metals totaling 1339.87

Table 4. Comparison of metal ions in beer and water with WHO guideline values for drinking water.

Metal	WHO guideline value (mg/L)	Mafefooane (bore hole)		Mahlanyeng (tap water)		Sea Point (tap water)		Mafikeng (bore hole)	
		Water	Beer	Water	Beer	Water	Beer	Water	Beer
As	0.01 mg/L (provisional) 10 µ/L	0.377±0.011*	0.038±0.002*	0.383±0.049*	0.013±0.003	0.037±0.010*	0.059±0.003*	0.033±0.010*	0.052±0.004*
Cd	0.003 mg/L	0.260±0.004*	0.036±0.006*	0.211±0.035*	0.036±0.006*	0.024±0.017	0.034±0.005*	0.023±0.005*	0.036±0.004*
Co	No limit listed	0.378±0.008	0.161±0.002	0.384±0.017	0.133±0.007	0.354±0.049	0.150±0.044	0.365±0.087	0.138±0.010
Cr	0.05 mg/L provisional	0.385±0.015*	0.072±0.007*	0.391±0.020*	0.053±0.003	0.364±0.001*	0.066±0.006*	0.351±0.004*	0.058±0.002*
Cu	2 mg/L	0.395±0.001	0.004±0.001	0.400±0.036	0.057±0.001	0.392±0.005	0.010±0.004	0.349±0.100	0.022±0.003
Fe	No limit listed 1–3 mg/L acceptable	3.929±0.007*	0.607±0.091	4.054±0.072*	0.381±0.000	3.590±0.200*	0.589±0.010	0.342±0.100	0.070±0.001
Hg	0.006 mg/L	0.210±0.013*	0.147±0.007*	0.381±0.001*	0.151±0.000*	0.021±0.003*	0.148±0.001*	0.201±0.011*	0.131±0.044*
Ni	0.07 mg/L	0.381±0.065*	0.004±0.001	0.384±0.010*	0.010±0.000	0.351±0.020*	0.032±0.005	0.135±0.042*	0.026±0.004
Pb	0.01 mg/L	0.373±0.003*	0.030±0.000*	0.400±0.009*	0.021±0.001*	0.400±0.011*	0.021±0.002*	0.332±0.100*	0.025±0.005*
Se	0.01 mg/L	0.420±0.010*	0.646±0.051*	0.409±0.007*	0.943±0.039*	0.371±0.000*	0.766±0.001*	0.407±0.003*	0.757±0.050*
Zn	No limit listed >3 mg/L unacceptable	3.746±0.100*	0.318±0.021	4.019±0.176*	0.773±0.003	3.916±0.224*	0.775±0.006	3.743±0.764	6.343±0.018*

*Concentration ($\bar{x} \pm$ S.D.) higher than the WHO guideline values for drinking water. WHO (2008).

± 10.84 ppb) and stout (14 metals totaling 464.85 ± 46.74 ppb) (Hague et al., 2008). The levels of 0.004 mg/L for Cu and Ni to 0.942 and 6.337 mg/L for Se and Zn, respectively, are equivalent to 0. 4 µg/L for Cu and Ni to 942.00 µg/L and 6337.00 µg/L for Se and Zn, respectively. Thus the metal ions concentrations in the sorghum beers in the present study were between 400 and 100000 times their concentrations in the wines and beverages in the two studies cited above. The concentration of, 6337.00 µg/L for zinc in the present study, for instance, was 48746.2 times the concentration of 0.130 µg/L for zinc in the Argentinean wines. Diets are known to account for the largest exposure to metals and metals play widespread roles in health and disease which range from the requirement for intake of essential trace elements to toxicity associated with metal overload (Hague et al.,

2008). Many of the toxic effects associated with metals are still under investigation, especially for low concentrations and for lifetime exposure, therefore the upper safe limits for many metal ions are unavailable (Lagerkvist and Oskarsson, 2007). However, the World Health Organization (WHO, 2008), has provided guideline values (mg/L) of the concentrations of some metals in drinking water. The concentrations of Co, Hg, Pb and Se in the beer samples were stronger than the WHO guideline values for drinking water while the concentrations of Ni and Cu were lower than the WHO guideline values. The concentrations of As and Cr in the beers were higher than their WHO guideline values for beers from three sources but not the Mahlanyeng source. No WHO guideline values was available for concentrations of Co, Fe and Zn. Many metal ions are associated with enhanced oxidative stress, inflammation and

cancer (U.S. EPA, 1989; Valko et al., 2006). Epidemiological evidence indicates that As is associated with cancers of the skin and internal organs, as well as with vascular disease (Luster and Simeonova, 2004).

Symptoms of short-term exposures of animals to cadmium include muscle cramps, sensory disturbances, liver injury, convulsions, shock and kidney failure (Rogers, 1996). The addition of cobalt salts as foam improvers in beer in the 1950s led to deaths in North America (Priest and Stewart, 2006). Chromium toxicity is very dependent on the species and oxidation states present and is still subject to considerable debate (Fisher and Naughton, 2005; Seenivasan et al., 2008). Some reported effects of high doses of copper include decreased levels of superoxide dismutase, a key enzyme for protection against oxidative damage which may be enhanced by

Table 5. Comparison of the alcohol content of the beers (% by volume).

Beer source	Total alcohol		Methanol		Ethanol	
	Mean	S.D	Mean	S.D	Mean	S.D
Mafefooane	6.36	0.666	0.2	0.021	6.4	0.37
Mahlanyeng	5.5	0.566	0.18	0.055	5.3	0.472
Sea Point	7.42	0.634	0.08	0.016	7.4	0.437
Mafikeng	7.42	0.634	0.2	0.019	6.4	0.337
One-way ANOVA result	13.146*		16.296*		22.15*	

n = sample size (5); * = statistically significant difference in alcohol concentrations in the beers from the different sources.

Table 6. Summary table of the incidence of MNPCE and frequency of PCEs in total erythrocytes in mouse bone marrow of 8-week-old males inbred NIH mice after 35 days of sorghum beer *ad libitumin* or water as negative control.

Treatment (% alcohol)	PCE scored	PCE (Data of five animals per group)			NCE scored of 5000 erythrocytes/group	PCE/ (PCE + NCE) (individual and Mean ±S.D)
		MNPCE (individual and total)	MNPCE/ 1000 PCE Individual data	Group Mean ± S.D		
0	10000	3; 2; 2; 3; 4; (14)	1.5; 1;1;1.5; 2	1.40 ± 0.40	1966	0.54; 0.75; 0.48; 0.52; (0.61 ± 0.13)
25	10000	6; 2; 2; 0; 4; (14)	3; 1; 1; 0; 2	2.8 ± 1.14	2176	0.54; 0.61; 0.53; 0.63; 0.52; (0.57 ± 0.05)
50 (†)	10000	4; 5; 2; 4; 0; (15)	2; 2.5; 1; 2; 0	3.0 ± 1.0	2829	0.48; 0.43; 0.46; 0.31; 0.49; (0.43 ± 0.07)
100	10000	0; 2; 2; 0; 2; (6)	0; 1; 1; 0; 1	0.60 ± 0.55	2621	0.42; 0.49; 0.43; 0.52; 0.52; (0.48 ± 0.05)
CP († and *)	10000	7; 15; 10; 8; 23 (63)	3.5; 7.5; 5; 4; 11.5	6.3 ±3.29	2888	0.36; 0.47; 0.37; 0.49; 0.42; (0.422± 0.058)

PCE = Polychromatic erythrocytes; NCE = Normochromatic erythrocyte; MNPCE = Micronucleated PCE; CP = Cyclophosphamide; * = Clastogenic; † =Toxic.

labile copper, which may initiate or exacerbate inflammatory disorders (Valko et al., 2006). High levels of ingested iron causes impaired oxidative phosphorylation and mitochondrial dysfunction particularly in the liver, which can result in cell death (Zacharski et al., 2004). Mercury toxicity is very dependent on the species and oxidation states present and is still subject to considerable debate (Fisher and Naughton, 2005; Seenivasan et al., 2008). Nickel has been shown to cause tissue injury and genotoxicity, to be cytotoxic to T-lymphocyte cells and to cause DNA damage

(Sunderman et al., 1976; environmental protection agency (EPA), 1995; Angela et al. 2006). The primary affects of lead are on the peripheral and central nervous system, kidney function, blood cells, and the metabolism of vitamin D and calcium. Lead can also cause hypertension, reproductive toxicity, and developmental effects (ATSDR, 1992).

Health effects of Selenium are very much dependent on dose. High blood levels of selenium is known to cause selenosis with symptoms that include gastrointestinal upsets, hair loss, white

blotchy nails, garlic breath odor, fatigue, irritability, and mild nerve damage (Goldhaber, 2003).

Zinc has numerous reported beneficial and detrimental effects, being essential components of many enzymes (Cherny et al., 2001; Alessio et al., 2007). Too little zinc can cause slow wound healing and skin sores, decreased sense of taste and smell, loss of appetite and damage in immune system. According to the Agency for toxic substances and disease registry (ATSDR) however, exposure to large amount of zinc for a long time can cause anemia, pancreas damage and low levels of high density lipoprotein cholesterol (ATSDR, 2005).

The results of the present study (Table 1) showed that the water samples from the four sources contained significantly different ($p < 0.05$) concentrations of As, Cd, Cr, Fe, Hg, Ni and Se but not significantly different concentrations of Co, Cu, Pb and Zn. The concentrations (mg/L) of the metal ions were in the ranges: As (0.033 – 0.383), Cd (0.023 – 0.260), Co (0.354 – 0.384, Cr (0.033 – 0.383), Cu (0.349 – 0.400), Fe (0.342 – 4.054), Hg (0.021 – 0.381), Ni (0.135 – 0.384), Pb (0.332 – 0.400), Se (0.371 – 0.420) and Zn (3.743 – 4.019).

According to the number of metal ions with the highest concentration, the water samples were ranked as follows: Mahlanyeng > Mafefooane > Sea Point > Mafekeng.

In a study of the presence of some metal ions in drinking water samples in Turkey, the metal ions concentrations varied as follows; Cu 0.17-1.19 µg/L, Fe 16.11-79.30 µg/L, Pb 0.18-0.99 µg/L and Mn 0.15-2.56 µg/L (Soylak et al., 2002).

Another study in Ghana of metal ions in drinking water demonstrated the presence of metal ions (mg/L) and the variation of their concentrations in water from different sources for Cu (1.19-2.75), Fe (0.05-0.85), Zn (0.04-0.15), Mn (0.003-0.011) and Al (0.05-0.15) (Akoto and Adiyiah, 2007). Whereas the concentrations of Cu in the present study were lower than their concentrations in the waters in the Ghana study, the concentrations of Zn were higher than their concentrations in the waters in the Ghana study. The concentrations of Fe in the waters in the present study were in the same ranges as their concentrations in the waters in the Ghana study. The concentrations of Cu, Zn and Fe in the waters of the present study and the waters in the Ghana study were much higher (>2000x), than their concentrations in the waters in the Turkey study. These observations could be attributed to differences in geology of the different regions. Furthermore, the concentrations of As, Cr, Hg, Ni, Pb and Se in the water samples analyzed in the present study were higher than the WHO guideline values for drinking water. The concentrations of Cd were also higher than the WHO guideline values in the waters from Mafefooane, Mahlanyeng and Mafikeng while the concentrations of Cu were lower than the WHO guideline values. The sources of the water at the 4 sites were as follows, Mafefooane (Bore hole), Mafikeng (Bore hole), Mahlanyeng (Tap water), Sea Point (Tap water).

The concentrations of Co, Cr, Cu, Fe, Ni and Pb in the beer samples were significantly lower ($p < 0.05$) than their concentrations in the water from each of the 4 water sources which suggested that the brewing process resulted in a lowering of the concentrations of these metal ions in the beers. However, the

concentrations of Se in the beers from Mafefooane and Mahlanyeng, As, Cd, Hg and Se ions in the beers from Sea Point and As, Cd, Se and Zn ions in the beers from Mafikeng, were significantly higher ($p < 0.05$), than their concentrations in the waters used for brewing. The observations suggested that these metal ions were introduced into the beers from extraneous sources during the brewing process. The extraneous source(s) of the metal ions in the beer samples in the cases where their concentrations were higher than their concentrations in the water used for brewing was not certain. One possible extraneous source of the metal ions could be the metal drums in which the beers were brewed. However, according to Larsson et al. (1998), as competition between brewers for customers set in, brewers invented brewing techniques that produced beer at lower and more affordable prices to consumers. Some of the new techniques, it is claimed, involved the introduction of radio batteries, brake fluid and car battery acids into the beer to make it stronger (that is, more intoxicating). Batteries are made of many different metal elements/compounds and, depending on the brand, may contain aluminum (Al), copper (Cu), mercury (Hg), arsenic (As), zinc (Zn), lead (Pb), nickel (Ni) and cadmium (Cd) (Dhar, 1973).

The beer samples contained measurable concentrations (percentage of alcohol by volume) of ethanol and methanol only, whose concentrations in the beers from the different sources varied significantly and with higher concentrations of ethanol than methanol in the beers from all the sources. The concentrations (percentage of alcohol by volume) of total alcohol in the beer samples were between 5.30 and 7.42, equivalent to 53,000 to 74,200 mg/L. The concentrations (in percentage of alcohol by volume) of methanol ranged from 0.08 to 0.2, equivalent to 800 to 2000 mg/L and for ethanol, 5.30 to 7.4, equivalent to 53,000 to 74,000 mg/L. The total alcohol contents in the beers from the different sources were thus different, which suggested that different brewers employed different methods. The average concentration of 6.36% alcohol in the beers was different from a reported concentration of 3% for *Joala*, another traditional beer that is brewed and consumed in Lesotho (Turner, 2001).

The United States Environmental Protection Agency, in their multimedia environmental goals for environmental assessment recommends a minimum acute toxicity concentration of methanol in drinking water at 3.9 ppm (3.9 mg/L), with a recommended limit of consumption below 7.8 mg/day (Cleland and Kingsbury, 1977). Death from consumption of the equivalent of 6 g of methanol has been reported the U.S. Department of Health, Education, and Welfare (HEW) (HEW, 1976; Wimer et al., 1983). The 1400 mg/L average content of methanol in the beers is equivalent to 1.4 mg/ml. Given that the daily volumes of traditional beer that were consumed on a typical drinking day ranged from 340 to 10 000 ml with a mean (SD) of 2634 (± 1737 ml) (Siegfried, 2001), a drinker who consumed the minimum of 340 ml/day, would have ingested 476 mg of methanol, which far exceeds the recommended limit of consumption of below 7.8 mg/day. For an average adult, the fatal ingested dose is approximately 1 L (approximately 2 pints) of 40 to 55% ethanol (the percentage found in whiskey, gin, rum, vodka, or brandy) consumed within a few minutes (Gosselin, 1984). In the present study, the beers contained on average, 6.36% total alcohol, equivalent to 36.125 g alcohol/pint (568 ml), made up of about 99% ethanol. Western beers contain, on average, 5% alcohol (Turner, 2001).

In the present study, the beer samples did not induce chromosomal aberrations in mice bone marrow but induced erythropoietic cell toxicity (decrease in the PCE/NCE ratio) at the 50% concentration of beer only. In one reported study in which rodents were exposed to ethanol *in vivo*, the ethanol did not induce micronuclei in mice, but conflicting results were obtained in rats. Ethanol induced sister chromatid exchanges in mouse embryos exposed *in vivo* and, in another study, chromosomal aberrations in rat embryos exposed *in vivo* (IARC, 1988). Statistically significant sister chromatid exchanges were observed in the bone marrow of male mice when 10 or 20% ethanol (approximately 20000 or 40000 mg/kg/day) was given as the only liquid source for 3 to 16 weeks (Obe et al., 1979). In the present study however, the mean alcohol concentration of the beers was much lower (6.36%) than, and the methods of administration of the beers different from, those of the studies previously cited.

Conclusions

The results showed that the beers from the four different sources contained significantly different (p< 0.05) concentrations (mg/L) of As, Cr, Fe, Ni Co, Cu, Pb, Zn and Se but similar (p> 0.05) concentrations of Cd and Hg. The concentrations (mg/L) of total alcohol (53,000 to 74,200); methanol (800 to 2000) and ethanol (53,000 to 74,000) differed significantly (p < 0.05) which suggested different brewing methods. Concentrations of Se, As, Cd, Hg, Se As, Cd, Se and Zn in the beers were higher (p< 0.05), than in the brewing water which suggested that these metals were introduced from extraneous sources.

Beer from only one source was assessed for genotoxicity. It was not clastogenic but induced erythropoietic cell toxicity at the 50% concentration only (p < 0.05).

The high alcohol and metal ions concentrations in the beers and water have implications for public health and highlights the need for public awareness, regulation and standardization of production methods and consumption.

Further research is necessary to ascertain the source(s) of the metal ions.

ACKNOWLEDGEMENT

Authors would like to express their appreciation to Mr. Andreas Thakaso for his help with the animal husbandry during this work.

REFERENCES

Aísa R, Pueyo F (2004). Endogenous longevity, health and economic growth: a slow growth for a longer life? Economics Bulletin, 9(3): 1-10.http://www.economicsbulletin.com/2004/volume9/EB–03I10002 A.pdf.

Akoto O, Adiyiah J (2007). Chemical analysis of drinking water from some communities in the Brong Ahafo region. Int. J. Environ. Sci. Technol., 4 (2): 211-214.

Alessio L, Campagna M, Lucchini R (2007). From lead to manganese through mercury: mythology, science, and lessons for prevention. Am. J. Ind. Med., 50: 779-787.

Ang HH, Lee KL, Kiyoshi M (2005). Determination of lead in Smilax luzonensis herbal preparations in Malaysia. Int. J. Toxicol., 24: 165-171.

Angela AU, Jinny HA, Hernandez M, Polotsky A, Hungerford DS, Frondoza CG (2006). Nickel and vanadium metal ions induce apoptosis of T-lymphocyte Jurkat cells. J. Biomed. Mater. Res. Pt A, 79(3): 512-521.

AOAC. Association of Official Analytical Chemists (1995). Official Methods of Analysis. 16th Edition. Washington DC, 28: 2-3.

Asita, AO, Dingann ME, Magama S (2008). Lack of modulatory effect of asparagus, tomato, and grape juice on cyclophosphamide-induced genotoxicity in mice. Afr. J. Biotechnol., 7 (18): 3383-3388.

ATSDR. Agency for Toxic Substance and Disease Registry (1992). Case Studies in Environmental Medicine: Lead Toxicity.

ATSDR. Agency for Toxic Substances and Disease Registry (2005). Toxicological Profile for Zinc (Update), Atlanta, GA: U.S. Department of Public Health and Human Services, Public Health Service.

Berger S, Braun S (2004). 200 and More NMR Experiments. Second Edition, Wiley-VCH, Germany, pp. 315-317.

Bonner P (1990). "Desirable or Undesirable Women? Liquor, Prostitution, and the Migration of Basotho Women to the Rand, 1920-1945," In: Cherryl Walker (Ed). Women and Gender in Southern Africa to 1945. Cape Town: David Philip; London: J. Currey.

Bourgeault-Tassé I (2011). Taking stock: Alternative indicators of national well-being. http://www.researchsea.com/html/article.php/aid/5840/cid/6/researc h/taking_stock__alternative_indicators_of_national__wellbeing.html ?PHPSESSID=fg9e65e3qp443iair1f79s8ds1. Accessed on 2011 07-10

Cherny RA, Atwood CS, Xilinas ME, Gray DN, Jones WD, McLean CA, Barnham KJ, Volitakis I, Fraser FW, Kim YS, Huang XD, Goldstein LE, Moir RD, Lim JT, Beyreuther K, Zheng H, Tanzi RE, Masters CL, Bush AI (2001). Treatment with a copper-zinc chelator markedly and rapidly inhibits β-amyloid accumulation in Alzheimer's disease transgenic mice. Neurology, 30: 665-676.

Cleland JG, Kingsbury GL (1977). Multimedia Environmental Goals For Environmental Assessment. U.S. Environmental Protection Agency: EPA-600/7-77-136b, E-28, November.

Dhar SK (1973). Metal Ions in Biological Systems: Studies of Some Biochemical and Environmental Problems In: Sanat K, Dhar (Editor) Plenum Press.

Fisher A, Naughton DP (2005). Therapeutic chelators for the twenty first Century: New treatments for iron and copper mediated inflammatory and neurological disorders. Curr. Drug. Deliv. 2: 261-

268.

Gimmler-Luz MC, Cardoso VV, Sardiglia CU, Widholzer DD (1999) Transplacental inhibitory effect of carrot juice on the clastogenicity of cyclophosphamide in mice. Genet. Mol. Biol., 22(1): 1-13.

Goldhaber SB (2003). Trace element risk assessment: essentiality vs. toxicity. Regulatory Toxicology and Pharmacology. 38: 232-42.

Gosselin RE (1984). Ethyl alcohol. In: Gosselin, RE, Smith RP, Hodge HC (Eds). Clinical Toxicol. Commercial Products. 5th ed. Baltimore: Williams and Wilkins, pp. III-267.

Hague T, Petroczi A, Andrews PLR, Barker J, Naughton DP (2008). Determination of metal ion content of beverages and estimation of target hazard quotients: a comparative study. Chem. Cent. J., 2: 13.

Halliwell B, Gutteridge JM (2007). Editors. Free radicals in biology and medicine. 4. Oxford University Press, UK.

HEW. US Department of Health, Education, and Welfare (1976). Occupational Exposure to Methyl Alcohol, HEW Pub. No. (NIOSH) (March), pp. 76-148. http://web.deu.edu.tr/geomed2010/2007/Sandal.pdf - Accessed on July 14, 2011.

IARC. International Agency for Research on Cancer (1988). Monographs on the evaluation of carcinogenic risks to humans. Vol. 44. Alcohol drinking. World Health Organization. 1988: 35.

Lagerkvist B, Oskarsson A (2007). Vanadium. Handbook on the Toxicology of Metals. Third Edition, pp. 905-923.

Lara R, Cerutti S, Salonia JA, Olsina RA, Martinez LD (2005). Trace element determination of Argentine wines using ETAAS and USN-ICP-OES. Food Chem. Toxicol., 43(2): 293-297.

Larsson A, Mapetla M, Schlyter A (1998). Changing Gender Relations in Southern Africa: Issues of Urban Life. The Nordic Africa Institute. 99911-31-21-3.

Luster MI, Simeonova PP (2004). Arsenic and urinary bladder cell proliferation. Toxicol. Appl. Pharmacol., 198: 419-423.

Navarro-Blasco I, Alvarez-Galindo JI (2005). Lead levels in retail samples of Spanish infant formulae and their contribution to dietary intake of infants. Food Addit. Contam., 22: 726-734.

Obe G, Natarajan AT, Meyers M, Hertog AD (1979). Induction of chromosomal aberrations in peripheral lymphocytes of human blood in vitro, and of SCEs in bone-marrow cells of mice in vivo by ethanol and its metabolite acetaldehyde. Mutat. Res. 68: 291-294.

OECD (2006). Life Expectancy vs. Alcohol Consumption by Country. Organisation for Economic Co-operation and Development: Health Data.

Priest FG, Stewart GG (2006). Handbook of Brewing (Editors: 2nd Edition) Taylor & Francis, Publishers.

Rogers JM (1996). The developmental toxicity of cadmium and arsenic with notes on lead In: L.W. Chang, Editor, Toxicology of Metals, Lewis Publishers, Boca Raton, pp. 1027–1045.

Salvadori DM, Ribeiro LR, Oliveira MD, Pereira CA, Becak W (1992). The protective effect of beta-carotene on genotoxicity induced by cyclophosphamide. Mutat. Res., 265(2): 237-244.

Seenivasan S, Manikandan N, Muraleedharan N (2008). Chromium contamination in black tea and its transfer into tea brew. Food Chem., 106: 1066-1069.

Senaoana MP, Turner SD, van Apeldoorn GJ (1984). Research on rural non-farm employment in Lesotho: results of a baseline survey. Roma: Institute of Southern African Studies, National University of Lesotho: Research report 6.

Sharma RK, Agrawal M (2005). Biological effects of heavy metals: An overview. J. Environ. Biol., 26: S301-313.

Siegfried N, Parry CD, Morojele NK, Wason D (2001). Profile of drinking behaviour and comparison of self-report with the CAGE questionnaire and carbohydrate-deficient transferrin in a rural Lesotho community. Alcohol and Alcoholism, 36(3): 243-248.

Soylak M, Armagan F, Aydin A, Saracoglu S, Elci L, Dogan M (2002). Chemical Analysis of Drinking Water Samples from Yozgat, Turkey. Polish J. Environ. Studies, 11(2): 151-156.

Sunderman FW, Jr Kasprzak K, Horak E, Gitilitz P, Onkelinx C (1976). Effects of TETA upon the metabolism and toxicity of $NiCl_2$ in rats. Toxicol. Appl. Pharmacol., 38:177–188.

Templer DI, Griffin PR, Hintze J (1993). Gender life expectancy and alcohol: an international perspective. Int. J. Addict., 28(14): 1613-1620.

Turner S (2001). Livelihoods in Lesotho. CARE Lesotho, 5 April.

U.S. EPA . Environmental Protection Agency (1989). Guidance manual for assessing human health risks from chemically contaminated, fish and shellfish, U.S. Environmental Protection Agency, Washington, D.C. EPA-503/8-89-002.

U.S. EPA. Environmental Protection Agency (1995). Nickel. Integrated Risk Information System (IRIS), Environmental Criteria and Assessment Office, Office of Health and Environmental Assessment, Cincinnati, OH.

Valko M, Rhodes CJ, Moncol J, Izakovic M, Mazur M (2006). Free radicals, metals and antioxidants in oxidative stress-induced cancer. Chem. Biol. Interact. 160: 1-40.

Varian AAS Handbook, Australia, 1988

VFAASAM. Varian Flame Atomic Absorption Spectrometry Analytical Methods (1989) Australia.

Wang X, Sato T, Xing B, Tao S (2005). Health risks of heavy metals to the general public in Tianjin, China via consumption of vegetables and fish. Sci. Total Environ., 350: 28-37.

WHO (2008). Who Guidelines for drinking-water quality [electronic resource]: Incorporating 1st and 2nd addenda, Vol.1, Recommendations. – 3rd ed.

WHO. World Health Organization (2011) Lesotho SOCIOECONOMIC CONTEXT.

WHO. World Health Organization (2011). WORLD HEALTH STATISTICS, 2011 http://www.who.int/whosis/whostat/EN_WHS2011_Full.pdf.

Wimer WW, Russell JA, Kappplan HL (1983). Alcohols Toxicology. Park Ridge New Jersey, Noyes Data Corporation.

Zacharski L, Chow B, Howes P, Lavori P, Shamayeva G (2004). Implementation of an iron reduction protocol in patients with peripheral vascular disease: VA cooperative study no. 410: The Iron (FE) and Atherosclerosis Study (FEAST)*1. American Heart J. 148(3): 386-392.

Zheng N, Wang QC, Zhang XW, Zheng DM, Zhang ZS, Zhang SQ (2007). Population health risk due to dietary intake of heavy metals in the industrial area of Huludao city, China. Sci. Total Environ., 387: 96-104.

16

Ameliorative effect of cabbage extract on cadmium-induced changes on hematology and biochemical parameters of albino rats

F. C. Onwuka[1], O. Erhabor[2] *, M. U. Eteng[3] and I. B. Umoh[3]

[1]Department of Biochemistry, Faculty of Basic Medical Sciences, University of Port Harcourt PMB 5323 Port Harcourt, River State, Nigeria
[2]Department of Haematology College of Health Sciences , University of Port Harcourt P. M. B. 5323, Port Harcourt, Rivers State, Nigeria.
[3]Department of Biochemistry, Faculty of Basic Medical Sciences, University of Calabar, P. M. B. 1115 Calabar, Cross River State, Nigeria.

The effect of dietary supplement containing cabbage on cadmium – induced toxicity was studied in wistar rats. The effect of cadmium was investigated in 3 animals groups. Group A, rats were fed normal basal diet only. Group B rats were placed on normal basal, mixed with cadmium chloride (3 mg/kg body weight daily) while group C rats were fed with basal diet mixed with Cadmium chloride (3 mg/kg body weight) and 0.5 kg dry cabbage pellets daily. Rats were monitored for 28 days. At the end of the treatment, the animals were sacrificed using chloroform vapor. Oral rat LD50 for Cadmium Chloride, Anhydrous is 88 mg/kg. The effect of cadmium treatment alone and combined cadmium-cabbage treatment on lipid peroxidation, as measured by malondialdehyde levels in testes and kidney, serum activities of acid phosphatase (ACP) and prostatic acid phosphatase (PAP) and alkaline phosphatase (ALP) were investigated alongside testicular and kidney organ weight and assessment of some hematological indices. The result showed that cadmium induced a significant increase in both testicular and kidney malondialdeyhde (MDA); but dietary cabbage seem to have a beneficial effect on lipid peroxidation. Cadmium also induced a 75% increase in ACP, 98% in PAP and 22% increase in ALP, but cabbage supplementation tended to produce a reduction in the activities of these enzymes (p = 0.001). Result of organ weight analysis in Cd –exposed rats showed a decrease in testes and kidney weight. Comparatively rats whose diet contained cadmium with cabbage supplementation showed an increase in organ weight. Administration of combined treatments of cadmium and cabbage may provide beneficial effects against cadmium-induced changes on the testicular and kidney weight, malondialdeyhde, liver enzymes; alkaline phosphatase, acid phosphatase and prostatic acid phosphatase levels by reducing cadmium –associated oxidative stress.

Key words: Cabbage, cadmium, toxicity, amelioration.

INTRODUCTION

Cadmium is a heavy metal, which is widely used in industry, affecting human health through occupational and environmental exposure. In mammals, it exerts multiple toxic effects and has been classified as a human carcinogen by the International Agency for Research on Cancer. Cadmium affects cell proliferation, differentiation, apoptosis and other cellular activities. Cadmium is particularly a concern in environmental pollution because it is said to accumulate in the human body causing renal dysfunction, pulmonary emphysema, kidney damage and osteoporosis (El-Demerdash et al., 2004).

Cadmium (Cd) is an environmental toxicant and an

endocrine disruptor in humans and rodents. Several organs are affected by Cd and recent studies have illustrated that the testis is exceedingly sensitive to Cd toxicity. More importantly, Cd and other toxicants, such as heavy metals ; lead, mercury, and estrogenic-based compounds like bisphenols may account for the recent declining fertility associated with reduced sperm count and testis function in men in developed countries (Siu et al., 2009)˙ Cadmium induced toxicity has been shown to be alleviated by antioxidants; L -ascorbic acid (Shirashi et al., 1993), broccoli (Zoyne et al., 2008), natural anti-oxidant Garlic (Kumar et al., 2009) and naringenin (Renugadevi et al., 2009), which is naturally occurring citrus flavonone.

Previous study (Ola-Mudathir et al., 2008) demonstrated that aqueous extracts of onion and garlic could proffer a measure of protection against Cd-induced testicular oxidative damage and spermiotoxicity by possibly reducing lipid peroxidation and increasing the antioxidant defense mechanism in rats. The hepatoprotective effect of onion and garlic extracts on cadmium (Cd)-induced oxidative damage in rats has also been reported (Hilman, 2009). Anti cancer properties of cabbage have been documented *in vivo and vitro* (Michael et al., 1999). These including the inhibition of the formation of free radicals, support of endogenous radical scavenging mechanism, enhancement of cellular antioxidant enzymes (superoxide dismutase, catalase, glutathione peroxidase), inhibition of low density lipoprotein from oxidation by free radicals and inhibition of the activation of the oxidant induced transcription factor and nuclear factor kappa (NP-KB) (Borek, 2001).

Cabbage is a member of the cruciferous family. Other vegetables that have developed or evolved from the early strains of cabbage include; brussels sprouts, cauliflower, kale and kohlrabi. Cabbage is a hardy vegetable that is available in various shades of green, as well as red or purple varieties. Most varieties have smooth leaves, but some types have ruffled textured leaves. The most popular varieties are green, red, savoy and Chinese. Cabbage is usually shredded into salads or used as an ingredient in stews, soups or baked dishes. Scientists at the American Cancer Society have found convincing evidence that diets high in cruciferous vegetables lower the risks of many forms of cancer (Byers et al., 2002). The present study was undertaken to assess if dietary

---

*Corresponding author. E-mail: n_osaro@yahoo.com.

Abbreviation: **MDA,** Malondialdeyhde; **ALT,** alanine transaminase; **AST,** aspartate transaminase; **ALP,** alkaline phosphatase; **ACP,** acid phosphatase prostatic; **PAP,** acid phosphatase; **EDTA,** ethylene diamine tetra acid; **NAC,** n-acetyl cysteine; **MiADMS,** monoisoamyl 2, 3-dimercaptosuccinate; **Hb,** blood hemoglobin; **TEC,** total erythrocytic count; **PCV,** packed cell volume; **TLC,** total leukocyte count.

supplementation with cabbage has any ameliorating effect on acute cadmium toxicity in Wistar rats.

MATERIALS AND METHODS

Animals

Thirty albino rats of wistar strain weighing 140 – 220 g were purchased from the disease free stock of the Department of Biochemistry animal house, University of Port Harcourt and transferred to the Department of Biochemistry University of Calabar, where the study took place. They were acclimatized for two weeks on a commercial rat diet prior to experimentation. Permission and approval for the animal studies were obtained from the College of Medical Science Animal Ethics Committee of the University of Calabar. The rats were weighed and randomly assigned on the basis of their weights into three study groups (A, B and C) of ten rats each and were housed in plastic cages with wire screen tops. They were kept under adequate ventilation with room temperature and relative humidity of 29 ± 2ºC and 40 - 70%, respectively, with a 12 h natural light-dark cycle. Food and water was provided *ad libitum*, and good hygiene was maintained by constant cleaning and removal of feaces with spilled feed from cages daily.

Animals in group A were fed normal basal diet only. Group B rats were placed on normal basal diet mixed with cadmium chloride (3 mg/kg body weight) daily while group C rats were fed with basal diet mixed with Cadmium chloride (3 mg/kg body weight) and 0.5 kg dry cabbage pellets daily. Experimental rats were monitored for 28 days. At the end of the treatment period, the animals were sacrificed using chloroform vapor. Oral rat LD50 for Cadmium Chloride, Anhydrous is 88 mg/kg. Blood was collected by cardiac puncture from all experimental rats into two tubes (a gel tube without anticoagulant and a tube containing Ethylene Diamine Tetra Acid (EDTA).

Sample in the gel tube was allowed to clot, centrifuged to obtain serum sample used for the colorimetric and enzyme immunoassay analysis of malondialdehyde level (MDA), alkaline phosphatase (ALP), acid phosphatase (ACP), and Prostatic acid phosphatase (PAP) using Randox kits (Randox Clinical Diagnostic solutions, UK). The EDTA anticoagulated sample was used for the analysis of hematological indices of hemoglobin (Hb), packed cell volume (PCV), white blood cells count (WBC) and red cell counts (RBC). Kidney and testes were surgically removed, weighed and placed in an iceberg and later trimmed to 5 g weight and further used to ascertain the extent of lipid peroxidation.

Statistics

Statistical analysis was carried out using a Statistical Package for Social Sciences version 10 (SPSS Inc., Chicago, IL.). Data collected were expressed as mean ± SD and the student's t – test was used for analysis. Descriptive analysis of percentages of continuous variables was reported. Comparisons were assessed using mean and chi-square test. A p-value of < 0.05 was considered statistically significant in all statistical comparison.

RESULTS

Malondialdehyde level (MDA) was used as an index of lipid peroxidation in testes and kidney homogenates of controls, and experimental animals treated with cadmium and cabbage supplements as shown in Table 1. The mean ± SD values of MDA levels obtained for the testes

Table 1. Malondialdehyde levels in testes and kidney of controls and cadmium treated rats.

Treatment group	Testes µmol/L	Kidney µmol/L	P-value
Group A (control)	2.14 ± 0.03	2.10 ± 0.07	0.01
Group B (Cd-treated)	3.42 + 0.02	2.76 + 0.03	
Group C (Cd+cabbage)	1.45 ± 0.07	2.53 ± 0.03	

Values are mean ± SD and indicates a statistically significant difference in the malondialdehyde levels in the testes and kidney of Cd treated rats compared to rats treated with Cd and cabbage supplementation.

Table 2. Total ACP, PAP, and ALP activities in control and cadmium exposed Wistar rats.

Treatment group	ACP (U/l)	PAP (U/l)	ALP(U/l)	P-value
Group A (Control)	6.05 ± 0.26	5.10 ±0.04	59.80 ± 0.57	0.001
Group B (Cd–treated)	10.60 ± 0.07	9.10 ± 0.06	62.00 ± 0.14	
Group C(Cd+cabbage)	5.95+0.21	5.25+0.03	51.30+0.14	

Values are mean ± SD and indicates a statistically significant difference between enzyme activities in Cd treated rats compared to rats treated with Cd and cabbage supplementation.

Table 3. Effect of Cd exposure and cabbage supplementation on testes weights of Wistar rats.

Treatment group	Testes weight (g)	P-value
Group A (Control)	2.12 ± 0.14	0.001
Group B (Cd-treated)	1.90 ±0.16	
Group C (Cd+ cabbage)	2.87 ± 0.17	

Values are mean ± SD and indicates a statistically significant difference in the testes weight of Cd treated rats compared to rats treated with Cd and cabbage supplementation.

in various groups were 2.14 ± 0.03 for controls, 3.42 ± 0.02 for Cd–treated and 1.45 ± 0.07 for rats fed with Cd supplemented with cabbage. Also, in the kidney tissue homogenates, the MDA levels of control and experiment groups were 2.10 ± 0.07 for controls, 2.76 ± 0.03 for Cd-treated and 2.53 ± 0.06 nmol/L for rats fed with Cd supplemented with cabbage. Result shows that cadmium treatment increased the MDA level in the testes by 60% relative to the control that received placebo.

The result further indicated a significant decrease (p = 0.01) in MDA levels in cd-treated with cabbage supplemented groups. The result equally showed a 31% increase in kidney MDA levels for the groups that had cd-treatment alone. Comparatively, cabbage supplementation tended to have a beneficial effect associated with a decline in the level of kidney MDA. The mean ± SD values of ACP for the control, Cd-treated and Cd-treated with cabbage supplement were 6.05 ± 0.26, 10.60 ± 0.07 and 5.95 ± 0.21 U/l, respectively. The mean ± SD values for PAP activities for the control, Cd-treated and Cd-treated with cabbage supplement were 5.10 ± 0.04, 9.10 ± 0.06 and 5.25 ± 0.03 U/l, respectively. The values of ALP for control, Cd-treated and Cd-treated with cabbage

were 59.80 ± 0.57, 62.00 ± 0.14 and 51.3.0 ± 0.14, respectively. The result showed that cadmium exposure significantly increased ACP, PAP and ALP activities in the cadmium fed rats, while cabbage supplementation tended to have a beneficial effect associated with a decline in the level of these enzymes as shown in Table 2.

There was a statistically significant difference in the values of testes weight of Control group A, Cd-treated group B and Cd-treated with cabbage supplements (group C), (2.12 ± 0.14, 1.90 ± 0.16 and 2.87 ± 0.17 g respectively) p = 0.001 as shown in Table 3. The values for kidney weight although, marginally higher in the cabbage supplemented groups (1.33 ± 0.15) compared to the Cd-treated rats (1.26 ± 0.18 g) and control rats (1.75 ± 0.31). The difference was however, not statistically significant (p = 0.06) as shown in Table 4.

The haematological parameters assessed were HB, PCV and total WBC counts. The mean + SD values are outlined in Table 5. The values for HB, PVC and WBC count for Cd-treatment group alone showed a significant reduction in comparison to the cabbage supplemented group. The value for total WBC count for Cd-group showed a significant increase compared to the cabbage supplemented group as shown in Table 5.

DISCUSSION

Cadmium is a well-known human carcinogen and a potent nephrotoxin. This study showed that Cd treatment induced as much as 60% of lipid peroxidation in the testes of test rats compared to control. This agrees with the report of Grupta et al. (1997) which indicated that Cd is an inducer of cell oxidative stress. Pre-treatment with cabbage reduced testicular MDA, significantly in rats fed with Cd-supplemented with cabbage compared to the

Table 4. Effect of Cd exposure and cabbage supplementation on kidney weights of Wistar rats.

Treatment group	Kidney weight (g)	P-value
Group A (Control)	1.75 ± 0.31	0.06
Group B (Cd-treated)	1.26 ± 0.18	
Group C (Cd+ cabbage)	1.33 ± 0.51	

Values are mean ± SD and indicates a non- statistically significant difference in the testes weight of Cd treated rats compared to rats treated with Cd and cabbage supplementation.

Table 5. Effect of Cd-treatment and cabbage supplementation on hematological parameters of albino rats.

Treatment group	Haemoglobin value (g/dl)	Paced cell volume (%)	White cell count (x10^9/L)	P-value
Group A (Control)	10.68 ± 0.41	35 ± 0.22	5.8 ± 0.41	0.001
Group B (Cd-treated)	7.28 ± 0.21	22 ± 0.60	2.97 ± 0.29	
Group C (Cd+ cabbage)	9.50 ± 0.30	31 ± 0.31	7.3 ± 0.50	

Values are mean ± SD and indicates a statistically significant difference in the hematological values of Cd treated rats compared to rats treated with Cd and cabbage supplementation.

Cd-treated ones. Similarly, Cd-treatment alone enhanced lipid peroxidation in kidney tissues of Cd fed rats by as much as 31% compared to control. Pre-treatment with cabbage alongside with Cd-treatment reduced significantly the MDA levels compared to the Cd-treated ones. These findings are consistent with the report of Fleischer et al., 2001, which reported that cabbage may have some anti cancer activity. Also anti cancer and antioxidants properties of the cabbage –family vegetable have been documented *in vivo and in vitro* (Michael et al., 1999).

The effect of cabbage on serum ACP, PAP and ALP activities of rats fed with a basal diet mixed with cadmium chloride (3 mg Cd/kg) revealed that Cd -treatment can cause a significant increase in serum ACP, PAP and ALP activity in comparison to control values. Cd-treated rats showed 75% increased in ACP, 78% in PAP and 22% in ALP compared to the values for untreated controls. Cadmium accumulation in blood affects the renal cortex and causes renal failure. The increased activity of these enzymes may be as a result of leakage of these enzymes into blood stream due to the compromised integrity of the testes and kidney resulting from cadmium toxicity. This finding is consistent with the report of Fouad et al. (2009) which indicated that Cd-induced oxidative damage in rat liver is amenable to attenuation by high dose of onion and moderate dose of garlic extracts possibly via reduced lipid peroxidation and enhanced antioxidant defense system that is insufficient to prevent Cd-induced heaptotoxicity. Cd was found to cause a marked increase in the levels of lipid peroxidation and glutathione S-transferase in Cd fed rats. They also observed a decrease in hepatic activities of alanine transaminase (ALT), aspartate transaminase (AST) and alkaline phosphatase and a concomitant increase in the plasma activities of ALT and AST. We observed a reversal in the cadmium-induced increase in the activities of these

enzymes in the cabbage supplemented group. Supplementation with cabbage seems to play a protective role against cadmium-induced increases in ACP, PAP and ALP activities in rats. Cabbage seems to have antioxidant effect against cadmium induced toxicity.

Previous reports indicates that cadmium induced toxicity has been alleviated by antioxidants; L -ascorbic acid (Shirashi et al., 1993), broccoli (Zoyne et al., 2008) natural anti-oxidant Garlic (Zoyne et al., 2008; Kumar et al., 2009), naringenin (Renugadevi et al., 2009), a naturally occurring citrus flavonone. Similarly, aqueous extracts of onion and garlic has been shown to proffer a measure of protection against Cd-induced testicular oxidative damage and spermiotoxicity by possibly reducing lipid peroxidation and increasing the antioxidant defense mechanism in rats (Ola-Mudathir et al., 2008). Antioxidant properties of cabbage seem to have the ability to mop up free radicals. Cabbage is rich in antioxidant nutrients, which play an important role in health maintenance. They neutralize harmful chemicals called "free-radicals" that cause cell damage in the body. Antioxidants have been strongly linked to the protection from numerous diseases; heart disease, cancer, eye disease as well as the regulation of the immune system. In addition, cabbage contains beta-carotene, lutein and zeaxanthin, carotenoids that are a large class of natural plant pigments. They have chemo-protective effects, exhibit strong antioxidant properties and may reduce the risk of age-related macular degeneration and some types of cancer (Michael et al., 1999; Lynn et al., 2006).

Our study indicated that testes weight was significantly lower in Cd- treated rats in comparison to rats feed with Cadmium supplemented with cabbage extract. This finding is in agreement with previous reports (Ola-Mudathir et al., 2008; Fouad et al., 2009) which indicated that Cd caused a marked rise in testicular lipid peroxidation (LPO) and glutathione S-transferase (GST) levels. Cd intoxication significantly decreased epididymal sperm concentration and sperm progress, motility, increased percent total sperm abnormalities and live/dead count. They also demonstrated that aqueous extracts of onion and garlic could proffer a measure of protection against Cd-induced testicular oxidative damage and spermiotoxicity by possibly reducing lipid peroxidation and increasing the antioxidant defence mechanism in rats. The mechanism of cadmium- induced testicular toxicity is poorly understood. Previous studies focusing on cadmium-related changes in testicular histopathology have implicated testicular blood vessel damage as the main cause of cadmium toxicity (Jian-Ming et al., 2003).

Although, not statistically significant, our study indicated that kidney weight was significantly lower in Cd-treated rats in comparison to rats feed with cadmium supplemented with cabbage extract. We observed that, cabbage supplementation produced a beneficial effect on the kidney weight. The degree of cadmium toxicity might depend on the extent of free radical load within the testes and kidney. Previous study by Sinha et al., 2009 investigated the protective role of taurine against cadmium-induced oxidative impairment in murine liver and observed that cadmium- induced hepatic oxidative stress and necrosis was prevented by the prophylactic properties of taurine. Similarly, the influence of an antioxidant agent such as N-acetyl cysteine (NAC) or mannitol on the cadmium chelating ability of monoisoamyl 2, 3-dimercaptosuccinate (MiADMS) was investigated in cadmium pre-exposed rats by Tandon et al., 2003. The combined treatments also improved liver and brain endogenous zinc levels, which were decreased due to cadmium toxicity.

The results suggest that, the administration of an antioxidant during chelation of cadmium may provide beneficial effects by reducing oxidative stress. Similarly, the administration of thiamine during chelation therapy in cadmium poisoning has been shown to have a beneficial effect and more effective than thiol chelating agents alone (Tandon et al., 2000). Also, the influence of vitamin E supplementation on the burden of cadmium was investigated in Cd-exposed rats in a previous report (Tandon et al., 1999). The treatment with MFA-vitamin E or CaNa3 DTPA-vitamin E was more effective than either vitamin E or chelating agent alone, in depleting blood and tissue Cd. The combined treatment showed an advantage over the individual agent in restoring Cd-induced biochemical changes. The treatment with chelator-vitamin E concomitantly with the exposure to Cd was more effective than post-Cd exposure treatment. Other antioxidants; S-adenosylmethionine, lipoic acid, glutathione, selenium, zinc, N-acetylcysteine (NAC), methionine, cysteine, alpha-tocopherol, and ascorbic acid have also been shown to have specific roles in the mitigation of cadmium toxicity (Patrick, 2003).

The effect of Cd-treatment and Cd-treatment with

cabbage on haematological indices and pathophysiogical fitness of albino rats revealed a significant distortion of the haemopoetic parameters by Cd and a distinct alleviation of cadmium-associated distortion in the hematological indices by cabbage supplementation. The red blood cells showed marked hypochromasia, with burr, target and crenated red blood cells. The differential leukocyte count showed marked neutropenia and basophilia. This finding is consistent with previous study (El-Demerdash et al., 2004) which indicated that treatment with cadmium chloride caused a significant decrease in blood hemoglobin (Hb), total erythrocytic count (TEC) and packed cell volume (PCV), while total leukocyte count (TLC) increased. This report suggest that, the administration of combined treatments of cadmium and cabbage may provide beneficial effects against cadmium- induced changes in the testicular and kidney weight, malondialdeyhde, liver enzymes; alkaline phosphatase (ALP), acid phosphatase (ACP), and prostatic acid phosphatase (PAP) levels by reducing cadmium –associated oxidative stress.

REFERENCES

Borek C (2001). Antioxidant health effects of aged cabbage extract. J . Nutr. 133(3): 1010-1015.

Byers T, Nestle M, McTiernan A, Doyle C, Currie-Williams A, Gansler T, Thun M (2002). The American Cancer Society 2001 Nutrition and Physical Activity Guidelines Advisory Committee. Reducing the Risk of Cancer with Healthy Food Choices and Physical Activity. CA Cancer J. Clin., 52: 92.

El-Demerdash FM, Yousef MI, Kedwany FS, Baghdadi HH (2004). Cadmium-induced changes in lipid peroxidation, blood hematology, biochemical parameters and semen quality of male rats: protective role of vitamin E and beta-carotene. Food Chem. Toxicol. 42(10): 1563-1571.

Fleischer M, Sarofim AF, Fasset DW, Hammond P, Shacklette HJ, Nibet IC, Epstein S (2001). Environmental impact of cadmium; a review by the panel on hazardous trace substances. Environ Perspect. 5: 253-323.

Fouad AA, Qureshi HA, Al-Sultan AI, Yacoubi MT, Ali AA (2009). Protective effect of hemin against cadmium-induced testicular damage in rats. Toxicol., 257(3): 153-160.

Grupta P, Kar A (1997). Role of testosterone in ameliorating the cadmium induced inhibition of thyroid function in adult male mouse. Bull. Environ. Contam. Toxicol., 58(3): 422-428.

Hilman C (2009). Hepatoprotective potentials of onion and garlic extracts on cadmium-induced oxidative damage in rats. Biol. Trace Elem. Res., 129(1-3): 143-156.

Jian-Ming Y, Marc A, Qiong-Yu C, Xiang-Dong W, Bing P, Xue-Zhi J (2003). Cadmium-induced damage to primary cultures of rat Leydig cells. Reprod. Toxicol., 17(5): 553-560.

Kumar P, Prasad Y, Patra AK, Ranjan R, Swarup D, Petra RC, Pal S (2009). Ascorbic acid, garlic extract and taurine alleviate cadmium-induced oxidative stress in freshwater catfish (Clarias batrachus). Sci. Total Environ., 407(18): 5024-5030.

Lynn A, Collins A, Fuller Z, Hillman K, Radcliffe B (2006). Cruciferous vegetables and colo-rectal cancer. Proc . Nutr. Soc. 65: 135-144.

Michael W, Pius J, Beverley H, Detmar B (1999). Molecular and cellular mechanisms of cadmium carcinogenesis. Toxicol. 192(2-3): 95-117.

Ola-Mudathir KF, Suru SM, Fafunso MA, Obioha UE, Faremi TY (2008). Protective roles of onion and garlic extracts on cadmium-induced changes in sperm characteristics and testicular oxidative damage in rats. Food Chem. Toxicol. 46(12): 3604-3611.

Patrick L (2003). Toxic metals and antioxidants: Part II. The role of antioxidants in arsenic and cadmium toxicity. Altern. Med. Rev.

8(2):106-128.

Renugadevi J, Prabu SM (2009). Cadmium-induced hepatotoxicity in rats and the protective effect of naringenin. Exp. Toxicol. Pathol., [Epub ahead of print].

Shirashi N, Uno H, Waalkes MP (1993). Effect of L- ascorbic acid pretreatment on cadmium toxicity in the male Fischer rat. Toxicol., 85 (2-3): 85-100.

Sinha M, Manna P, Sil PC (2009). Induction of necrosis in cadmium-induced hepatic oxidative stress and its prevention by the prophylactic properties of taurine. Trace Elem. Med. Biol., 23(4): 300-313.

Siu ER, Mruk DD, Porto CS, Cheng CY (2009). Cadmium-induced testicular injury. Toxicol. Appl. Pharmacol. 238(3): 240-249.

Tandon SK, Singh S (1999). Influence of vitamin E on preventive or therapeutic effect of MFA and DTPA in cadmium toxicity. Biomed. Environ. Sci., 8(1): 59-64.

Tandon SK, Prasad S (2000). Effect of thiamine on the cadmium — chelating capacity of thiol compounds. Hum. Exp. Toxicol., 19(9): 523-528.

Tandon SK, Singh S, Prasad S, Khandekar K, Dwivedi VK, Chatterjee M, Mathur N (2003). Reversal of cadmium induced oxidative stress by chelating agent, antioxidant or their combination in rat. Toxicol. Letts. 145(3): 211-217.

Zoyne P, Yolanda M, Helinä H, Carmen C (2008). Protective Effect of Selenium in Broccoli (Brassica oleracea) Plants Subjected to Cadmium Exposure. J. Agric. Food Chem. 56(1): 266-271.

17

Attenuation of *t*-Butylhydroperoxide induced oxidative stress in HEK 293 WT cells by tea catechins and anthocyanins

Kerio L. C.[1], Bend J. R.[2], Wachira F. N.[1, 3]*, Wanyoko, J. K.[1] and Rotich M. K.[4]

[1]Tea Research Foundation of Kenya, P. O. Box 820, 20200, Kericho, Kenya.
[2]Department of Pathology, Schulich School of Medicine and Dentistry, University of Western Ontario, 1400 Western Road London, N6G 2V4, Canada.
[3]Biochemistry Department, Egerton University, P. O. Box 536, Egerton, Kenya.
[4]Chemistry Department, Egerton University, P. O. Box 536, Egerton, Kenya.

The health promoting properties of catechins and anthocyanins have been of great interest to researchers in the recent past due to their significant *in vitro* and *in vivo* antioxidant activities. Most research on anthocyanins has been based on berry anthocyanins. These potentially health enhancing pigments are also found in some Kenyan tea cultivars. An *in vitro* study was carried out to determine the effects of pure catechins (EGCG and EC) and tea anthocyanin extract from cultivar TRFK 306/1 on *t*-Butylhydroperoxide (*t*-BHP) induced oxidatively stressed HEK 293 cells. The effects of the catechins and tea anthocyanin extract on untreated cells (without *t*-BHP) cells and cells treated (with *t*-BHP, 500 µM) were determined by measuring cytotoxicity (lactate dehydrogenase (LDH) leakage from the cells) and depletion of cellular glutathione (GSH). Cells were preincubated with the antioxidants for 30 min before addition of *t*-BHP (500 µM) and additional incubation for 6 h. The results showed that epigallocatechin gallate (EGCG) and epicatechin (EC) as well as tea anthocyanin extract significantly (p<0.001) attenuated *t*-BHP induced LDH leakage in a concentration dependent manner in treated cells. One way ANOVA analysis showed significant (p<0.001) differences in the various effective concentrations of the catechins and tea anthocyanin extract used. Intracellular GSH content was also increased in a dose dependent manner. From these results, it is concluded that anthocyanin rich tea from selected Kenyan cultivars may have cytoprotective effects against oxidative stressors.

Key words: Anthocyanins, catechins, oxidative stress, lactate dehydrogenase, glutathione, *t*-BHP, attenuate, antioxidant, reactive oxygen species (ROS).

INTRODUCTION

Kenya is one of the most important producers of black tea in Africa. In fact, Kenya supplies 22% of the world's black tea and this crop is a major foreign exchange earner for the country (TBK, 2010), contributing about 26% of all foreign exchange earnings and 4% of the gross domestic product (GDP). In 2010, tea earnings in Kenya was Kshs 98 billion compared to Kshs 69 billion earned in 2009 (TBK, 2010). Tea production in 2010 was 399 million Kg compared to 314 million Kg in 2009, accounting for much of the increase in revenue. Despite high production, local consumption of tea in Kenya is very low. For example, the estimated national per capita consumption of tea was only 0.46 Kg made tea/person/year in 2009 (TBK, 2010). To spur local use of Kenyan tea, it may be necessary to add nutraceutical value to the existing black tea products or to diversify into other products that can use tea and tea products as a raw material. Some specialty tea products are particularly sought after because they are considered to be healthful,

*Corresponding author. Email: fwachira@yahoo.com.

often because they contain or are believed to contain a higher content of biologically active antioxidants. Alternatively, teas can be more appealing to consumers because they contain colour additives or flavours.

The pharmacological properties of the tea plant (*Camellia sinensis*) have historically been ascribed to its polyphenols particularly the catechins; (catechin (+)-C, epicatechin (EC), epigallocatechin (EGC), epigallocatechin gallate (EGCG), gallocatechin (GC) and epicatechin gallate (ECG) in green tea; and theaflavins (TFs) and thearubigins (TRs) in black tea. Tea catechins have been reported to exhibit the following pharmacological properties: radical scavenging of reactive oxygen species (ROS) and reactive nitrogen species (RNS) *in vitro* (Muzolf et al., 2008; Lee et al., 2008); metal chelation (Khokhar and Owusu, 2003; Tang et al., 2002); anticarcinogenic (Roy et al., 2005; Seely et al., 2005); antiinflammatory (Karori et al., 2008); antimicrobial (Almajano et al., 2008); antioxidant (Gramza et al., 2006; Luczaj and Skrzydlewska, 2005; Maurya and Rizvi, 2008; Karori et al., 2007); antidiabetic (Cabrera et al., 2006); pro-apoptotic in monocytes (Kawai et al., 2004) and protection of cells against oxidative stress-induced apoptosis (Yao et al., 2007). The most abundant and bioactive green tea catechin is EGCG (Cabrera et al., 2006). The antioxidant effectiveness of a tea therefore depends on the tea variety in concert with its content of EGCG (Katalinic et al., 2006). Though pharmacological properties of tea have traditionally been attributed to catechins in green tea and theaflavins in black tea (Leung et al., 2001), anthocyanins which occur in grapes, have also occur in some tea plants (Terahara et al., 2001; Kerio et al., 2011).

In the plant kingdom, anthocyanins occur ubiquitously conferring red, blue and purple colours to fruits, vegetables and some grains (Kong et al., 2003; Prior, 2003). Anthocyanins are most abundant in berries (*Vaccinium* sp.*)*, grapes, apples, purple cabbage, eggplant, black carrots, purple fleshed sweet potato and grains like black rice and purple corn. In plants anthocyanins are localized in the cell vacuole (epidermis and peripheral mesophyll cells), in the leaves, stems, roots, flowers and fruits. In nature, over 600 different anthocyanins have been identified to date, though only six are predominant; cyanidin, peonidin, petunidin, pelargonidin, delphinidin and malvidin. A recent study we carried out on anthocyanin rich tea germplasm revealed that malvidin was the most predominant in Kenyan processed black and green tea (Kerio et al., 2011). The different anthocyanidins found in nature differ in their substitution patterns in the B-ring (hydroxylation and methylation), which also affects their biological activities. Anthocyanins have been found to exhibit superior antioxidant activities to some other polyphenols although just like the catechins in the tea plant, they have been reported to have a wide range of biological effects which include antioxidant (Bae and Suh, 2007; Orak, 2007;

Choi et al., 2007), anti-inflammatory (Dai et al., 2007; Arli and Cau, 2007) antimicrobial (Viskelis et al., 2009; Heinonen et al., 2007), antiartherosclerotic (Mazza, 2007) and anticarcinogenic (Wang and Stoner, 2008) activities. Anthocyanins have also been implicated in induction of apoptosis in cancer cells (Hafeez et al., 2008; Lee et al., 2009) as well as with the chemoprotection of cells against oxidative stress-induced apoptosis (Elisia and Kitts, 2008), improvement of vision (Lee et al., 2005) and neuroprotective effects (Tarrozia et al., 2007). Anthocyanins have also been linked with the modulation of oxidative stress in cells which has long been recognized as a mediator of cell death either by necrosis or apoptosis (Yao et al., 2008; Elisia and Kitts, 2008). Anthocyanins do this by preventing the accumulation of intracellular ROS that cause loss of mitochondrial membrane potential ($\Delta\Psi m$) (Haidara et al., 2002; Gogvadze et al., 2006). Apoptosis is an important physiological process of cell death that plays a critical role in tissue homeostasis (Krysko et al., 2008) but that also has been implicated in a number of pathological conditions for example Parkinson's Syndrome and Alzheimer's disease (Jellinger and Bancher, 1998; Erdem et al., 1998).

Though numerous potential pharmacological properties have been described for anthocyanins, little work has been carried out to establish the bioactivity of tea-derived anthocyanins. A cell culture assay based on human embryonic kidney (HEK 293) cells was used in this study to determine the ability of a tea anthocyanin extract and catechins to attenuate oxidative stress. Oxidative stress in cultured cells results in damage to three major categories of essential macromolecules, causing lipid peroxidation (Duthie et al., 2005) protein oxidation (Halliwell and Whiteman, 2004) and DNA damage due to formation of oxidized nucleotide derivatives (Lazze et al., 2003). To detoxify ROS, cells largely depend on innate antioxidants like glutathione (GSH), dietary antioxidants such as Vitamin C and E and enzymic antioxidants including superoxide dismutase, catalase, and the peroxiredoxins. The innate antioxidant glutathione (GSH), the oxidized form (glutathione disulphide) of which is a marker of oxidative stress, is a tripeptide (γ-glutamyl-cysteinyl-glycine) responsible, in significant part, for maintaining the redox balance of most cells, where it is the most abundant non-protein thiol (Lapenna et al., 1998). GSH can scavenge ROS directly or act as a co-factor for enzymatic defenses (Franco and Cidlowski, 2009). Glutathione is found almost exclusively in a reduced state in cells and the enzyme glutathione reductase, which is constitutively active and inducible upon oxidative stress, converts glutathione disulphide (GSSG) to GSH (Shih et al., 2005). The ratio of GSH: GSSG within cells is often used as a measure of cellular oxidative stress (Atakisi et al., 2010). If oxidative stress results in irreparable cell damage, then apoptosis (programmed cell death) are activated but if cells are exposed to extremely high levels of ROS, they also become necrotic (Krysko et al., 2008).

In the present study, HEK 293 cells were treated with t-BHP (500 µM) for 6 h to induce oxidative stress, to evaluate the ability of pretreatment with pure tea catechins and tea anthocyanin extract to attenuate the t-BHP-mediated cytotoxicity and GSH depletion.

MATERIALS AND METHODS

Anthocyanin extraction and purification

Five grams of ground steamed tea leaf from tea clone TRFK 306/1 samples were weighed into 250 ml conical flasks covered with foil and mixed with 50 ml methanol-1.5 M hydrochloric acid MeOH/HCl; (99/1v/v) and magnetically stirred at 900 rpm for 4 h at room temperature. The resultant solution was filtered and evaporated to dryness using a Rotavapour (Buchi Rotavapour R-300, Switzerland) under reduced pressure at 35°C. The extract was dissolved in 10 mL distilled water and passed through a 0.45 µm membrane filter and kept in an ice bath until the next phase of isolation.

The extracts were passed through reverse phase (RP) C_{18} solid phase extraction (SUPELCO, SPE) cartridges (Sigma-Aldrich, USA) previously activated with 10% MeOH/HCl and washed with 0.01% HCl in distilled water. Anthocyanins were adsorbed onto the column while sugars, acids and other water-soluble compounds were eluted from the column with 0.01% HCl in distilled water. Anthocyanins were then recovered using methanol acidified with 10% v/v formic acid. The cartridges were subsequently eluted with ethyl acetate (Fischer Scientific, UK) to remove phenolic compounds other than anthocyanins. The purified extracts were stored at -20°C until further analysis. The extracted anthocyanins crystallized upon solvent removal.

Catechins

The catechins (-)-epicatechin and (-)-epigallocatechin gallate were purchased from Sigma Aldrich, USA.

Incubation of HEK 293 cells

HEK 293 cells (Eton Bioscience, USA) were cultured in 75 cm^3 flasks in minimum essential medium MEM (Invitrogen, Canada) and 10% FBS containing 0.5% gentamycin until confluent. Then 4×10^5 cells/mL were seeded in 60 mm Petri dishes for treatment prior to assay for GSH/GSSG or Western Blot or 4×10^5 cells/ml were seeded in 96 well plates for treatment prior to assay for lactate dehydrogenase release. The cells were incubated overnight at 37°C in a humidified atmosphere containing 95% air and 5% CO_2 prior to treatment.

LDH leakage

Overall cell damage was determined by LDH leakage into the extracellular medium according to the method of Korzeniewski and Callewaert, (1983). Cells were first pre-treated with varying concentrations of either the catechins (10-50µM) or the tea anthocyanin extract (10-100 µM) or vehicle (blank) for 30 min prior to incubation with 500 µM t-BHP (Sigma-Aldrich, Canada) for 6 h. Released LDH was analyzed at 490 nm using a Titertek Multiscan Microplate Reader (Phoenix AZ, USA). Cell damage was expressed

as the fraction of LDH released into the culture medium relative to the total cellular LDH activity determined after lysis of cells with 0.1% v/v Triton X-100 (Sigma-Aldrich, USA).

Analysis of intracellular GSH

The intracellular concentrations of GSH were determined according to the method of Habig et al. (1974). Cells were plated overnight in 2 ml of minimum essential medium (MEM) (Invitrogen, Canada) and treated with the antioxidants (catechins or anthocyanin) for 30 min after adding fresh media. After incubation with 500 µM t-BHP for 6 h, the cells were harvested from the wells by gentle scraping, placed in ice, and centrifuged at 500 x g for 5 min. Cell pellets were re-suspended in 250 µl lysis buffer (10 mM Tris, 130 mM NaCl, 10 mM NaF, 10 mM NaH_2PO_4, 10 mM $Na_4P_2O_7$, and 2 mM EDTA) and quickly frozen in liquid nitrogen. The remaining aliquots were centrifuged for 2 min at 14,000 x g, and the supernatants were used for GSH content measurement based on a GST enzymatic method of Habig et al., (1974). Sample supernatant (50 µl) was added to wells of a 96-well plate and made up to 190 µl with 0.1 M potassium phosphate buffer (pH 7.6). To each sample reaction, 5 µl of 40 mM DNCB, and 0.2 unit of glutathione S-transferase were added, followed by incubation in the dark at room temperature for 20 min. The formation of S-2, 4-dinitrophenyl glutathione was measured by monitoring the absorbance at wavelength 340 nm with a Titertek Multiskan Microplate Reader (Phoenix AZ, USA). GSH content was expressed as nmol GSH/1000 cells.

Statistical analysis

The LDH assays were carried out in triplicate with three different batches of cultured cells and the data were subjected to one way analysis of variance. Student's t-test was used to determine significance between treatments. Differences between treatments were considered to be statistically significant if $p<0.05$. The GSH assays were carried out twice with different batches of cells and the means of the two experiments are reported here.

RESULTS

Effect of tea catechins and anthocyanins on LDH leakage from HEK 293 cells

Tea catechins and anthocyanins were compared for their ability to attenuate t-BHP induced oxidative stress in HEK 293 cells. Epicatechin (EC) and epigallocatechin gallate (EGCG) were used as representative tea catechins in the assays while the tea anthocyanin extract was obtained from steamed leaves of the tea clone TRFK 306/1, as described above. The molarity of the tea anthocyanin extract solution was obtained from the Relative Molecular Mass (RMM) of anthocyanins (449.2 g) as Cyanidin-3-glucoside equivalents. HEK 293 cells were subjected to oxidative stress by treatment with 500µM t-BHP, a cell-permeant organic oxidant that generates ROS and increased the amount of LDH released from these cells by disrupting cell membrane integrity. The effects of catechins and tea anthocyanin extract in attenuating the t-BHP induced oxidative stress was evaluated by determining by comparing the release of LDH to the medium cells treated with the tea constituents and t-BHP vs cells treated only with t-BHP.

Lactate dehydrogenase leakage is recognized as a sensitive measure of plasma membrane damage and cytotoxicity caused by reactive oxygen species (ROS) which can eventually lead to cell death. Pre-treatment of the cells for 30 min with the catechins (-)-EC alone or before addition of 500 µM t-BHP and incubation for 6 h resulted in lowering of LDH leakage to 4.71% (p<0.001), 1.57% (p<0.001) and 0.16% (p<0.001) for 10, 25 or 50 µM (-)-EC, respectively without t-BHP compared to 500 µM t-BHP while the cells that were oxidatively stressed after pre-treatment showed a reduction of LDH leakage to 14.6% (p>0.05), 9.1% (p>0.05) and 5.2% (p<0.05) compared to 500 µM t-BHP (Figure 1). Treatment of cells with 500 µM t-BHP for 6 h caused a rapid release of LDH into the cytosol which was attenuated in a dose dependent manner by pre-treatment of cells with increasing Epicatechin levels. One way ANOVA results showed significance (p<0.001) between the various concentrations of epicatechin used showing that EC did protect the cells under oxidative stress. To examine the attenuation of LDH leakage by EGCG, a similar treatment was carried out. It was observed that LDH leakage in the controls (without 500 µM t-BHP) was lowered in a dose dependent manner from 2.58% to 1.67% (p<0.01), 1.11% (p<0.01) and 0.11% (p<0.01) while for the cells that were treated (with 500 µM t-BHP) LDH leakage was lowered from 15.1% to 12.2% (p>0.05), 10.9% (p>0.05) and 2.8% (p<0.01), respectively compared to t-BHP treatment (Figure 2). One way ANOVA results showed significant cytoprotective effects (p<0.001) against oxidative stress between the various concentrations of EGCG tested.

In an assay to evaluate the effect of tea anthocyanin extract on LDH leakage in HEK 293 cells, the first group of cells was treated with anthocyanins only and the second group was pre-treated with increasing anthocyanin concentrations for 30 min before induction of oxidative stress with t-BHP (500µM) (Figure 3). LDH leakage in the controls (without 500 µM t-BHP) was lowered to 9.35% (p<0.001), 3.7% (p<0.001), 2.0% (p<0.000) and 0.23% (p<0.0001) for 10, 25, 50 and 100 µM, respectively. In the oxidatively stressed cells (with 500 µM t-BHP), LDH leakage was dose dependently lowered from 20.63% to 16.79% (p>0.05), 13.92% (p<0.01), 7.23% (p<0.001) and 2.41% (p<0.0001) for the same tea anthocyanin extract concentrations, respectively. ANOVA results showed significant differences (p<0.001) between the various concentrations of tea anthocyanin extract used. These results demonstrated that the anthocyanins were effective cytoprotectants against oxidative stress.

Effect of catechins and tea anthocyanins on intracellular GSH content

The effect of catechins and tea anthocyanin extract on

reduced glutathione (GSH) content of HEK 293 cells was also determined in this study. GSH is an important high capacity cellular antioxidant, widely distributed among living cells. It is also involved in many biological functions such as amelioration of t-BHP-induced rat hepatotoxicity (Hwang et al., 2005). though not statistically significant, EGCG marginally increased cellular GSH content in t-BHP treated cells from 0.043 GSH nmol/mg protein for t-BHP only to 0.058, 0.067 and 0.074 GSH nmol/mg protein, respectively for 10, 25 and 50 µM EGCG (Figure 4). Tea anthocyanin extract at concentrations of 10, 50 and 100 µM similarly, marginally increased GSH concentration in the t-BHP treated cells. (Figure 5) but not significantly. The data presented for EGCG and tea anthocyanin extract were obtained from the mean of two independent experiments that were repeated with similar results.

DISCUSSION

Interest in the protective function of naturally occurring antioxidants in biological systems has risen in the recent past. For a long time, tea has been used as a healthy beverage owing to its pharmacological properties which are largely ascribed to the catechins contained. Catechins are known to quench free radicals, a characteristic associated with the hydroxyl groups in the B-ring. Evidence has shown that increased hydroxylation increases the antioxidant power of the catechins with EGCG in green tea being the most potent of all (Saffari and Sadrzadeh, 2004). Epigallocatechin-3-gallate (EGCG) is the most abundant compound found in green tea and has been associated with various health benefits as nutritional supplements for various ailments. The potential health benefits ascribed to EGCG include cancer chemoprevention (Hsuuw and Chan, 2007), antioxidant (Fu and Koo, 2006), improving cardiovascular health (Hirai et al., 2007) and neuroprotective (Mandel et al., 2004) among others. However, increasing evidence has also shown that EGCG is a pro-oxidant in higher concentrations and enhances production of ROS which subsequently lead to cell death (Bandele and Osheroff, 2008; Lambert et al., 2010; Rohde et al., 2011).

Recently, anthocyanin containing tea cultivars have been developed and characterized (Kerio et al., 2011). The anthocyanins/anthocyanidins identified in the tea cultivars were cyanidin-3-O-galactoside, cyanidin-3-O-glucoside, cyanidin, delphinidin, peonidin, pelargonidinand malvidin (Kerio et al., 2011) but the most predominant in the TRFK 306 cultivars is malvidin. Anthocyanins are pigmented plant polyphenols, whose colour in a plant depends on the pH of the cell sap. One purpose of this preliminary study was to compare the antioxidant potency of a mixed extract of tea anthocyanins with the two pure tea catechins, EC and EGCG. All three treatments attenuated the oxidative stress induced by 500 µM t-BHP in HEK 293 cells by

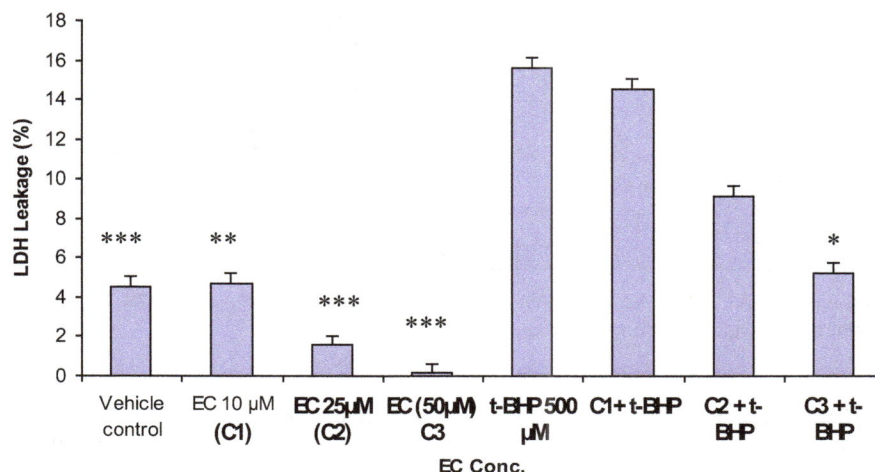

Figure 1. The effect of pre-incubating HEK 293 cells with 0-50 µM epicatechin for 30 min on LDH leakage following incubation with or without t-BHP (500µM) for 6 h. DMSO is the vehicle control (0 µM epicatechin); C1, C2 and C3 contain 10, 25 or 50 µM epicatechin (EC), respectively. Each column represents the mean and SD of three independent experiments. *p<0.05, **p<0.01, ***p<0.001 compared with t-BHP treatment alone (student's t-test).

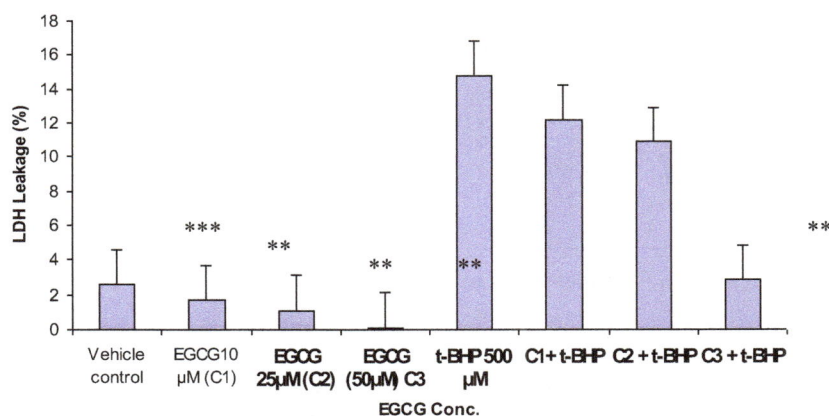

Figure 2. The effect of pre-incubating HEK 293 cells with 0-50 µM Epigallocatechin gallate (EGCG) for 30 min on LDH leakage following incubation with or without *t*-BHP (500 µM) for 6 h. DMSO is the vehicle control (0 µM epicatechin); C1, C2 and C3 contain 10 µM, 25 µM or 50 µM EGCG, respectively. Each column represents the mean and SD of three independent experiments. **p<0.01 and ***p<0.001 compared with *t*-BHP treatment alone.

reducing LDH leakage, an assay for cell damage. The integrity of cellular membranes is critical for normal cell function. Peroxidative decomposition of the membranes leads to impairment of normal cellular functions. Others have shown that t-BHP enhances lipid membrane peroxidation that contributes to cytotoxicity of HEK 293 cells (Kim et al., 2011; Hwang et al., 2005; Wang et al., 2000).

The results of this study revealed that catechins (EC and EGCG) as well as tea anthocyanin extract were able to partially protect the cells from the oxidative damage caused by *t*-BHP (500 µM). The effectiveness of the anthocyanidin extract might be attributed to its major constituent, malvidin. Malvidin is a dimethylated compound that is non-polar in nature. It is postulated that this anthocyanidin will have increased partitioning into the lipid phase of the membrane thereby increasing the antioxidant capacity of this tea anthocyanin. In addition, anthocyanin antioxidants are unique in their mode of action which may be attributed to their ability for electron

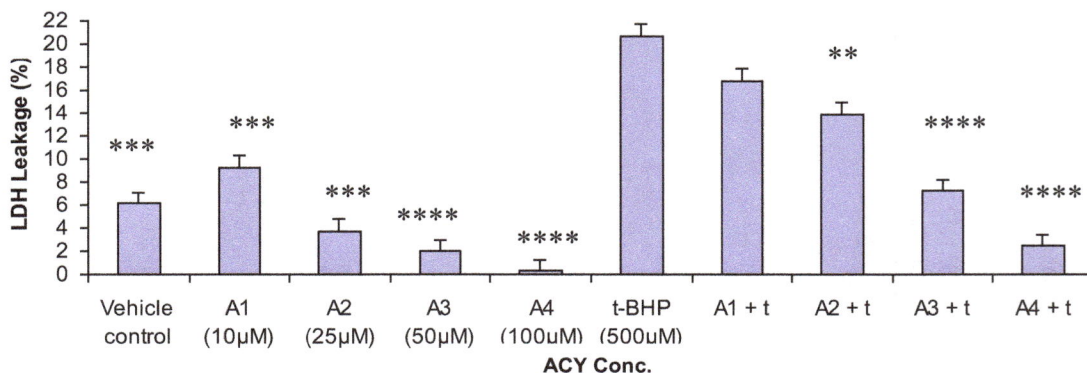

Figure 3. The effect of pre-incubating HEK 293 cells with 0-100 μM tea anthocyanin extract for 30 min on LDH leakage following incubation with or without t-BHP (500 μM) for 6 h. DMSO is the vehicle control (0 μM tea anthocyanin extract); A1, A2 and A3 contain 10, 25, 50 or 100 μM tea anthocyanin extract, respectively. Each column represents the mean and SD of three independent experiments. **p<0.01, ***p<0.001 and ****p<0.0001, compared with t-BHP treatment alone.

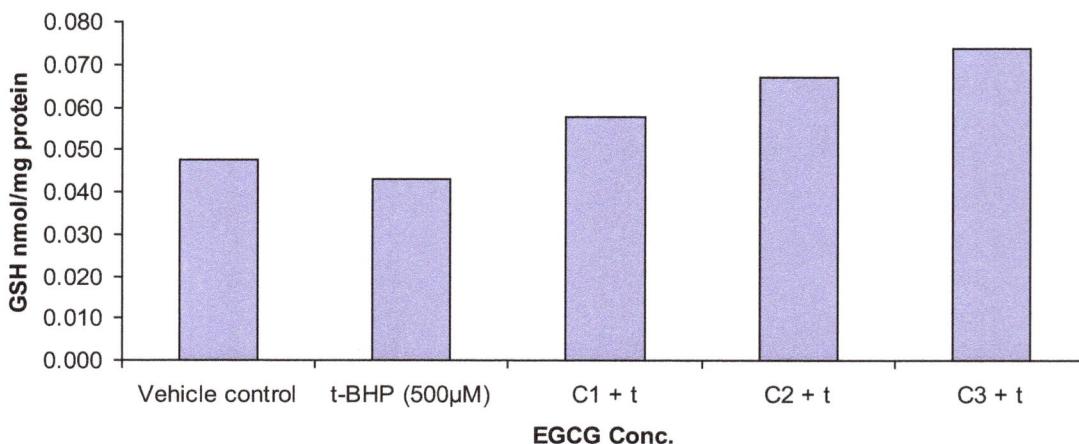

Figure 4. Effect of EGCG on the concentration of GSH in the treated cells and on the depletion of GSH induced by t-BHP (500 μM, 6 h) in HEK 293 cells. C1, C2 and C3 contain 10, 25 or 50 μM EGCG, respectively. Each column represents the mean of two independent experiments.

delocalization and to form resonating structures following changes in pH (Bagchi et al., 2006). Anthocyanins have been shown to be novel antioxidants and potent inhibitors of lipid peroxidation in comparison with other classic antioxidants. For example, grape seed proanthocyanidin extract was found to have superior antioxidant efficacy compared to vitamins C, E and β-carotene (Bagchi et al., 2003).

GSH has been established as a very important endogenous molecule for protection against the cytotoxicity of reactive electrophilic metabolites by converting them to covalent S-glutathione conjugates (Habig et al., 1974). GSH is also a very important antioxidant, essential for the maintenance and balance of thiol (-SH) groups on intracellular proteins. Antioxidants

(vitamin C, catechins and other polyphenols) have been shown to increase the GSH pool and decrease GSSG content as well as attenuate hydroperoxide induced oxidative stress by stimulating the activity of two enzymes involved in GSH/GSSG balance that is glutathione peroxidase (GPx) and glutathione reductase (GRx) (Wijeratne et al., 2005). Increase in GSH content and decrease in cellular hydroperoxides (oxidants) are related phenomena. Dietary antioxidants for example catechins in tea and lately anthocyanins in the same plant, assist the cells to quench the oxidant evaluated in this study caused by t-BHP, complementing cellular antioxidants such as GSH and preventing their depletion. In cells, antioxidants often cause cyclic stimulation of glutathione peroxidase (GPx) and glutathione reductase (GRx), the

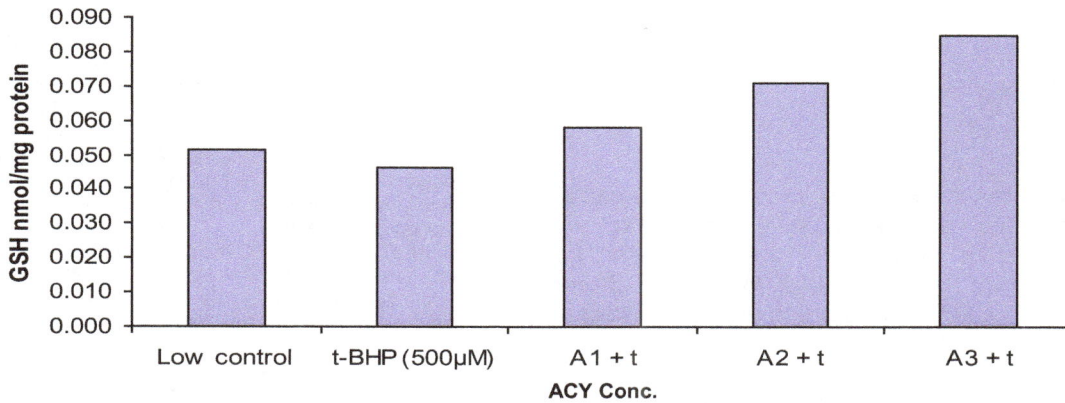

Figure 5. Effect of tea anthocyanin extract on the concentration of GSH in the treated cells and on the depletion of GSH induced by t-BHP (500 μM, 6 h) in HEK 293 cells. A1, A2 and A3 contain 10 μM, 50 μM or 100 μM tea anthocyanin extract. Each column represents the mean of two independent experiments.

latter which regenerates GSH from GSSG to be used in other antioxidant processes in the mitochondria or they improve mitochondrial function by detoxifying hydroperoxides (Shih et al., 2007). For Example, EGCG has been shown to modulate GSH metabolism in rat astrocytes (Ahmed et al., 2002). A key finding in this study is that anthocyanins and EGCG not only maintain a good redox status in basal conditions for GSH, but also do so in the t-BHP treated cells that have undergone oxidative insult. Our t-BHP treatment resulted in depletion of GSH possibly as a result of increased the levels of cellular hydroperoxides and high amounts of ROS responsible for oxidative stress (Alia et al., 2005). The 500 μM t-BHP used in this study caused oxidation of cellular GSH and subsequently resulted in an increase in the levels of GSSG, an inhibitor of GPx that can saturate GRx. (Yang et al., 2006). t-BHP-induced ROS formation affects GRx activity indirectly by oxidation of the pyridine nucleotide NADPH, a limiting cofactor for GRx. On the positive side, anthocyanins and catechins behave as electron donors to the mitochondrial electron transport chain or as mitochondrial respiring substrates that support the reduction of GSSG formed during oxidative stress (Weissel et al., 2006). The ability of dietary antioxidants including catechins and anthocyanins to function as direct free radical scavengers is related to their electron donating ability which is a direct function of the number of hydroxyl groups in the B-ring and the structural orientation of the compound (Castaneda-Ovando et al., 2009). Studies on the chemistry of EGCG as an antioxidant have shown that the trihydroxyphenyl B-Ring is the more active site of the antioxidant reaction than is the galloyl moiety (Nagle et al., 2006). In anthocyanins, the ring orientation has been found to determine the ease by which a hydrogen atom from a hydroxyl group can be donated to a free radical and its ability to support a free electron by delocalization of its own electrons (Castaneda-Ovando et al., 2009).

A study on t-BHP induced hepatotoxicity in rats demonstrated the protective effects of anthocyanins extracted from dried hibiscus flowers, primarily delphinidin-3-glucoside and cyanidin-3-glucoside (Wang et al., 2000) which decreased cytotoxicity (LDH release) and lipid peroxidation at concentrations of 0.1 and 0.2 mg/mL. Similar attenuation of oxidative stress-mediated toxicity was found *in vivo* after the oral administration of 100 or 200 mg/kg anthocyanins to rats for 5 days before a single dose of 0.2 mmol/kg of t-BHP. Although the catechin EGCG, also has cytoprotective properties it can also causes oxidative stress, as demonstrated by the effect of ECGC on GSH metabolism in cultured rat astrocytes (Ahmed et al., 2002). Similarly, several other studies have shown that EGCG, at high concentrations has pro-oxidative activities. In pancreatic beta cells, treatment with EGCG reduced cell viability and increased apoptotic cell death, H_2O_2 and ROS production, particularly at 50 and 100 μM although enhanced cell death was found at concentrations as low as 5 μM after 48 h (Suh et al., 2010). Recently, there was a case reported in Denmark of liver toxicity in an individual who drank 4-6 cups of green tea per day for six months (Rohde et al., 2011) which was ascribed to EGCG, the major catechin in green tea. Cell culture studies have revealed that EGCG could inhibit intercellular communication via gap junctions in normal rat liver epithelial cells (Kang et al., 2008). Indeed, ECGC has been shown to poison both isomers of topoisomerase II by a redox-dependent mechanism similar to that for 1, 4-benzoquinone (Bandele and Osherof, 2008).

The lowered levels of LDH leakage and apparent increase in cellular GSH content (due to attenuated depletion) in cells treated with the tea anthocyanin extract in this study reveal potential benefits of these compounds in the normal body cells. The tea anthocyanin extract was able to partially attenuate t-BHP-induced oxidative stress in the HEK 293 cells, effectively demonstrating the cytoprotective properties of these antioxidants in cultured cells.

Conclusion

Oxidative stress results from an excess of intercellular ROS and RNS free radicals which are associated with several degenerative diseases, including cancer and chronic inflammation (Halliwell, 2002; Wang and Stoner, 2008; Tang et al., 2007). Therefore, it seems reasonable that consumption of reasonable amounts of anthocyanin rich teas, containing both anthocyanins and catechins, is likely to attenuate the adverse effects of oxidative stress in the body by decreasing oxidation of essential cellular proteins, lipids and nucleic acids while sparing GSH depletion.   In this study, it is shown that the tea anthocyanin extract at the highest concentrations tested (100 μM), did not cause any obvious oxidative stress on the cells. This shows the potential of the tea anthocyanins for cytoprotection against oxidative stressors. It should however be stressed that although regardless of the class of chemical antioxidants selected for this purpose, ingestion of high concentrations over time is likely to be associated with negative effects, as has been reported in experimental animals and humans for EGCG, particularly when taken on an empty stomach which results in enhanced absorption. Additional work is however required to prove that anthocyanin rich tea cultivars have less potential for negative effects because of the presence of both anthocyanin and catechins antioxidants.

ACKNOWLEDGEMENTS

Authors thank the Graduate Student Exchange Program organized by the Canadian Bureau for International Education (CBIE) funded by the Department of Foreign Affairs and International Trade Canada (DFAIT) for funding this work. We also thank The Tea Research Foundation of Kenya for providing the tea samples used in this study and for permission to publish these results.

REFERENCES

Alia M, Ramos S, Mateus R, Bravo L, Goya L (2005). Response of the antioxidant defense system to *tert*-butyl hydroperoxide and hydrogen peroxide in a human Hepatoma cell line (HepG2). J. Bioch. Mol. Toxicol. 19: 119-128.

Ahmed I, John A, Vijayasarathy C, Robin MA, Raza H (2002). Differential modulation of growth and glutathione metabolism in cultured rat astrocytes by 4-hydroxynonenal and green tea polyphenol, epigallocatechin-3-gallate. Neurotoxicology, 23: 289-300.

Almajano PM, Rosa C, Jimenez L, Gordon H (2008). Antioxidant and antimicrobial activities of tea infusions. Food Chem., 108: 55-63.

Arli G, Cau NO (2007). Determination of naturally occurring antioxidant and anti-inflammatory compounds in fresh fruits by HPLC. Toxicol. Lett., 172: S221-S222.

Bae SH, Suh HJ (2007). Antioxidant activities of five different mulberry

cultivars in Korea. Food Sci. Technol., 40: 955-62.

Bagchi D, Roy S, Patel V, He G (2006). Safety and whole body antioxidant potential of a novel anthocyanins-rich formulation of edible berries. Mol. Cell. Bioch., 281: 197-209.

Bagchi D, Sen CK, Ray SD, Das DK, Bagchi M, Preuss HG Vinson JA (2003). Molecular mechanisms of cardioprotection by a novel grape seed proanthocyanidin extract-A review. Mutation Res., 524: 87-89.

Bandele OJ, Osheroff N (2008). (-)-epigallocatechin gallate, a major constituent of green tea, poisons type II topoisomerases. Chem. Res. Toxicol., 21: 936-943.

Cabrera C, Artacho R, Gimenez R (2006). Beneficial effects of green tea-A review. J. Am. Coll. Nutr., 25: 79-99.

Castaneda-Ovando A, Pacheco-Hernandez MDL, Paez-Hernandez ME, Rodriguez JA, Galan-Vidal CA (2009). Chemical studies of anthocyanins: A review. Food Chem., 113: 859-871.

Choi E, Chang H, Cho J, Hyan S (2007). Cytoprotective effects of anthocyanins against doxorubicin-induced toxicity in H9c2 cardiomyocytes in relation to their antioxidant activities. Food Chem. Toxicol., 45: 1873-1881.

Dai J, Patel JD, Mumper RJ (2007). Characterization of blackberry extract and its antiproliferative and anti-inflammatory properties. J. Med. Food, 10: 258-265.

Duthie SJ, Gardner PT, Morrice PC, Wood SG, Pirie L, Bestwick CC, Milne L, Duthie GG (2005). DNA stability and lipid peroxidation in vitamin E deficient rats *in vivo* and *in vitro*. Modulation by the dietary anthocyanin cyanidin-3-glycoside. Eur. J. Nutr., 44: 195-203.

Elisia I, Kitts DD (2008). Anthocyanins inhibit peroxyl radical induced apoptosis in CaCO-2 cells. Mol. Cell Biochem 312: 139-145.

Erdem S, Mendell JR, Sahenk Z (1998). Fate of Schwann cells in CMT1A and HNPP: Evidence for apoptosis. J. Neur. Exp. Neur., 57: 635– 642.

Franco R, Cidlowski JA (2009). Apoptosis and glutathione; beyond an antioxidant. Cell Death Diff., 16: 1303-1314.

Fu Y, Koo MW (2006). EGCG protects HT-22 cells against glutamate-induced oxidative stress. Neur. Res., 10: 23-30.

Gogvadze V, Orrenius S, Zhivotovsky B (2006). Multiple pathways of cytochrome *c* release from mitochondria in apoptosis. Biochim. Biophys. Acta., 1757: 639-647.

Gramza A, Khokhar S, Yoko S, Swiglo AG, Hes M, Korczak J. (2006). Antioxidant activity of tea extracts in lipids and correlation with polyphenol content. Eur. J. Lipid Sci. Tech.,108: 351-362.

Habig WH, Pabst MJ, Jakoby WB (1974). Glutathione S-transferases. The first enzymatic step in mercapturic acid formation. J. Biol. Chem. 249: 7130-7139.

Haidara K, Morel I, Abalea V, Barre MG, Denizeau F (2002). Mechanism of *tert*-butylhydroperoxide induced apoptosis in rat hepatocytes: involvement of mitochondria and endoplasmic reticulum. Biochim. Biophys. Acta., 1542: 173-185.

Hafeez BB, Siddiqui IA, Asim M, Malik A, Afaq F, Adhami VM, Saleem M, Din M, Mukhtar H (2008). A dietary anthocyanidin delphinidin induces apoptosis of human prostrate cancer PC3 cells *in vitro* and *in vivo*: Involvement of nuclear factor-κB signalling. Cancer Res. 68: 8564-8572.

Halliwell B, and Whiteman M. (2004). Measuring reactive species and oxidative damage *in vivo* and in cell culture- how should you do it and what do the results mean? British J. Pharm., 142: 231-255.

Halliwell B (2002). Effect of diet on cancer development: Is oxidative DNA damage a biomarker? Free Rad. Biol. Med., 32: 968-974.

Heinonen M (2007). Antioxidant activity and antimicrobial effect of berry phenolics- A Finnish perspective. Mol. Nutr Food Res. 51: 684-691.

Hirai M, Hotta Y, Ishikawa N, Wakida Y, Fukuzawa Y, Isobe F, Nakano A, Chiba T Kawamura N (2007). Protective effects of EGCG or GCG, a green tea catechin epimer against postischemic myocardial dysfunction in guinea-pig hearts. Life Sci., 80: 1020-1032.

Hsuuw Y, Chen W (2007). Epigallocatechin gallate dose-dependently induces apoptosis or necrosis in Human MCF-7 cells. Ann. N.Y. Acad. Sci., 1095: 428-440.

Hwang JM, Wang CJ, Chou FP, Tseng TH, Hsieh YS, Hsu JD, Chu, CY (2005). Protective effect of baicalin on *tert*-butyl hydroperoxide-induced rat hepatotoxicity. Arch. Toxicol., 79: 102-109.

Jellinger KA, Bancher C (1998). Neuropathology of Alzheimer's disease: A critical update. J. Neural Trans., 54: 77–95.

Kang NJ, Lee KM, Kim JH, Lee BK, Kwon JY, Lee KW, Lee HJ (2008). Inhibition of gap junctional intercellular communication by the green tea polyphenol (-)-epigallocatechin gallate in normal rat liver epithelial cells. J. Agric. Food Chem., 56: 10422-10427.

Karori SM, Ngure RM, Wachira FN, Wanyoko JK, Mwangi JN (2008). Different types of tea products attenuate inflammation induced in *Trypanosoma brucei* infected mice. Parasit Int., 57: 325-333.

Karori SM, Wachira FN, Wanyoko JK Ngure RM (2007). Antioxidant capacity of different types of tea products. Afr. J. Biotech., 6: 2287-2296.

Katalinic V, Milos M, Kulisic T, Jukic M (2006). Screening of 70 medicinal plant extracts for antioxidant capacity and total phenols. Food Chem., 94: 550–557.

Kerio LC, Wachira FN, Wanyoko JK, Rotich MK (2011) Characterization of anthocyanins in Kenyan teas: Extraction and identification. Food Chem., 131(1): 31-38.

Kim SM, Kang K, Jho EH, Jung YH, Nho CW, Um BH, Pan CH (2011). Hepatoprotective effects of flavonoid glycosides from *Lespedeza cuneata* against oxidative stress induced by *tert*-butyl hydroperoxide. Phytotherapy Res. (Epub ahead of print).

Khokhar S, Owusu RK (2003). Iron binding characteristics of phenolic compounds: some tentative structure-activity relations. Food Chem., 81: 133-140.

Kong JM, Chia LS, Goh NK, Chia TF, Brouillard R (2003). Analysis and biological activities of anthocyanins. Phytochemical, 64: 923 –933.

Korzeniewski C, Callewaert DM (1983). An enzyme-release assay for natural cytotoxicity. J. Immun. Methods 64: 313-20.

Krysko D, Berghe TV, D'Herde K, Vandenabeele P (2008). Apoptosis and necrosis: Detection, discrimination and phagocytosis. Methods 44: 205-221.

Lambert JD, Kennet MJ, Sang S, Reuht KR, Ju J, Yang CS (2010). Hepatotoxicity of high oral dose (-)-epigallocatechin-3-gallate in mice. Food Chem.Toxicol., 48: 409-416.

Lapenna D, de Goia S, Ciofani G, Mezzetti A, Ucchino S, Calafiore AM, Napotaliano AM, Di Ilio C. Cuccurulo F (1998). Glutathione-related antioxidant defenses in human artherosclerotic plaques. Circulation 97: 930.

Lazze M, Pizzala R, Savio M, Stivala L, Prosperi E, Bianchi RL (2003). Anthocyanins protect against DNA damage induced by *t*-BHP in rat smooth muscle and hepatoma cells. Mutation Res., 535: 103-115.

Lee V, Chen C, Liao Y, Tzen J, Chang C (2008). Structural determination and DPPH radical scavenging activity of two acylated flavonoids tetraglycosides in oolong tea (*Camellia sinensis*). Chem. Pharm. Bull., 56: 851-853.

Lee J, Lee HK, Kim CY, Hong Y, Choe CM, You T, Seong GJ (2005). Purified high dose of anthocyanoside oligomer administration improves nocturnal vision and clinical symptoms in myopia subjects. British J. Nutr., 93: 895-899.

Lee S, Park S, Park JH, Shin, DY, Kim GY, Rya, CH, Choi YH (2009). Induction of apoptosis in human leukaemia U937 cells by anthocyanins through downregulation of Bcl-2 and activation of caspases. Interntl. J. Onc., 34: 1077-1083.

Leung LK, Su Y, Chen R, Zhang Z, Huang Y, Chen ZY (2001). Theaflavins in black tea and catechins in green tea are equally effective antioxidants. J. Nutr. 131: 2248-2251.

Luczaj W, Skrzydlewska E (2005). Antioxidative properties of black tea- A review. Prev. Med. 40: 910-918.

Mandel S, Weinreb O, Amit T, Youdim MBH (2004). Cell signalling pathways in the neuroprotective actions of the green tea polyphenol (-)-epigallocatechin-3-gallate: Implications for neurodegenerative diseases. J. Neurochem., 88: 1555-1569.

Maurya PK, Rizvi SI (2008). Protective role of tea catechins on erythrocytes subjected to oxidative stress during human aging. Nat. Prod. Res., pp. 1-8.

Mazza G. (2007). Anthocyanins and heart health. Ann. Ist. Super San. 43: 369-374.

Muzolf M, Szymusiak H, Anna G, Rietjens MCM, Tyrakowska B (2008). pH-dependent radical scavenging capacity of green tea catechins. J. Agric. Food Chem., 56: 816-832.

Nagle DG, Ferreira D, Zhou Y (2006). Epigallocatechin-3-gallate (EGCG): Chemical and biomedical perspectives-Molecules of interest. Phytochemical, 67: 1849-1855.

Orak HH (2007). Total antioxidant activities, phenolics, anthocyanins, polyphenol oxidase activities of selected red grape cultivars. Scientia Hort., 111: 235-41.

Prior RL (2003). Fruits and vegetables in the prevention of cellular oxidative damage. Am. J. Clin. Nutr., 78: 570S–578S.

Rohde J, Jacobsen C, Kromann-Andersen H (2011). Toxic hepatitis triggered by green tea. Ugeskr Laeger, 173: 205-206

Royr AM, Baliga MS, Katiyar SK (2005). Epigallocatechin-3-gallate induces apoptosis in estrogen receptor-negative human breast carcinoma cells via modulation in protein expression of p53 and Bax and caspase-3 activation. Mol. Cancer Ther., 4: 81 - 90.

Saffari Y, Sadrazeh SMH (2004). Green tea metabolite EGCG protects membranes against oxidative damage in vitro. Life Sci., 74: 1513-1518.

Seely D, Mills EJ, Wu P, Verma S, Guyatt, GH (2005). The effects of green tea consumption on incidence of breast cancer and recurrence of breast cancer: A systematic review and meta-analysis. Integr. Cancer Ther. 4: 144 - 155.

Shih P, Yeh C, Yen G (2007). Anthocyanins induce the activation of Phase II enzymes through the antioxidant response element pathway against oxidative stress induced apoptosis. J. Agric. Food Chem., 55: 9427-9435.

Shih PH, Yeh CT, Yen GC (2005). Effects of anthocyanidin on the proliferation and induction of apoptosis in human gastric endocarcinoma cells. Food Chem.Toxicol., 55: 1557-1566.

Suh KS, Chon S, Oh S, Kim JW, Kim YS, Woo JT (2010). Pro-oxidative effects of green tea polyphenol (-)-epigallocatechin-3-gallate on the HIT-T15 pancreatic beta cell line. Cell Biol. Toxicol., 26: 189-199.

Tang Y, Zhao DY, Elliott S, Zhao W, Curriel TJ, Beckman BS, Burrow ME (2007). Epigallocatechin-3-gallate induces growth inhibition and apoptosis in human breast cancer cells through surviving suppression. Internat. J. Oncol., 31: 705-711.

Tang SZ, Kerry JP, Sheehan D, Buckley DJ (2002). Antioxidative mechanisms of tea catechins in chicken meat systems. Food Chem., 76: 45-51.

Tarozzia A, Morroni F, Hrelia S, Ageloni C, Machesi A, Cantelli FG, Hrelia P. (2007). Neuroprotective effects of anthocyanins and their *in vivo* metabolites in 5H-SY5Y cells. Neurosci. Lett., 424: 36-40.

Tea Board of Kenya (2010). Tea industry performance in 2009. The Tea Board of Kenya.

Terahara N, Takeda Y, Nesumi A, Honda T (2001). Anthocyanins from red-flower tea (Benibana-cha) *Camellia sinensis*. Phytochem. 56: 359-361.

Viskelis P, Rubiskiene M, Jasutiene I, Sarkinas A, Daubaras R, Cesoniene L (2009). Anthocyanins, antioxidative and antimicrobial properties of American Cranberry (*Vaccinium macrocarpon* Ait) and their press cakes. J. Food Sci., 74: 157-161.

Wang C, Wang J, Lin W, Chu C, Chou F, Tseng T (2000). Protective effects of Hibiscus anthocyanins against *tert*-butyl hydroperoxide-induced hepatotoxicity in rats. Food Chem. Toxicol., 38: 411-416.

Wang L, Stoner GD (2008). Anthocyanins and their role in cancer prevention. Cancer Lett., 269: 281-290.

Weissel T, Baum M, Eisenbrand G, Dietrich H, Will F, Stockis JP, Kulling S, Rufer C, Johannes C, Jackowski C (2006). An anthocyanins/phenolic-rich fruit juice reduces oxidative DNA damage and increases glutathione levels in healthy probands. Biotechnol. J., 1: 388-397.

Wijeratne SS, Cuppet SL, Schlegel V (2005). Hydrogen peroxide induced oxidative stress damage and antioxidant enzyme response in Caco-2 human colon cells. J. Agric Food Chem., 53: 8768-8774.

Yang MS, Chan HW, Yu LC (2006). Glutathione peroxidase and glutathione reductase activities are partially responsible for determining the susceptibility of cells to oxidative stress. Toxicology, 226: 126-130.

Yao K, Ye P, Zhang L, Tang X, Zhang Y (2007). Epigallocatechin gallate protects against oxidative-stress induced mitochondria-dependent apoptosis in human lens epithelial cells. Mol. Vis., 14: 217-223.

18

Mycotoxins in animals: Occurrence, effects, prevention and management

Manal M. Zaki[1]*, S. A. El-Midany[2], H. M. Shaheen[3] and Laura Rizzi[4]

[1]Department of Veterinary Hygiene and Management, Faculty of Veterinary Medicine, Cairo University, Giza 11221 Egypt.
[2]Department of Hygiene and Preventive Medicine, Faculty of Veterinary Medicine Kafr El-Sheikh Univ., Egypt.
[3]Department of Pharmacology, Faculty of Veterinary Medicine South Valley University, Qena., Egypt.
[4]Department of Morphophysiology and Animal Production (DIMORFIPA), University of Bologna, 40064, Italy.

Globalization of the trade in agricultural commodities has contributed significantly to the discussion about potential hazards involved and has increased in particular the awareness of mycotoxins. Safety awareness in food and feed production has also risen due to the simple fact that methods for testing residues and undesirable substances have become noticeably more sophisticated and more available at all points of the supply chain. Mycotoxins comprise of a family of fungal toxins, many of which have been implicated as chemical progenitors of toxicity in man and animals. There are four classes of mycotoxins of major concern namely aflatoxins, zearalenone, ochratoxins, and fumonisins. Formation of mycotoxins varied between species as well as within a given species. A variety of physical, chemical, and biological methods to counteract the mycotoxin problem have been reported, but large-scale, practical, and cost-effective methods for detoxifying mycotoxin-containing feedstuffs are currently not available. Detoxification strategies for the contaminated foods and feeds should be done to reduce or eliminate the adverse actions of mycotoxin to improve food safety and prevent economic losses. The most recent approach to the problem has been the addition to the animal's diet of nonnutritive sorbents that sequester mycotoxins, reduce their gastrointestinal absorption and avoiding their toxic effects on livestock and toxin carryover into animal products. This review comments on the potential hazards of several mycotoxins together with prevention strategy for fungal and mycotoxin contamination.

Key words: Mycotoxins, detoxification, aflatoxins, zearalenone, ochratoxins, prevention.

INTRODUCTION

Toxic substances are almost ubiquitous in the environment. Thus, they are also present in ingredients for animal feed. Adequate risk management depends on knowledge of absorption, metabolism, carry-over and toxicological profile of these substances and on practical measures to reduce them. Generally, toxic substances are metabolized before or after absorption through the intestinal tract (Kan and Meijer, 2007). Depending on their physico-chemical characteristics, some substances are metabolized into naturally occurring and generally harmless constituents. Most veterinary drugs and feed additives fall into this group (Kan and Meijer, 2007). Some mycotoxins were heat stable up to as much as 400°C. As a result, they may also be of relevance in processing operations (Mayer et al., 2008).

Molds (fungi) develop from spores that are found ubiquitously in the environment. Mold growth on grain under field conditions or during storage can occur at moisture levels above 16% and at temperatures above freezing. The growth of molds on grain can affect the nutritional quality of grain in several ways (Marguardt, 1996). First, they decrease the nutritional value of the commodity as they consume fats, protein and carbohydrates

*Corresponding author. E-mail: drmanalmoustafa2008@yahoo.com.

Table 1. Examples of fungal species and mycotoxins of economical significance in animal agriculture (D'Mello and Macdonald, 1997).

Fungal species	Mycotoxins
Aspergillus flavus and *A. parasiticus*	Aflatoxins
A. Ochraceus, Penicillium viridicatum, and *P. cyclopium*	Ochratoxin A
Fusarium culmorum, F .graminearum, and *F. sportrichoides*	Deoxynivalenol
F. sporotrichoides and *F. poae*	T-2 toxin
F. sporotrichoides, F. graminearum, and *F. poae*	Diacetoxyscirpenol
F. culmorum, F. graminearum, and *F. sporotrichoides*	Zearalenone
F. proliferatum, F. verticillioides	Fumonisins
Acremonium coenophialum	Ergopeptine
Alkaloids	
A. lolii	Lolitrem alkaloids

that are present in the grain. Thus these nutrients are no longer available to the animal. Secondly, some species of mold are able to produce highly toxic compounds called mycotoxins.

These toxins can adversely affect animal health and production and can cause harmful effects to humans if transmitted into foods. The combined presence of mold and mycotoxins may cause decreased feed intake, decreased feed efficiency, decreased rate of gain, and increased risk of infection as well as reproductive problems (Marguardt, 1996).

Mycotoxins are metabolized in the liver and the kidneys and also by microorganisms in the digestive tract. Therefore, often the chemical structure and associated toxicity of mycotoxin residues excreated by animals or found in their tissues are different from the parent molecule (Ratcliff, 2002). No region of the world escapes the problem of mycotoxins and according to Lowlor and Lynch (2005) mycotoxins are estimated to affect as much as 25% of the world's crop each year.

Moulds and associated mycotoxins are important factors adversely affecting foods produced using contaminated plant products or animal products derived from animals fed on contaminated feeds (Robens and Cardwell, 2003). Mycotoxins are toxic to humans and animals, which explains the major concern of food and feed industries in preventing them from entering the food chain (Pierre, 2007). Toxin-producing moulds may invade plant material in the field before harvest, during post-harvest handling and storage and during processing into food and feed products. Thus, toxigenic fungi have been roughly classified into two groups (i) field fungi; (ii) storage fungi (Pierre, 2007).

Mycotoxins are secondary metabolites produced by filamentous fungi that cause a toxic response (mycotoxicosis) when ingested by higher animals. Cereal plants may be contaminated by mycotoxins in two ways: fungi growing as pathogens on plants or growing saprophytically on stored plants (Glenn, 2007). However,

not all fungal growth results in mycotoxin formation and detection of fungi do not imply necessarily the presence of mycotoxins. Consumption of a mycotoxin-contaminated diet may induce acute and long-term chronic effects resulting in teratogenic, carcinogenic, and oestrogenic or immune-suppressive effects. Direct consequences of consumption of mycotoxin-contaminated animal feed include: reduced feed intake, feed refusal, poor feed conversion, diminished body weight gain, increased disease incidence (due to immune-suppression), and reduced reproductive capacities (Fink-Gremmels and Malekinejad, 2007; Morgavi and Riley, 2007; Pestka, 2007; Voss and Haschek, 2007) which leads to economic losses (Huwig et al., 2001; Wu, 2004, 2006).

Due to modern laboratory methods and a growing interest in this field of research, more than 300 different mycotoxins have been differentiated thus far. However, for a practical consideration in the feed-manufacturing process only a small number of toxins are of relevance, with aflatoxins, trichothecenes, zearalenone, ochratoxins and fumonisins (Table 1) being of particular interest, although it has to be mentioned that the extent of each toxin impairment is highly species-dependant (Erber and Binder, 2004). The Fusarium genus, e.g. *Fusarium verticillioides* (formerly *Fusarium moniliforme*), *Fusarium roseus*, *Fusarium tricinctum* and *Fusarium nivale*, are ubiquitous soil organisms, which may infect cereals directly in the field thereby, increasing fumonisins, trichothecene, and zearalenone levels (depending on the species) during growth, ripening of grain and at harvesting.

Although the scientific literature offers a broad variety of information on the effects of individual mycotoxins in various animal species, concurrent exposure to multiple mycotoxins is more likely in the livestock industry (Table 2). For example, aflatoxin and fumonisin B1, as well as deoxynivalenol (DON) or other trichothecenes (one or even more of them) and zearalenone frequently occur

Table 2. Geographic occurrence of mycotoxins.

Location	Mycotoxins
Western Europe	Ochratoxin, Vomitoxin, Zearalenone
Eastern Europe	Zearalenone, Vomitoxin.
North America	Ochratoxin, Vomitoxin, Zearalenone, Aflatoxins
South America	Aflatoxins, fumonisins, Ochratoxin, Vomitoxin, T-2 toxin.
Africa	Aflatoxins, fumonisins, Zearalenone.
Asia	Aflatoxins
Australia	Aflatoxins, fumonisins

Source: Devegowda et al. (1998).

together in the same grain. Additionally, in the feed manufacturing process, various batches of different raw materials are mixed together thus producing a totally new matrix with a new risk profile. Poor livestock performance and/or disease symptoms observed in commercial operations may be due to the synergistic interactions between multiple mycotoxins. Scientific reports on synergistic effects of mycotoxins at acute toxicity levels describe combinations of aflatoxins with various trichothecenes, as well as with ochratoxins and fumonisins, but also combinations of fumonisins plus DON. Nevertheless it has to be pointed out that far more work has to be done in this particular field of research, especially in the sub-acute contamination range as well as with combinations of more than two toxins (CAST, 2003; Erber and Binder, 2004).

Several of the major mycotoxins exert their effects through different organ systems and different biological pathways. Aflatoxin, ochratoxin, and T-2 toxin all interfere with protein formation, but each does so in a different manner; aflatoxin binds to both RNA and DNA and blocks transcription (Kan and Meijer, 2007). T-2 toxin blocks initiation of translation, and ochratoxin blocks phenylalanine-t RNA synthetase, and thus blocks translation. Some scientists assume complete absorption of these noxious substances, as a worst-case scenario to predict residues in animal products from those in feed (Kan and Meijer, 2007). By doing so, they ignore the physiological processes occurring during transit through the intestine and after absorption into the general circulation as well as intermediary metabolism. Furthermore this approach does not take advantage of existing knowledge to identify or implement possible control points for reduction of levels of residues in animal products (Kan and Meijer, 2007).

Multi-toxin occurrence may be one important explanation for divergences in effect levels described in the scientific literature, where defined, mostly purified mycotoxins are used in most studies. In field outbreaks, naturally contaminated feeds may contain multiple mycotoxins and thus apparently lower contamination levels of a single specific mycotoxin can be associated with more severe effects.

Analytical methods for separation of mycotoxins included thin-layer chromatography (TLC), gas chromatography (GC), and High-performance liquid chromatography (HPLC). The few TLC methods had been used for screening and not for quantification (Abbas et al., 2004; Benedetti et al., 2006). GC with high-resolution MS had also been used for analysis of some mycotoxins (Fernandez et al., 2007; Ikonomou et al., 2008). However, a labour-intensive derivatization step was often indispensable prior to GC analysis. Therefore, HPLC–MS–MS had increasingly become the method of choice for mycotoxin analysis (Abbas et al., 2004; Benedetti et al., 2006; Fernandez et al., 2007).

Mycotoxin losses and costs of mycotoxin management are overlapping areas of concern. Costs of mycotoxin management include research production practices, testing and research necessary to try to prevent the toxins from appearing in food and feed products of affected commodities (Robens and Cardwell, 2003). Mycotoxin losses result from lowered animal production (Robens and Cardwell, 2003) and any human toxicity attributable to the presence of the toxin (Council for Agricultural Science and Technology (CAST, 1989) in the affected commodity which lowers its market value, as well as secondary effects on agriculture production and agricultural communities (CAST, 1989).

Due to the multiple possible origins of fungal infection, any prevention strategy for fungal and mycotoxin contamination must be carried out at an integrative level all along the food production chain (Robens and Cardwell, 2003). Three steps of intervention have been identified. The first step in prevention should occur before any fungal infestation; the second step is during the period of fungal invasion of plant material and mycotoxin production; the third step is initiated when the agricultural products have been identified as heavily contaminated. Such hazard analysis has some similarity with the HACCP management system of food safety (Degirmencioglu et al., 2005), mainly with the principles 2 (Determination of critical control points) and 3 (Establish critical limits). Most of the efforts must be concentrated on the two first steps since, once mycotoxins are present, it is difficult to eliminate them in a practical way.

Approaches to prevent mycotoxicoses include pre- and post-harvest strategies; the latter are often categorized into physical, chemical and biological methods (Jouany, 2007). The best way would be the prevention of mycotoxin formation in the field of its first place, which is supported by proper crop rotation and fungicide administration at the right time. In case of toxin manifestation, measures are required that act specifically against certain types and groups of toxins. The most prevalent approach counteracting mycotoxins in the feed industry is to include sorbent materials into the feed, for more or less selective removal of toxins by means of adsorption within the route of the gastrointestinal tract, or to add enzymes or microbes capable of detoxifying certain mycotoxins or toxin groups (Leibetseder, 2005).

FACTORS THAT PROMOTE FUNGAL GROWTH AND MYCOTOXIN PRODUCTION

Besides the presence of nutrients, the most important factors for growth and mycotoxin production are temperature, water activity (aw) and oxygen. Often contamination of food by fungi may vary due to different origins of contamination, especially storage buildings, bins or underground pits (Christensen and Sauer, 1982). Often, fungi invade only a minor fraction of feed particles with appropriate condition for a growth such as enough water content, aeration, etc.

Substrates differ in their ability to support fungal growth due to differences in their physical and chemical characteristics, which include water activity, oxygen availability and surface area, while chemical characteristic include carbohydrates, fat, protein, trace elements and amino acid composition (Russell et al., 1991). While some substrates are susceptible to colonization, other environmental conditions increase the vulnerability of the fungi to the substrate. The conditions include temperature, water activity, pH and atmospheric air (oxygen) (Moss, 1991).

Temperature

It has been shown that *Penicillium* species have a lower minimum temperature range than *Aspergillus* species. The optimal temperature for *Penicillium* and *Aspergillus* is 25 to 30°C and 30 to 40°C, respectively. The maximal temperature is 28 to 30°C for *Penicillia* and 37 to 47°C for most *Aspergilli*. Various *Fusarium* species can also be regarded as psychrophilic, because of their low optimal temperature of 8 to 15°C (Moss, 1991).

Water activity

Water activity (aw) is a measure of unbound water in the food available for the growth of the mould. Values for water activity appreciation vary between 0.61 and 0.91. Most storage fungi grow at aw<0.75 (Moss, 1991). It is important to note that ambient conditions (temperature and humidity) do not only influence the rate at which chemical changes may take place, but also the growth of fungi and insect pests (Francis and Wood, 1982). This is because high temperatures and relative humidity provide ideal conditions for growth and development of moulds with possible production of mycotoxins (Pitt and Hocking, 1997).

Smith and moss (1985) reported that moisture determines whether microbes can colonize a substrate or not. These factors enable moulds to break down complex macromolecular compounds and utilize them for growth and metabolism. In the process, they produce and secrete toxic secondary metabolite, which are "mycotoxins" (Moss, 1996). Excessive moisture in the field and in storage, temperature extremes, humidity, drought, variations in harvesting practices and insect infestations are major environmental factors that determine the severity of mycotoxin contamination (Hussein and Brassel, 2001).

pH

At high water activities, fungi compete with bacteria as food spoilers (Wheeler et al., 1991). Most fungi are little affected by pH over a broad range, commonly 3 to 8 (Wheeler et al., 1991), however, the pH of a medium may exercise important control over a given morphogenic event without remarkably influencing the overall growth of a fungus (Pitt and Hocking, 1997).

Oxygen

Oxygen is essential for the growth of fungi, but certain species can also grow under anaerobic conditions with the formation of ethanol and organic acids. Oxygen also influences production of mycotoxins. The production of patulin and penicillic acid decrease sharply at low oxygen concentrations, while fungal growth is not noticeably influenced (Northolt, 1979). *Aspergillus* growth is restricted at an oxygen concentration of less than 1% (Pitt and Hocking, 1997) (Tables 1 and 2).

Aflatoxins

Aflatoxins, a family of closely related, biologically active mycotoxins, have been known as a prominent cause of animal disease for many years. The toxins occur naturally on several key animal feeds, including corn, cottonseed, and peanuts. Occurrence of aflatoxins on some field crops tends to spike in years when drought and insect damage facilitate invasion by the causative organisms, *Aspergillus flavus* and *Aspergillus parasiticus,* which abound in the crop's environment. Aflatoxins B1 was present in contaminated peanut containing feed that caused a mass death of turkeys in England in 1960; an incidence that triggered much subsequent mycotoxin research.

Acute aflatoxicosis causes a distinct overt clinical disease marked by hepatitis, icterus, hemorrhage, and death. More chronic aflatoxins poisoning produces very variable signs that may not be clinically obvious; reduced rate of gain in young animals is a sensitive clinical register of chronic aflatoxicosis. The immune system is also sensitive to aflatoxins, and suppression of cell-mediated immune responsiveness, reduced phagocytosis, and depressed complement and interferon production are produced. Acquired immunity from vaccination programs may be substantially suppressed in some disease models. In such cases the signs of disease observed are those of the infectious process rather than those of the aflatoxins that predisposed the animal to infection. Of considerable potential economic consequence is the fact that aflatoxins can suppress the immune system of young animals by in utero-transfer across the placenta of the pregnant dam (Pier et al., 1985). In these cases the affected newborn animals lack resistance to infection and cannot respond well to vaccines. These are reactions of considerable consequence in colonized animals in which we rely on elective vaccination procedures in disease prevention.

Aflatoxins in feedstuffs

Aflatoxins, primarily aflatoxin B1, occur in a number of important animal feeds. Growth of toxigenic strains of *A. flavus* and *A. parasiticus* on corn, cottonseed, and peanuts often results in injurious levels of aflatoxin B1, the most biologically active member of the aflatoxin family (Cheeke and Shull, 1985). These three feedstuffs are the most important sources of aflatoxin in animal feeds (Cheeke and Shull, 1985). The causative molds may occasionally colonize small cereal grains (barley, oats, and wheat) and produce low to moderate levels of aflatoxin. Soybeans do not support appreciable levels of aflatoxin B1 production (Lillehoj et al., 1991).

The moisture content promotes the growth of the toxigenic molds and grinding of the kernel destroys the natural barrier to infestation. Moisture content of the feed must be $\geq$ 15% to support growth of the molds. The fungus must gain access to susceptible parts of the plant (e.g., the corn kernel, cotton seed, etc.) before it grows and elaborates aflatoxins. Seasonal peaks in aflatoxin content are seen in key years when drought-damaged plants or insect-damaged crops are rendered more susceptible to fungal invasion. Wet harvest seasons also may contribute to high levels of aflatoxin in certain crops.

Aflatoxin sometimes develops in crops stored at levels of moisture content > 15% or properly dried crops stored in leaky bins. Development of aflatoxin can be prevented in stored grains by good management practices (Christensen and Meronuck, 1986); the occurrence of aflatoxin in field crops, however, is largely a matter of uncontrollable natural events. In these events careful use of blending with clean crops or detoxification through ammonization, with close attention to existing rules and regulations, may be possible to reduce the toxin content in animal feeds to safe levels (Park et al., 1988; FDA, 1989). Recent information suggests that binding agents fed with aflatoxins may reduce the availability of the toxins and thereby reduce their effects in some animal species (Harvey et al., 1989). In the absence of one of these control procedures the feed should be withheld from animal use.

Zearalenone

Zearalenone (previously known as F-2 toxin) was produced by some Fusarium species *Fusarium graminearum* (Gibberella zeae), *Fusarium culmorum, Fusarium cerealis, Fusarium equiseti, Fusarium crookwellense* and *Fusarium semitectum*. These fungi infected contaminants of cereal crops worldwide (Bennett and Klich, 2003). The concentration of accumulated Zearalenone (ZEA) in cereals depended on several factors such as the substrate, temperature, duration of Fusarium growth and strain of fungal species. Moreover, the humid tropical climate promoted microbial proliferation on food and feedstuffs and finally mycotoxin biosynthesis (Nuryono et al., 2005).

Toxicity of ZEA and its metabolites was related to the chemical structure of the mycotoxins, similar to naturally occurring estrogens (Gromadzka et al., 2009). ZEA was heat-stable, which made it difficult to remove and/or decomposed from food (Kuiper-Goodman et al., 1987). Additionally, it was observed that during food and feed processing (e.g. milling, extrusion, storage and heating) ZEA was not decomposed (Yumbe-Guevara et al., 2003).

Zearalenone imitates the effect of female hormone oestrogen and at low doses, increases the size or early maturity of mammary glands and reproductive organs. At higher doses, Zearalenone interferes with conception, ovulation, implantation, fetal development and the viability of new born animals (Zinedine et al., 2007). Zearalenone causes estrogenic responses in dairy cattle and large doses of this toxin are associated with abortions. Other responses of dairy animals to zearalenone are reduced in feed intake, decreased milk production, vaginitis, increase vaginal secretions, poor reproductive performance and mammary gland enlargement in heifers. It is recommended that zearalenone should not exceed 250 ppb in the total diet (Zinedine et al., 2007).

Ochratoxins

The ochratoxins are metabolites produced by certains species of genera Aspergillus and penicillium (Wood, 1992). Ochratoxins A was discovered in 1965 by South African Scientists as a toxic secondary metabolite of *Aspergillus ochraceus* (Van der Merwe et al., 1965). Other species of *A. ochraceus* group and several Penicillium species, including *Penicillium viridicatum*, have been shown to form ochratoxin A.

Ochratoxin A is the major metabolite of toxicological significance and it is mainly a contaminant of cereal grains (corn, barely, wheat and oats). It has also been found beans (soyabeans, coffee, cocoa) and peanuts and meat in some countries (Krogh, 1987). Ochratoxin A is teratogenic in rat, hamster and chick embryo and is an inhibitor of hepatic mitochondrial transport to cause damage to the liver, gut, lymphoid tissue and renal tubular damage (Krogh, 1987).

Fumonisins

The fumonisins are a group of compounds originally isolated from *Fusarium moniliforme* (Gelderblom et al., 1988). Six different fumonisins (FA_1, FA_2, FB_1, FB_2, FB_3 and FB_4) have been reported, the A series are amides and the B series have a free amine (Gelderblom et al., 1991).

In most animals tumonisin impairs immune function, causes liver and kidney damage, decreases weight gains, and increases mortality rates. The fumonosins (FB_1 and FB_2) were recently isolated from *F. moniliforme* cultures and found to promote cancer in rats (Gelderblom et al., 1988). These toxins occur naturally in corn and have been associated with equine leukoencephalomalacia (Ross et al., 1990).

Fumonisins are stable during food processing: they are not degraded during corn fermentation (Scott and Lawrence, 1995); they are heat stable (Marasas, 1997) and resistant to canning and baking processes (Castelo et al., 1998), although in corn the nixtamalization process reduces fumonisin B1 levels, a five-fold more toxic product with respect to the original level (Bullerman and Bianchini, 2007; Hendrich et al., 1993; Voss et al., 1996).

The use of natural bioactive substances for control of postharvest fungal infections has gained attention due to problems associated with chemical agents. These include the development of fungal species resistant to chemical treatments, which increases food-borne pathogenic microorganisms, in addition to increasing the number of pesticides under observation or regulation (Rabea et al., 2003). Also, essential oils of cinnamon (Cinnamomum zeylanicum Blume) and oregano have shown fungicidal activity *in vitro* against *A. flavus* Link: Fr. (García-Camarillo et al., 2006).

In addition, Sánchez et al. (2005) reported the inhibition of both growth and mycotoxin production by *A. flavus* and *A. parasiticus* Speare when exposed to ethanolic, methanolic, and aqueous extracts of Agave species. For that reason, it is possible that native plants such as *Larrea tridentata*, *Baccharis glutinosa*, *Ambrosia confertiflora* DC, and *Azadirachta indica* A. Juss. can be used as source of natural preservative compounds for the control of filamentous fungi like *Fusarium verticillioides*.

T-2 toxin

The T-2 toxin, produced mainly by *Fusarium tricinctum*, was the first trichothecene to be found as a naturally occurring grain contaminant in the United States (Hsu et al., 1972). It was associated with a lethal toxicosos in dairy cattle that had consumed moldy corn in Wisconsin. This mycotoxin rarely associated with disorders in animals or humans in other countries (Mroch et al., 1983). Yoshizawa et al. (1981), stated that the chance of finding T-2 toxin as a residue in edible tissue is remote because it is rapidly metabolized *in vivo*.

In dairy cattle T-2 toxin has been associated feed refusal, production losses, gastroenteritis, intestinal hemorrhages and death. T-2 has also been associated with reduced immune response in calves. In poultry, T-2 toxin has been implicated to cause mouth and intestinal lesion as well impair the bird's immune response, causing decreased in egg production, decreased feed consumption, weight loss and altered feather patterns (Mroch et al., 1983).

Vomitoxin

Vomitoxin also called deoxynivalenol is stable, survive processing, milling and does occur in food products and feeds prepared from contaminated corn and wheat. The most common producer of vomitoxin is *F. graminearum* (Marasas et al., 1984). Corn contaminated with *F. graminearum* was shown to contain the trichothecene vomitoxin (3, 7, 15-trihydroxy- 12, 13-epoxytrichothec-9-en-8-one). Vomitoxin is perhaps, the most commonly detected *Fusarium* mycotoxin. Vomitoxin has been associated with reduced milk production in dairy cattle, vomiting by swine contaminated feed or their refusal to eat feed containing the toxin, and inhibiting reproductive performance and immune function in several animal species (Marasas et al., 1984).

Toxicology and syndromes

In common with other physiologically active compounds, the Fusarium mycotoxins are capable of inducing both acute and chronic effects. The effects observed are often related to dose levels and duration of exposure. Although

Table 3. Mycotoxins and their effects on different species of livestock.

Mycotoxins	Species susceptability	Effects
Aflatoxins	All domestic animals and poultry	Hepatoxic, carcinogenic, immnosuppressive
Zearalenone	Mainly pigs and dairy animals	Estrogenic and reproductive disorder
Vomitoxin	Mainly pigs and dairy animals	Dermatotoxic, feed refusal
Ochratoxin	Mainly pigs and poultry	Nephrotoxic, gout
T-2 Toxin	Mainly pigs and poultry	Mouth lesions, loss of appetite
Fumonisins	Mainly pigs and horses	Neurological disorders, liver damage.

Source: Ratcliff, 2002.

acute and chronic effects in farm livestock are readily demonstrated under experimental conditions, similar manifestations have been reported in natural outbreaks of Fusarium mycotoxicoses in Europe, Asia, New Zealand and South America (Fazekas and Bajmocy, 1996; Prathapkumar et al., 1997; Kramer et al., 1997; Galhardo et al., 1997).

Chronic exposure of farm animals to DON is a continuing hazard in Canada, the USA and continental Europe. In Japan, several cases of mycotoxicoses in animals have been attributed to consumption o f cereals contaminated with DON and NIV (Yoshizawa, 1991). A number of specific syndromes in farm livestock have now been positively linked with exposure to certain trichothecenes, ZEN, and fumonisins. These include feed refusal, emesis and anorexia; oral and gastro-intestinal lesions; ill-thrift; reproductive dysfunction; equine leukoencephalomalacia; and porcine pulmonary edema. In addition, Duodenitis/ proximal jejunitis and acute mortality syndrome have tentatively been linked with particular Fusarium mycotoxins.

Effect of mycotoxins on animals

Acute primary mycotoxicoses are produced if high to moderate amounts of mycotoxins are consumed. Specific, overt, acute episodes of disease ensue, which include hepatitis, hemorrhage, nephritis, necrosis of oral and enteric epithelium, and death. These effects belonged to the target organs usually affected by specific mycotoxins. Chronic primary mycotoxicoses, resulting from moderate to low levels of mycotoxin intake, often cause reduced productivity in the form of slower rate of growth, reduced production and inferior market quality. These effects often occur without the production of an overt, primary mycotoxicosis syndrome (Table 3).

Consumption of low levels of mycotoxins through the feeds do not cause overt mycotoxicoses, but often predisposes to various infectious diseases and especially to secondary bacterial infections or to a heavy progression of some often encountered parasitic diseases (Stoev et al., 2000; Koynarski et al., 2007), because of the suppression in both humoral and cell-

mediated immune response in such animals (Stoev et al., 2000). Suppression of the cellular immune system is a known result after ingestion of several mycotoxins. Cheeke (1998) confirmed that in monogastrics, variable immune responses have been observed after ingestion of these mycotoxins. Various degrees of mycotoxicoses from natural sources occur in different animal species because of the wide range of feed ingredients used and the differences among and within species (Hussein and Brassel, 2001).

Mycotoxins have several effects in poultry (Figure 1). Early investigations concerning the sudden death of 100, 000 turkey poults consuming groundnuts in England linked A. flavus to acute hepatic necrosis and hyperplasia of the bile ducts of intoxicated birds (Newberne and Butler, 1969). High levels of aflatoxins (from 0.2 to 1 mg/kg) in combination with other mycotoxins (OTA and / or trichothecenes) in poultry feed may cause diseases such as hepatitis and can lead to the development of salmonelosis, coccidiosis, and infectious bursal disease (Ratcliff, 2002). Chickens have been shown to bruise and haemorrhage from AF (Table 4 and Figures 2 to 4b).

De-contamination and amelioration

A number of de-contamination procedures have been investigated, broadly divisible into physical and chemical principles (Placinta et al., 1999). Physical methods include milling which has been shown to be highly effective for DON, and density segregation which has resulted in reduced levels of trichothecenes and ZEN. Super activated charcoalis partially effective at reducing the incidence of oral lesions in broilers fed T-2 toxin, but mortality remains unaffected (Edrington et al., 1997). Furthermore, amelioration of oral lesions was not consistent between experiments. Chemical methods tested include calcium hydroxide monomethylamine, sodium bisulphite and ammonia.

The commercial potential of these de-contamination procedures, however, has yet to be determined. Antioxidants such as vitamin E have been considered as dietary supplements to counteract the effect of T-2 toxin.

A partial beneficial effect, in terms of reduced in vivo

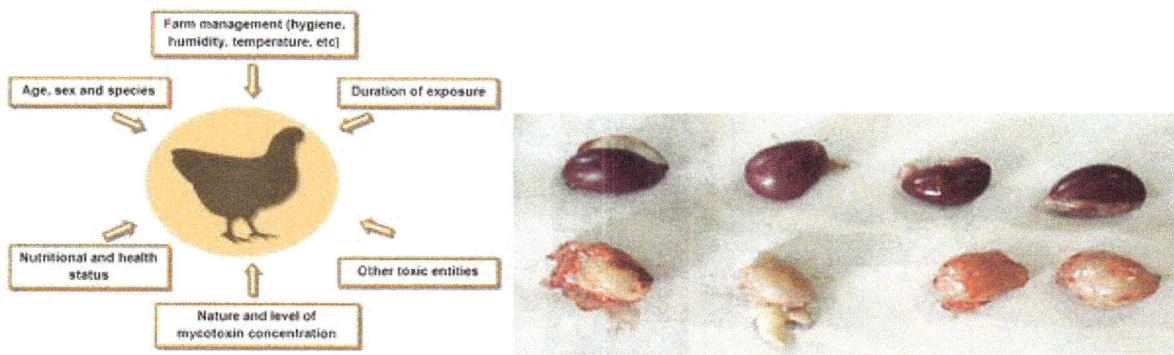

Figure 1. Mycotoxins effects on poultry (The poultrysite.com, mycotoxin.com).

Table 4. Mycotoxins and their effects on different species of livestock.

Mycotoxins	Species susceptability	Effects
Aflatoxins	All domestic animals and poultry	Hepatoxic, carcinogenic, immnosuppressive
Zearalenone	Mainly pigs and dairy animals	Estrogenic and reproductive disorder
Vomitoxin	Mainly pigs and dairy animals	Dermatotoxic, feed refusal
Ochratoxin	Mainly pigs and poultry	Nephrotoxic, gout
T-2 Toxin	Mainly pigs and poultry	Mouth lesions, loss of appetite
Fumonisins	Mainly pigs and horses	Neurological disorders, liver damage.

Source: Ratcliff, 2002.

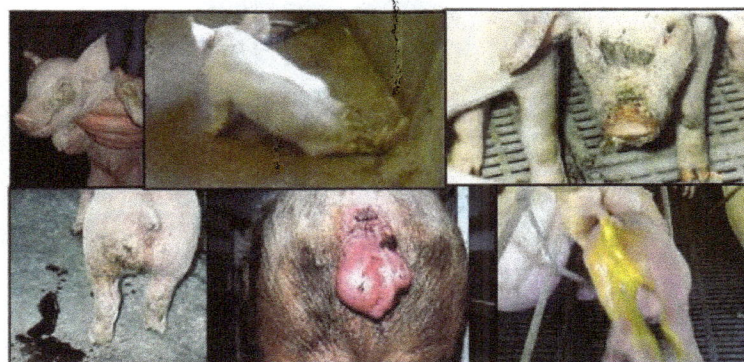

Figure 2. Mycotoxins effects on pigs (en.engormix.com).

lipid peroxidation, has been reported in one study with chickens (Hoehler and Marquardt, 1996). Vitamin C was ineffective in this respect.

The general harmony now prevailing is that preventive measures offer greater potential than remedial procedures (Figure 5). With ZEN, a feeding strategy for breeding ewes has been suggested, based on the use of chicory pastures containing inherently low levels of the mycotoxin (Kramer et al., 1997). However, selection of cultivars of cereal and forage plants that are resistant to infection by toxigenic species of Fusarium pathogens is likely to be the long-term objective of any effort to control contamination with the associated mycotoxins.

ALawadi and AL-Jedabi (2000) proved an inhibitory and antibiotic activity of camel urine against the growth of Candida albicans (yeast), Aspergillus niger, Fusarium oxysporum even after it is boil to 100°C. The effect of camel urine and milk on the growth properties of such fungi or on the efficiency of aflatoxins as inhibitors to Bacillus subtilus growth is seen as a primary step fined

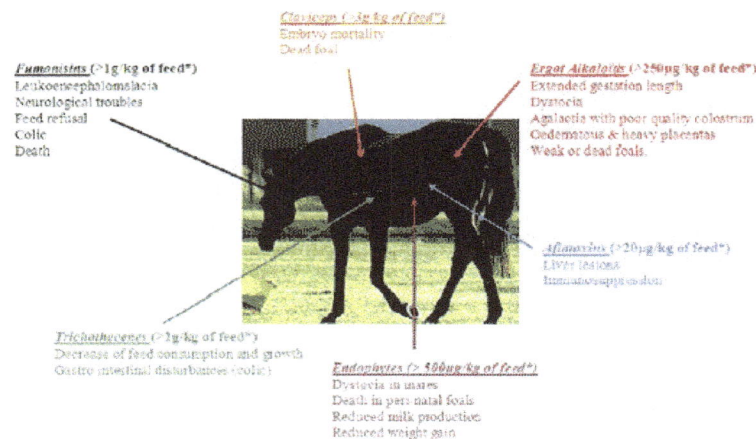

Figure 3. Mycotoxin effects in Equine (en.engormix.com).

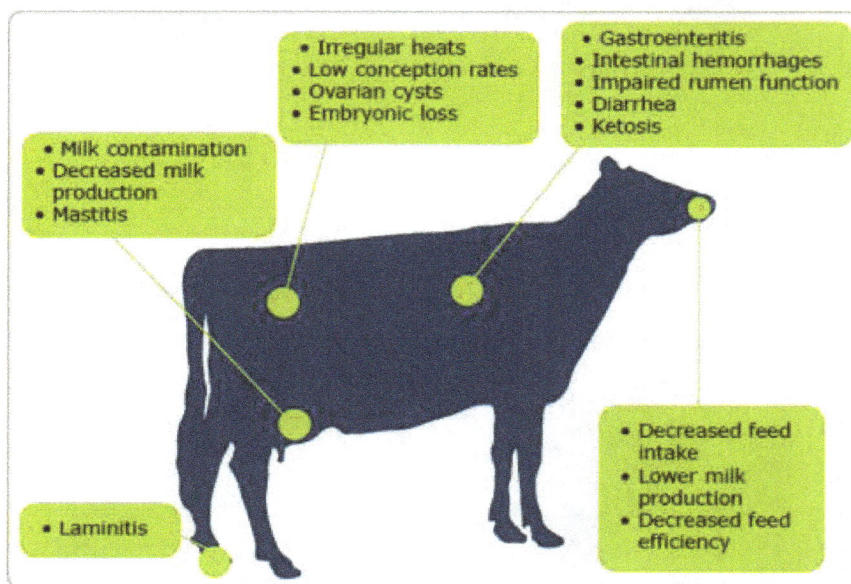

Figure 4a. Mycotoxins effects on ruminants (en.engormix.com).

away to get rid of fungal toxins. The chemical and organic constituents of urine proved to have inhibitory properties against fungal and bacterial growth (Ghosal et al., 1974; Varley et al., 1980; Mura et al., 1987; Amer and Hendi, 1996).

When Amer and Al-hendi (1996) analyzed urine of mature camels of between 5 to 10 years old, they found that its relative density ranged from 1.022 to 1.07, while pH values varied to be either acidic or alkaline. Urea level ranged from 18 to 36 g/dl. Keratin recorded 0.2 to 0.5 g/L. Microscopical analysis proved the presence of phosphorus and calcium oxalate and ammonium urate; some epithelial and granular cells appeared. Al-Attas

(2008), using neutron activation analysis, estimated some essential elements within milk and urine of camels, and discovered that it contains large amount of Na and K substituting the loss of such elements in the case of diarrhea. Also it contains large amount of Zn which assists in the cure of the infection due to diarrhea.

Prevention and management of mycotoxins in food and feed

When contamination cannot be prevented at pre-harvest or during the post-harvest stage, decontamination/

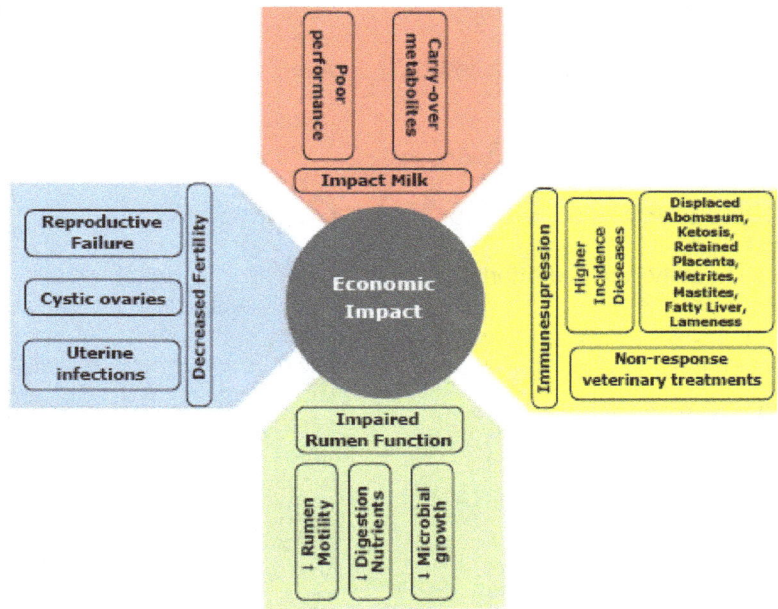

Figure 4b. Mycotoxins effects on ruminants (en.engormix.com).

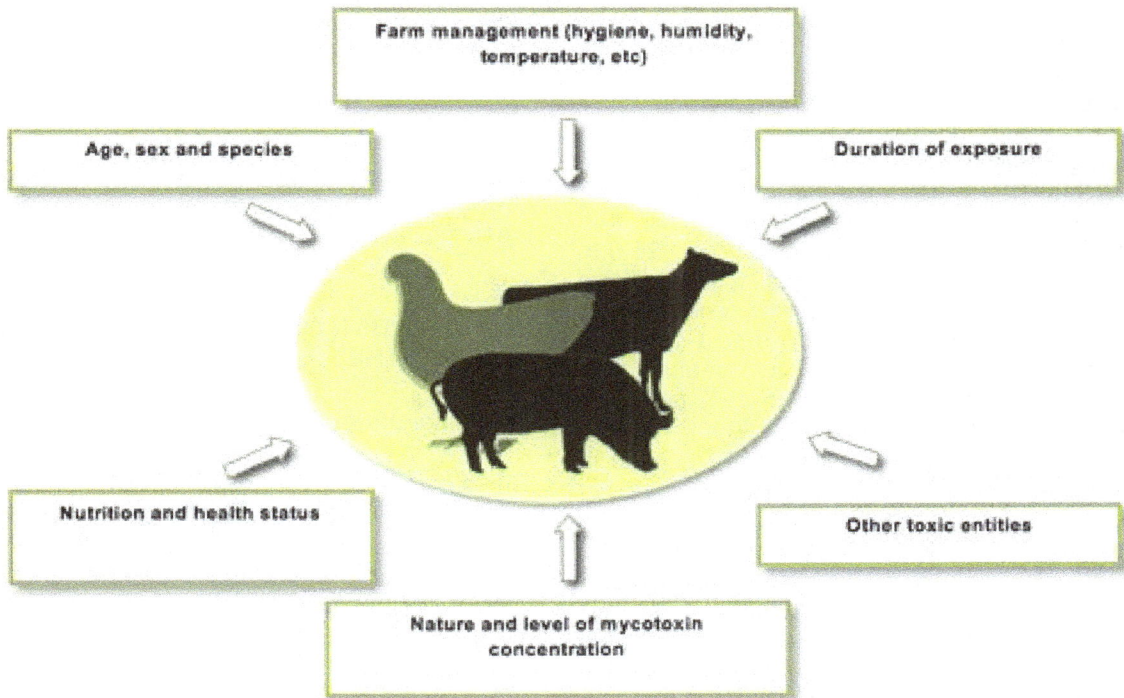

Figure 5. Farm management for prevention and control of mycotoxin (en.engormix.com).

detoxification procedures played an important role in helping prevent exposure to the toxic and carcinogenic effect of mycotoxins through the physical separation and physical, chemical and biological inactivation and/or

removal of the toxin (Kabak et al., 2006). Any detoxification procedure to reduce the toxic
and economic impact of mycotoxins needs the following basic criteria (Jemmali, 1979):

1) It must destroy, inactivate or remove the mycotoxins in foods and feeds.
2) It must not produce or leave toxic and/or carcinogenic residues in the final products.
3) It should not alter significantly the nutritional and technological properties of the product.
4) It must be capable of destroying fungal spores and mycelia in order to avoiding new toxin forming under favourable conditions.
5) It had to be technically and economically feasible.

REMOVAL OF MYCOTOXINS FROM CONTAMINATED COMMODITIES

Several methods were reported for the removal of mycotoxins from contaminated commodities, including physical separation, extraction with solvents and adsorption.

Physical separation

Since detoxification of mycotoxins by chemical applications was not an acceptable practice in some regions, physical separation of contaminated crops was a very important option for the producer (Kabak et al., 2006). Cleaning grains removed kernels with extensive mold growth, broken kernels and fine materials, which reduced mycotoxin concentration (Bullerman and Bianchini, 2007). Cleaning of the maize removed 26.6 to 69.4% of the fumonisins (Sydenham et al., 2004), while a 40 to 80% reduction in aflatoxin levels were reported after physical cleaning and separation of mould-damaged kernels and seeds (Park, 2002). However, cleaning was not effective in removing DON; only 6 to 19% reduction was achieved in wheat by cleaning (Abbas et al., 1985).

Washing procedures, using distilled water, resulted in 65 to 69% reductions of DON and 2 to 61% reductions of ZEA in barley and maize, whereas using 1 M sodium carbonate solution for the first wash reduced DON by 72 to 74% and ZEA by 80 to 87% (Trenholm et al., 1992). This process might be a useful treatment before wet milling and brewing; otherwise, the cost of seed drying would be prohibitive. Such approaches are also capable of reducing patulin levels in the final juiced products (Acar et al., 1998).

Extraction with solvents

Extraction with a variety of solvents including ethanol, aqueous isopropanol, methanol–water, and acetonitril–

water removed aflatoxins from contaminated commodities such as cottonseed and peanuts. On the other hand, high cost and problems related to disposal of the toxic extracts restrict its use for large scale application (Rustom, 1997).

Adsorption

Two of the most potent adsorbents for removal of mycotoxins were activated carbon (AC) and bentonite. When phosphate-buffered saline (PBS) and wine samples contaminated with 5 ng/ ml OTA were treated with 1 mg/ ml AC, 100 and 87% of the available toxin were absorbed by the sorbent respectively (Var et al., 2008). In relation to other mycotoxins, AC was shown to considerably decrease patulin levels in apple juice (Artuk et al., 1995). Bentonite, which had a negative charged surface, for its part showed a very poor affinity for OTA (Var et al. ,2008), DON and NIV (Avantaggiato et al. ,2004), while Diaz et al. (2002) observed that bentonite was effective in removing AFB1 in the range 95 to 98.1%. Yeasts were focused on the removal of mycotoxins in liquids in recent years.

Cecchini et al. (2006) demonstrated that the percentage of OTA removal during fermentation was between 46.83 and 52.16% in white wine and between 53.21 and 70.13% in red wine, depending on the yeast strain used. Similarly, Caridi et al. (2006) reported that the removal of OTA in wines by 20 different *Saccharomyces sensu stricto* strains, using a naturally and spiked OTA-containing grape must (1.58 and 7.63 ng/ ml respectively), after 90 days of fermentation was between 39.9 and 92.1% and between 67.9 and 83.4% respectively.

INACTIVATION OF MYCOTOXINS IN CONTAMINATED COMMODITIES

Physical methods

Physical strategies including thermal processing (cooking, boiling, baking, frying, roasting, microwave heating, extrusion) and irradiation were applied for inactivation of the toxin or to reduce its content in foods and feeds.

Thermal treatment

Most mycotoxins were heat-resistant within the range of conventional food processing temperatures (80 to 121°C), so little or no reduction in overall toxin levels occurred as a result of normal cooking conditions such as boiling and frying, or even following pasteurization. The initial level of contamination, type of mycotoxin and its concentration, heating temperature and time, and the degree of heat penetration, as well as the moisture content, pH and ionic strength of food, among other factors, played a significant role in the achievement of toxin degradation (Samarajeewa et al., 1990; Rustom, 1997).

Chemical methods

A variety of chemicals, including acids, bases, oxidizing reagents, reducing agents, chlorinating agents, and miscellaneous reagents were tested to detoxify mycotoxins. The success of detoxification process by chemical treatments highly depends on the type of food and/or feed. The use of chemicals in combination with physical treatments such as thermal processing for the detoxification of food products contaminated with mycotoxins increased the efficacy of mycotoxins degradation.

Acid treatment

It was clear from the accumulated evidence that treatment of aflatoxins with strong acids destroyed the biological activity of AFB1 and AFG1 by converting them to the hemiacetal forms AFB2a and AFG2a respectively, due to acid-catalysed addition of water across the double bond in the furan ring (Heathcote and Hibbert, 1978). Treatment with HCl (pH 2) reduced AFB1 levels by 19.3% within 24 h (Doyle et al., 1982).

Treatment with bases

Among bases and other chemicals, ammoniation was proved to be an effective method for detoxifying aflatoxin-contaminated agricultural products and animal feeds. The ammoniation process, using either ammonium hydrochloride or gaseous ammonia (NH_3), was equally effective in the detoxification of aflatoxins in maize, and was shown in some cases to decrease aflatoxin levels by more than 75% (Burgos-Hernández et al., 2002). Ammoniation caused a 79% reduction of FB1 in contaminated maize (Park et al., 1992). It was reported that ammoniation almost completely decomposed OTA in maize, wheat and barley (Scott, 1996).

The ammoniation process did not leave toxic metabolites of mycotoxins in feed (Scott, 1998), but the relatively long period of aeration and its cost, which increased the price of the product by 5 to 20%, restricted its use in animal feeds (Peraica et al., 2002). In addition, some undesirable effects in the sensory and nutritional quality of the feed, such as brown colour of the treated feed, a decrease in lysine and sulphur-containing amino acids, cannot be overlooked (Piva et al., 1995; Scott, 1998).

Oxidizing agents

It was well-known that aflatoxins such as AFB1, AFG1 and AFM1 which had a terminal double bond in the dihydrofuran ring were more susceptible to attack by Ozone (O_3) and other oxidizing agents than AFB2, AFG2 and AFM2, which lack this double bond (McKenzie et al., 1997). Ozone was reported to reduce AFB1 and AFG1 levels by 77 and 80% respectively in peanuts after treatment at 75°C for 10 min, while the maximum degradation was 51%, occurring for AFB2 and AFG2 in peanuts, regardless of the exposure times (Proctor et al., 2004). In another study, the reductions of AFB1 in paprika were 80 and 93% after exposures to 33 mg/L O_3 and 66 mg/L O_3 for 60 min respectively (Inan et al., 2007).

However, limited experiments with other mycotoxins showed that patulin, CPA, OA, FB1 and ZEA were effectively degraded after treatment with O_3 at 10% for 15 s (McKenzie et al., 1997). H_2O_2, one of the oxidizing agents, was used on a commercial scale to detoxify aflatoxin. Treatment of figs with H_2O_2 at 0.2% caused a 65.5% reduction in AFB1 levels following 72 h storage (Altuğ et al., 1990). Additionally, citrinin can be completely detoxified by H_2O_2 at 0.05% for 30 min at room temperature, whereas OTA was not detoxified by treatment with 0.05 to 0.1% H_2O_2 (Fouler et al., 1994). Abd Alla (1997) revealed that ZEA was degraded by 83.9% when using 10% H_2O_2 at 80°C for 16 h.

Reducing agents

Sodium bisulfite ($NaHSO_3$) was shown to destroy mycotoxins, primarily AFB1 in maize (Doyle et al., 1982) and dried figs (Altuğ et al., 1990). Additionally, $NaHSO_3$ solutions reduced DON level (85%) in contaminated maize (4.4 mg/ kg) and form a DON-sulfonate conjugate when the treatment was performed at 80°C for 18 h (Young et al., 1987). Also, sodium metabisulfite at 10 g/ kg was reported to be an effective tool for overcoming the depressing effects of DON on feed-intake in piglets (Dänicke et al., 2005). Alternatively, the reaction of FB1 with reducing sugars such as D-glucose, D-fructose at 65°C for 48 h blocked the primary amino group of FB1, and prevented FB1-induced toxicity on cell tissue cultures on rats and swine (Fernandez-Surumay et al., 2005).

Biological methods

An alternative approach to remove the toxic and carcinogenic potential of mycotoxins was the biological detoxification, intended as enzymatic degradation or modifying of toxins that led to less toxic products. Studies in this area were dramatically increased with the recent advances in molecular biology, genetic engineering and microbial genomics, coupled with the discovery of the catabolic capabilities of microbial populations. Detoxification of mycotoxins by microorganisms was reviewed extensively by Bata and La´sztity (1999) and Karlovsky (1999).

Many species of bacteria were reported to degrade mycotoxins. Earlier work by Ciegler et al. (1966) identified *Flavobacterium aurantiacum* NRRL B-184, which could irreversibly remove AFB1 from a variety of food products including milk, oil, peanut butter, peanuts and maize without leaving toxic by-products. On the other hand, the bright orange pigmentation associated with *F. aurantiacum* restricted its use in food and feed fermentations (Line et al., 1994). Apart from *F. aurantiacum*, a variety of lactic acid bacteria originating from fermented products were reported to inhibit mutagenic activity of AFB1 (Park and Rhee, 2001). Earlier work demonstrated that more than 99% of patulin (50 mg/L) removed during alcoholic fermentation of apple juice, while only 10% decrease was observed in the control sample (Stinson et al., 1978). Later, three commercial cider strains of *S. cerevisiae* degraded patulin during active fermentative growth (Moss and Long, 2002).

With respect to other mycotoxins, fermentation by *S. cerevisiae* of wort containing ZEA resulted in conversion of 69% of the toxin to b-zearalenol and 8.1% to a-zearalenol (Scott et al., 1992). Similarly, cultures of *Candida tropicalis*, *Torulaspora delbrucki*, *Zygosaccharomyces rouxii*, and seven *Saccharomyces* strains were able to convert ZEA to a- and b-zearalenol (Boswald et al., 1995). In another study, OTA, FB1 and FB2 at the levels of 0.19, 0.95 and 0.95 mg/ ml respectively were degraded in the range of 87 to 91% by three strains of *S. cerevisiae* during fermentation of worth at 25°C for 8 days (Scott et al., 1995). Additionally, some losses (<40%) of OTA occurred during fermentation (Baxter et al. 2001), while alcoholic fermentation of malt by *S. cerevisiae* resulted in an average of 53% decrease in the initial contamination level of DON and T-2 toxin (Garda et al., 2005).

ACKNOWLEDGEMENTS

Authors want to thank all the scientists that have contributed significantly to existing research attempts that are reviewed in this article and/or who could serve as contacts for more detailed information in their particular areas of research. Also, we are grateful to Dr. Wael Anwar, Professor of Veterinary Hygiene and Management at the Faculty of Veterinary Medicine, Cairo University for his endless support and advice during entire study.

REFERENCES

Abbas HK, Mirocha CJ, Pawlosky RJ, Pusch DJ (1985): Effect of cleaning, milling, and baking on deoxynivalenol in wheat. Appl. Environ. Microbiol., 50: 482-486.

Abbas HK, Zablotowicz RM, Weaver MA, Horn BW, Xie W, Shier WT (2004). Can J. Microbiol., 50: 193-199.

Abd Alla ES (1997): Zearalenone: incidence, toxigenic fungi and chemical decontamination in Egyptian cereals. Nahrung, 41: 362-365.

Acar J, Gokmen V, Taydas EE (1998): The effects of processing technology on the patulin content of juice during commercial apple juice concentrate production. Z Lebensm Unters Forsch A, 207: 328-331.

Al-Attas AS (2008). Determination of Essential Elements in Milk and Urine of Camel and in Nigella sativa Seeds. Arabian J. Chem., 1(2): 123-129.

Alawadi A, AL-Jedabi A (2000). Antimicrobial agents in cemel's Urine (9B) Microbiol. Viruses, 8(11): 265-281.

Altug T, Yousef AE, Marth EH (1990). Degradation of aflatoxin B1 in dried figs by sodium bisulfite with or without heat, ultraviolet energy or hydrogen peroxide. J. Food Prot., 53: 581-582.

Amer HA, Al-Hendi AB (1996). Physical, biochemical and microscopically analysis of camel urine. J. Camel Practice Res., 3(1): 17-21.

Artuk N, Cemeroglu B, Aydar G, Saglam N (1995): Elma suyu konsantresinde aktif ko¨mu¨r kullanm u¨zerinde aras¸trmalar. Tr. J. Agric. For., 19: 259-265.

Avantaggiato G, Havenaar R, Visconti A (2004). Evaluation of the intestinal absorption of deoxynivalenol and nivalenol by an *in vitro* gastrointestinal model, and the binding efficacy of activated carbon and other adsorbent materials. Food Chem. Toxicol., 42: 817-824.

Bata A, La´sztity R (1999): Detoxification of mycotoxin-contaminated food and feed by microorganisms. Trends Food Sci. Technol., 10: 223-228.

Baxter ED, Slaiding IR, Kelly B (2001): Behavior of ochratoxin A in brewing. J. Am. Soc. Brew Chem., 59: 98-100.

Benedetti R, Nazzi F, Locci R, Firrao G (2006): Biodegradation, 17: 31-38.

Bennett JW, Klich M (2003). Mycotoxins. Clin. Microbiol. Rev., 16: 497-516.

Boswald C, Engelhardt G, Vogel H, Walnofer PR (1995) Metabolism of the Fusarium Mycotoxins zearalenone and deoxynivalenol by yeast strains of technological relevance. Nat Toxins, 3: 138-144.

Bullerman LB, Bianchini A (2007): Stability of mycotoxins during food processing. Int. J. Food Microbiol., 119: 140-146.

Burgos-Hernandez A, Price RL, Jorgensen-Kornman K, Lopez-Garcia R, Njapau H, Park DL (2002). Decontamination of aflatoxin B1-contaminated corn by ammonium persulphate during fermentation. J. Sci. Food Agric., 82: 546-552.

Caridi A, Galvano F, Tafuri A, Ritieni A (2006). Ochratoxin A removal during winemaking. Enzyme Microb Technol., 40: 122-126.

Castelo MM, Sumner SS, Bullerman LB (1998). Occurrence of fumonisins in corn-based food products. J. Food Protection, 61: 704-707.

CAST Report (2003). Mycotoxins: risks in plant, animal, and human systems. In: J.L. Richard, G.A. Payne (Eds.), Council for Agricultural Science and Technology Task Force Report No. 139, Ames, Iowa, USA. ISBN 1-887383-22-0

Council for Agricultural Science and Technology (CAST), 1989) Mycotoxins.Economic and health risks. Task Force Rep. No. 116. November 1989. Council for Agric. Sci. Technol., Ames, IA.

Cecchini F, Morassut M, Moruno EG, Di Stefano R (2006): Influence of yeast strain on ochratoxin A content during fermentation of white and red must. Food Microbiol., 23: 411-417.

Cheeke PR (1998). Natural Toxicants in Feeds and Poisonous Plants, 2nd ed Interstate Publishers, Danville, United States, pp. 116-122.

Cheeke PR, LR Shull (1985). Natural Toxicants in Feeds and Poisonous Plants. AVI Publishing Com- pany, Westport, CT, pp. 393-476.

Christensen CM, Sauer DB (1982). Microflora. In: Storage of cereal grains and their products (Christensen CM ed), American Association of Cereal Chemists, St. Paul, pp. 219- 240.

Christensen CM, RA Meronuck (1986). Quality maintenance in stored grains and seeds. University of Minnesota Press, Minneapolis.

Ciegler A, Lillehoj EB, Peterson RE, Hall HH (1966): Microbial detoxification of aflatoxin. Appl. Microbiol., 14: 934–939.

Danicke S, Valenta H, Gareis M, Lucht HW, Reichenbach H (2005). On the effects of a hydrothermal treatment of deoxynivalenol (DON)-contaminated wheat in the presence of sodium metabisulfite (Na2S2O5) on DON reduction and on piglet performance. Animal Feed Sci. Technol., 118: 93–108.

Degirmencioglu N, Esecali H, Cokal Y, Bilgic M (2005). From safety feed to safety food: the application of HACCP in mycotoxin control Arch. Zootech., 8: 19-32.

Devegowda G, MVLN Raju, and HVLN Swamy (1998). Mycotoxins: Novel solutions for their counteraction. Feedstuffs, pp. 12-15.

Diaz DE, Hagler WM Jr, Hopkins BA, Whitlow LW (2002). Aflatoxin binders I: In vitro binding assay for aflatoxin B1 by several potential sequestering agents. Mycopathologia, 156: 223–226.

D'Mello JPF, Porter JK, Macdonald AMC, Placinta CM (1997) Fusarium mycotoxins. In: Handbook of Plant and Fungal Toxicants. Edited by JPF D'Mello, CRC Press, Boca Raton, FL, USA, p. 287.

Doyle MP, Applebaum RS, Brackett RE, Marth EM (1982). Physical, chemical and biological degradation mycotoxins in foods and agricultural commodities. J. Food Prot., 45: 964–971.

Edrington TS, Kubena LF, Harvey RB, Rottinghaus RE (1997). Influence of superactivated charcoal on the toxic effects of aflatoxin or T-2 toxin in growing broilers. Poult. Sci., 76: 1205–1211.

Erber E, Binder EM (2004). Managing the risk of mycotoxins in modern feed production. In: The 5th Korea Feed Ingredient Association International Symposium, Korea Feed Ingredient Association, Seoul, Korea, July 16, pp. 21–45.

Fazekas B, Bajmocy E (1996). Occurrence of equine leukoencephalomalacia caused by fumonisin-B1mycotoxin in Hungary.Magy.Allatorv. Lap., 51: 484-487.

Fernandez MP, Ikonomou, MG, Buchanan I (2007). Sci Total Environ., 373: 250–269

Fernandez-Surumay G, Osweiler GD, Yaeger MJ, Rottinghaus GE, Hendrich S, Buckley LK, and Murphy PA (2005): Fumonisin B-glucose reaction products are less toxic when fed to swine. J. Agric. Food Chem., 53: 4264–4271.

Fink-Gremmels J, Malekinejad H (2007). Biochemical mechanisms and clinical effects associated with exposure to the mycoestrogen zearalenone. In: Morgavi, D.P., Riley, R.T. (Eds.), Fusarium and their toxins: Mycology, occurrence, toxicity, control and economic impact. Anim. Feed Sci. Technol., pp. 283-298.

Food and Drug Administration (1989). Corn shipped in interstate commerce for use in animal feeds; action levels for aflatoxin in animal feeds. Fed. Reg., 54: 100- 22622.

Fouler SG, Triverdi AB, Kitabatake N (1994): Detoxification of citrinin and ochratoxin A by hydrogen peroxide. J AOAC Int., 77: 631–636.

Francis BJ, JF Wood (1982). Changes in The Nutritive Content and Value of Feed Concentrate During Storage. In: M. Rechigl (Ed). Handbook of Nutritive Value of Processed Food, CRC Press, Florida.

Galhardo M, Birgel EH, Soares LMV, Furlani RPZ (1997). Poisoning by diacetoxyscirpenol in cattle fed citrus pulp in the state of Sao Paulo, Brazil. Braz. J. Vet. Res. Anim. Sci., 34: 90-91.

García-Camarillo EA, Y Quezada-Viay, R Mexicana (2006). Actividad Antifúngica de Aceites Esenciales de Canela (Cinnamomum zeylanicum Blume) y Orégano (Origanum vulgare L.) y su Efecto sobre la Producción deAflatoxinas en nuez pecancra {Carya illinoensis(F.A. Wangenh) K.Koch }. Revista Mexicana de Fitopatologia, 24: 8 – 12.

Garda J, Macedo RM, Faria R, Bernd L, Dors GC, Badiale-Furlong E (2005). Alcoholic fermentation effects on malt spiked with trichothecenes. Food Control, 16: 423–428.

Gelderblom WCA, Jaskiewicz K, Marasas WFO, Thiel PG, Horak RM, Vleggaar R, Kriek NPJ (1988). Cancer promoting potential of different strains of Fusarium moniliforme in a short-term cancer initiation/promotion assay. Carcinogenesis (Eynsham), 9: 1405-1409.

Gelderblom WCA, Kriek NPJ, Marasas WFO, Thiel PG (1991). Toxicity and carcinogenicity of the Fusarium moniliforme metabolite, fumonisin Bj, in rats. Carcinogenesis (Eynsham), 12: 1247-1251.

Ghosal AK, Appanna TC, Dwaraknath PK (1974). Seasonal variations in water compartment of the Indian camel. Brit. Vet. J., 130: 132.

Glenn AE (2007). Mycotoxigenic Fusarium species in animal feed. In: Morgavi DP, Riley RT (Eds.), Fusarium and their toxins: Mycology, occurrence, toxicity, control and economic impact. Anim. Feed Sci. Technol., pp. 213-240.

Gromadzka K, Waskiewicz A, Golinski P, Swietlik J(2009). Occurrence of estrogenic mycotoxin - Zearalenone in aqueous environmental samples with various NOM content. Water Res. (Oxford), 43(4): 1051-1059.

Harvey RB, LF Kubena, TD Phillips (1989). Prevention of aflatoxicosis by addition of hydrated sodium, calcium, aluminosilicate to the diets of growing barrows. Am. J. Vet. Res., 50: 416.

Heathcote JG, Hibbert JR (1978). Aflatoxin chemical and biological aspects. Elsevier Scientific, Amsterdam, p. 212.

Hendrich S, Miller KA, Wilson TM, Murphy PA (1993). Toxicity of Fusarium proliferatum -fermented nixtamalized corn-based diets fed to rats: Effect of nutritional status. J. Agric. Food Chem., 41: 1649-1654.

Hoehler D, RR Marquardt (1996). Influence of vitamins E and C on the toxic effects of ochratoxin A and T-2 toxin in chicks. Poult. Sci., 75: 1508-1515.

Hsu IC, Smalley EB, Strong FM, RIbelln WE (1972) Idcntific:lliun ofT•2 toxin in moldy corn associated with a lethal toxicosis in dairy cattle. Appl. Microbiol., 24: 682-690.

Hussein S, Brasel J (2001).Toxicity, metabolism, and impact of mycotoxinx on humans and animals. Toxicol., 167: 101–134.

Huwig A, Freimund S, Kappeli O, Dutler, H. (2001). Mycotoxin detoxification of animals feed by different adsorbents. Toxicol Lett., 122: 179–188.

Ikonomou MG, Cai SS, Fernandez MP, Blair JD, Fischer M (2008): Environ. Toxicol. Chem., 27: 243–251.

Inan F, Pala M, Doymaz I (2007). Use of ozone in detoxification of aflatoxin B1 in red pepper. J. Stored Prod. Res., 43: 425–429.

Jemmali M (1979): Decontamination and detoxification of mycotoxins. Pure Appl. Chem., 52: 175–181.

Jouany JP (2007). Methods for preventing, decontaminating and minimizing the toxicity of mycotoxins in feeds. In: Morgavi, D.P., Riley, R.T. (Eds.), Fusarium and their toxins: Mycology, occurrence, toxicity, control and economic impact. Anim. Feed Sci. Technol., pp. 342-362.

Kabak B, Dobson ADW, Var I (2006): Strategies to prevent mycotoxin contamination of food and animal feed: a review. Crit Rev. Food Sci. Nutr., 46: 593–619.

Kan CA, GAL Meijer (2007). The risk of contamination of food with toxic substances present in animal feed. Animal Feed Sci. Technol., 133: 84–108.

Pierre JJ (2007). Methods for preventing, decontaminating and minimizing the toxicity of mycotoxins in feeds. Animal Feed Science and Technol., 137(3-4): 342-362.

Karlovsky P (1999): Biological detoxification of fungal toxins and its use in plant breeding, feet, and food production. Nat Toxins, 7: 1–23.

Koynarski V, Stoev S, Grozeva N, T Mirtcheva, H Daskalov, J Mitev, P Mantle (2007). Experimental coccidiosis provoked by Eimeria acervulina in chicks simultaneously fed on ochratoxin A contaminated diet. Res. in Veterinary Sci,, 82(2): 225-231.

Kramer R, Keogh RG, Sprosen JM, McDonald MF (1997). Free and conjugated levels of zearalenone in ewes mated on grass-dominant pasture or chicory. Proc. N. Z. Soc. Anim. Prod., pp. 57-90.

Krogh P (1987): Ochratoxins in food. In: Krogh P, editor. Mycotoxins in food. London: Academic Press, pp. 97–121.

Kuiper-Goodman T, Scott PM, Watanabe H (1987): Risk assessment of the mycotoxin zearalenone. Regul. Toxicol. Pharmacol., 7: 253–306.

Lawlor PG, Lynch PB (2005). Mycotoxin management. Afr. Farming Food Process, 46: 12-13.

Leibetseder J (2005). Decontamination and detoxification of mycotoxins. In: Mosenthin R, Zentek J, Zebrowska T (Eds.), Biology of Nutrition in Growing Animals. Biology of Growing Animals Series, 4. Elsevier, Amsterdam.

Lillehoj EB, TE Cleveland, D Bhatnagar (1991). Mycotoxins in feedstuffs: Production in novel substrates. In: J. E. Smith and R. S. Henderson (Ed.) Mycotoxins in Animal Foods, pp. 399-413. CRC

Press, Boca Raton, FL.

Line JE, Brackett RE, Wilkinson RE (1994): Evidence for degradation of aflatoxin B1 by *Flavobacterium aurantiacum*. J. Food Prot., 57: 788–791.

Marasas WFO (1997). Risk assessment of fumonisins produced by *Fusarium moniliforme* in corn. Cereal Res. Commun., 25: 399–406.

Marasas WFO, Kriek NPJ, Fincham JE, van Rensburg SJ (1984). Primary liver cancer and oesophageal basal cell hyperplasia in rats caused by *Fusarium moniliforme*. Int. J. Cancer, 34: 383–387.

Marguardt RR (1996). Effects of molds and their toxins on livestock performance: A western Canadian perspective. Animal Sci. Technol., 58: 77-89.

Mayer S, Engelhart S, Kolk A, Blome H (2008): The significance of mycotoxins in the framework of assessing workplace related risks. Mycotoxin Res., 24(3): 151-164.

McKenzie KS, Sarr AB, Mayura K, Bailey RH, Millar DR, Rogers TD, Corred WP, Voss KA, Plattner RD, Kubena LF, Phillips TD (1997): Oxidative degradation and detoxification of mycotoxins using a novel source of ozone. Food Chem. Toxicol., 35: 820-807.

Morgavi D, Riley RT (2007). An historical overview of field disease outbreaks known or suspected to be caused by consumption of feeds contaminated with Fusarium toxins. In: Morgavi, D.P., Riley, R.T. (Eds.), Fusarium and their toxins: Mycology, occurrence, toxicity, control and economic impact. Anim. Feed Sci. Technol., pp. 201–212.

Moss MO (1991). Influence of agricultural biocides on mycotoxin formation in cereals. Chełkowski, J. Cereal grain: Mycotoxins, fungi and quality in drying and storage, pp. 281-295.

Moss MO (1996). Centenary review. Mycotoxins. Mycol. Res., 100: 513 -523.

Moss MO, Long MT (2002). Fate of patulin in the presence of yeast Saccharomyces cerevisiae. Food Addit Contam., 19: 387–399.

Mroch CJT, Pawlosky RA, Chatterjee K, Watson S, Hayes W (1983). Analysis for Fusarium toxins in various samples implicated in biological warfare in Southeast Asia. J. Ass. off. analytic. Chem., 66: 14-85.

Mura U, Osman AM, Mohamed AS, Di-Martino D, Ipata PL (1987). Purine salvage as metabolite and energy saving mechanism in Camelus dromedarius: the recovery of guanine. Comparative Biochemistry and Physiology Part B: Biochem. Molecular Biol., 87: 157-160.

Newberne PM, Butler WH (1969). Acute and Chronic Effects of Aflatoxin on the Liver of Domestic and Laboratory Animals-a Review. Cancer Res., 29: 236.

Northolt MD (1979). The Effect of Water Activity and Temperature on the Production of Some Mycotoxins. (PhD Dissertation). Bilthoven, Holland, pp. 170-174.

Nuryono N, Noviandi CT, Bohm J, Razzazi-Fazeli E (2005): A limited survey of zearalenone in Indonesian maize-based food and feed by ELISA and high performance liquid chromatography. Food Contr., 16: 65–71.

Park DL (2002): Effect of processing on aflatoxin. Adv Exp Med Biol 504:173–179.

Park DL, Lee LS, Price RL, Pohland AE (1988). Review of the decontamination of aflatoxins by ammoniation: Current status and regulation. J. Assoc. Off. Anal. Chem., 71: 685.

Park DL, Rua SM, Mirocha CJ Jr, Abd-Alla E, Weng CJ (1992): Mutagenic potentials of fumonisin contaminated corn following ammonia decontamination procedure. Mycopathologia, 117: 105-108.

Park HD, Rhee CH (2001). Antimutagenic activity of *Lactobacillus plantarum* KLAB21 isolated from kimchi Korean fermented vegetables. Biotechnol. Lett., 23: 1583–1589.

Peraica M, Domijan AM, Jurjevic Z, Cvjetkovic B (2002). Prevention of exposure to Mycotoxins from food and feed. Arh Hig Rada Toxicol., 53: 229-237.

Pestka JJ (2007). Deoxynivalenol: toxicity, mechanisms and health risks. In: D.P. Morgavi, R.T. Riley (Eds), Fusarium and their toxins: Mycology, occurrence, toxicity, control and economic impact. Anim. Feed Sci. Technol., pp. 283-298.

Pier AC, McLoughlin ME, Richard JL, Baetz AL, Dahlgren RR (1985). In utero transfer of aflatoxin an n d selected effects on neonatal pigs. In: J. Lacey (Ed.) Trichothecenes and Other Mycotoxins. John Wiley and Sons, New York. pp. 495-506.

Pitt JI, Hocking AD (1997). Fungi and Food Spoilage, 2nd ed. Aspen Publishers, Gaithersburg, MD, USA, pp. 209–220.

Piva G, Galvano F, Pietri A, Piva A (1995): Detoxification methods of aflatoxins. A review. Nutr. Res., 15: 767–776.

Placinta CM, D'Mello JPF, Macdonald AMC (1999). A review of worldwide contamination of cereal grains and animal feed with Fusarium mycotoxins. Anim. Feed Sci. Technol., 78: 21-37.

Prathapkumar SH, Rao VS, Paramkishan RJ, Bhat RV (1997). Disease outbreak in laying hens arising from the consumption of fumonisin-contaminated food. Br. Poult. Sci., 38: 475-479.

Proctor AD, Ahmedna M, Kumar JV, Goktepe I (2004): Degradation of aflatoxins in peanut kernels/flour by gaseous ozonation and mild heat treatment. Food Addit Contam., 21: 786–793.

Rabea EI, Badawy ME, Stevens CV, Smagghe G, Steurbaut W (2003). Chitosan as antimicrobial agent:applications and mode of action. Biomacromolecules, 4(6): 1457–1465.

Ratcliff J (2002). The role of mycotoxins in food and feed safety. Presented at AFMA (Animal Feed Manufacturers Association) on 16th August, 2002.

Robens J, Cardwell K (2003). The Costs of Mycotoxin Management to the USA: Management of Aflatoxins in the United States. Toxin Reviews, 22(2-3): 139-152.

Ross FF, Nelson PE, Richard JL, Osweiler GD, Rice LG, Plattner RD, Wilson TM (1990). Production of fumonisins by Fusarium moniliforme and Fusarium proliferation isolates associated with equine leukoencephalomalacia and a pulmonary edema syndrome in swine. Appl. Environ. Microbiol., 56: 3225-226.

Russell L, Cox DF, G Larsen, Bodwell K, CE Nelson (1991). Incidence of molds and mycotoxins in commercial animal feed mills in seven midwestern states. J Anim. Sci., 69: 5-12.

Rustom IYS (1997). Aflatoxin in food and feed: occurrence, legislation and inactivation by physical methods. Food Chem., 59: 57–67.

Samarajeewa U, Sen AC, Cohen MD, Wei CI (1990). Detoxification of aflatoxins in foods and feeds by physical and chemical methods. J Food Prot., 53: 489–501.

Sánchez E, Heredia N, Garcia S (2005). Inhibition of growth and mycotoxin production of *Aspergillus flavus* and *Aspergillus parasiticus* by extracts of Agave species. Int. J. Food Microbiol., 98: 271–279.

Scott PM (1996). Effects of processing and detoxification treatments on ochratoxin A: Introduction. Food Addit. Contam., 13: 19–21.

Scott PM (1998). Industrial and farm detoxification processes for mycotoxins. In: Le Bars J, Galtier P (eds) Mycotox'98 international symposium, 2–4 July, Toulouse, France, pp. 543–548.

Scott PM, Kanhere SR, Dailey EF, Farber JM (1992). Fermentation of wort containing deoxynivalenol and zearalenone. Mycotoxin Res., 8: 58–60.

Scott PM, Kanhere SR, Lawrance GA, Daley EF, Farber JM (1995): Fermentation of wort containing added ochratoxin A and fumonisins B1 and B2. Food Addit. Contam., 12: 31-40.

Scott PM, Lawrence GA (1995). Mycotoxin methodology Food Additives & Contaminants: Part A: Chemistry, Analysis, Control. Exposure & Risk Assessment, 12(3): 395-403.

Smith JE, Moss MO (1985). Mycotoxins, Formation, Analysis. and Signifiance. Chichester, Wiley & Sons, p. 148.

Stinson EE, Osman SF, Huhtanen CN, Bills DD (1978). Disappearance of patulin during alcoholic fermentation of apple juice. Appl. Environ. Microbiol., 36: 620–622.

Stoev SD, Anguelov G, Ivanov I, Pavlov D (2000): Influence of ochratoxin A and extract of artichoke on the vicinal immunity and health in broiler chicks. Exp. Toxicol. Pathol., 52(1) :43-55.

Sydenham EW, Van Der Westhuizen L, Stockenstrom S, Shephard GS, Thiel PG (2004): Fumonisin-contaminated maize: physical treatment for the partial decontamination of bulk shipments. Food Addit Contam., 11: 25–32.

Trenholm HL, Charmley LL, Prelusky DB, Warner RM (1992). Washing procedures using water or sodium carbonate solutions for the decontamination of three cereals contaminated with deoxynivalenol and zearalenone. J. Agric. Food Chem., 40: 2147–2151.

Van der Merwe KJ, Steyn PS, Fourie L, Scott DB, Theron JJ (1965). Ochratoxin A, a toxic metabolite produced by *Aspergillus ochraceus* Wilh. Nat., 205: 1112–1113.

Var I, Kabak B, Erginkaya Z (2008). Reduction in ochratoxin A levels in white wine, following treatment with activated carbon and sodium bentonite. Food Control, 19: 592–598.

Varley H, Gowenlock AH, Bell M (1980). Practical Clinical Biochemistry.5th ed. Vol.1, William Heinemann Medical Books, Ltd., London.Wen K, Seguim P, St-Arnaud M, Jabaji-Hare S (2005). Real-time quantitative RT-PCR of defense- associated gene transcripts of Rhizoctonia solani infected bean seedlings in response to inoculation, pp. 741–742.

Voss KA, Bacon C, Norred W, Chapin R, Chamberlain W, Plattner R, Meredith F (1996).Studies on the reproductive effects of *Fusarium moniliforme* culture material in rats and the biodistribution of [14C] fumonisin B1 in pregnant rats. Nat. Toxins, 4: 24–33.

Voss KA, Smith GW, Haschek WM (2007). Fumonisins: toxicokinetics, mechanism of action and toxicity. In: Morgavi DP, Riley RT (Eds.), Fusarium and their toxins: Mycology, occurrence, toxicity, control andeconomic impact. Anim. Feed Sci. Technol., pp. 299–325.

Wheeler KA, Hurdman BF, Pitt JI (1991): Influence of pH on the growth of some toxigenic species of Aspergillus, Penicillium and Fusarium. Int. J. Food Microbiol. 12: 141-150.

Wood GE (1992). Mycotoxins in foods and feeds in the United States. J. Anim. Sci., 70: 3941–3949.

Wu F (2004). Mycotoxins risk assessment for the purpose of setting International Regulatory Standards. Environ. Sci. Technol., 38(15): 4049–4055.

Wu F (2006). Economic impact of fumonisin and aflatoxin regulations on global corn and peanut markets. In: Barug D, Bhatnager D, van Egmond HP, van der Kamp JW, van Osenbruggen WA, Visconti A (Eds.), The Mycotoxin Factbook. Food & Feed Topics. Wageningen Academic Publishers, The Netherlands, pp. 277–289.

Yoshizawa T (1991). Natural occurrence of mycotoxins in small grain cereals (wheat, barley, rye, oats, sorghum, millet, rice). In: Smith JE, Henderson RS. (Eds.), Mycotoxins and Animal Foods. CRC Press, Boca Raton, FL, pp. 301-324.

Yoshizawa T, Mlltoclla CJ, Behrens JC, Swanson SP (1981). t'oletabolic fate of T-2 toxin in a lactating cow. Fd Cosmet. Toxic., 19: 31.

Young JC, Trenholm HL, Friend DW, Prelusky DB (1987). Detoxification of deoxynivalenol with sodium bisulfite and evaluation of the effects when pure mycotoxin or contaminated corn was treated and given to pigs. J. Agric. Food Chem., 35: 259–226.

Yumbe-Guevara BE, Imoto T, Yoshizawa, T (2003): Effects of heating procedures on deoxynivalenol, nivalenol and zearalenone levels in naturally contaminated barley and wheat. Food Addit. Contam., 20: 1132–1140.

Zinedine A, Soriano JM, Moltó JC, Mañes J (2007). Review on the toxicity, occurrence, metabolism, detoxification, regulations and intake of zearalenone: An oestrogenic mycotoxin. Food Chem. Toxicol., 45(1): 1-18.

19

Paraphenylenediamine induces apoptosis of Murine myeloma cells in a reactive oxygen species dependant mechanism

Zineb Elyoussoufi[1,2], Norddine Habti[2], Said Motaouakkil[2,3] and Rachida Cadi[1]*

[1]Laboratory of Physiology and Molecular Genetics associated with CNRST, Department of Biology, Ain Chock Faculty of Sciences, Hassan II University, Casablanca, Morocco.
[2]Laboratory of Experimental Medicine and Biotechnology, Faculty of Medicine and Pharmacy, University Hassan II, Casablanca, Morocco.
[3]Medical intensive care unit, Ibn Rochd university hospital, Casablanca, Morocco.

Paraphenylenediamine (p-PD) is the main aromatic amine used in the formulation of hair dyes. Epidemiologic studies have suggested that the use of p-PD based hair dyes might be related to increase risk of human malignant tumors including multiple myeloma and hematopoietic cancers. However, the toxicity of p-PD on myeloma cells has not been well elucidated yet. Therefore, the association between the cytotoxicity of p-PD and the reactive oxygen species (ROS) generation on murine myeloma cells P3X63Ag8.653 (P3) was evaluated. Treatment with p-PD decreases cell viability in a dose and time dependent manner. In addition, p-PD markedly enhanced lipid peroxidation. This increase was accompanied with a decrease in both glutathione reductase and superoxide dismutase activities. Furthermore, Pre-treatment of P3 cells with antioxidants, reduced glutathione or manganese II chloride, significantly inhibited p-PD induced cytotoxicity and ROS generation. Based on these results, p-PD might induce apoptosis via the involvement of ROS.

Key words: Murine myeloma cells, p-Phenylenediamine, apoptosis, oxidative stress, ROS, antioxidants.

INTRODUCTION

Paraphenylenediamine (p-PD) a monocyclic arylamine, is frequently used as an ingredient of oxidative hair coloring products (Corbett and Menkart, 1973) and black henna dyes (Kang and Lee, 2006). Currently, p-PD is present in more than 1000 hair dye formulations marketed all over the world (Stanley et al., 2005). Epidemiologic studies demonstrated that workers in the textile dye and rubber industries, hair dye users and barbers incurred a high risk of bladder cancer, non-Hodgkin's lymphoma, multiple myeloma and hematopoietic cancers (Thun et al., 1994; Yu

et al., 1998; Gago-Dominguez et al., 2001; Rauscher et al., 2004). Carcinogens usually cause genomic damage to expose cells which may either undergo apoptosis or proliferation with genomic damage and potentially leading to transformation in cancerous cells (Steller, 1995; Thompson, 1995).

The apoptotic regulatory process involves activation of a cascade of molecular events that lead to cell death. Oxidants including reactive oxygen species (ROS), lipid hydroperoxides and nitrogen monoxide (NO) are believed to be widely involved in oxidative stress leading to the induction of apoptosis (Forrest et al., 1994). It was found that p-PD is able to induce oxidative stress in keratinocytes and other cells (Picardo et al., 1996; Rioux and Castonguay, 2000; Brans et al., 2007; Chen et al.,

---

*Corresponding author. E-mail:rachidacadi@gmail.com.

2010). This prompted us to investigate the impact of p-PD on the ROS pathway in murine myeloma cells P3X63Ag8.653. In this study we investigate more the role of ROS in p-PD induced apoptosis. Reduced glutathione (GSH) and manganese II chloride ($MnCl_2$) were used as antioxidant and Hydrogen peroxide (H_2O_2) was used as an oxidant to confirm the role of ROS in apoptosis.

MATERIALS AND METHODS

Cell culture

The murine myeloma cells P3X63Ag8.653 (P3) (European collection of cell cultures, grande bretagne) was maintained in RPMI 1640 medium (Sigma Aldrich, USA) supplemented with 5 or 10% fetal bovine serum (FBS), 100 UI/ml penicillin, 100 µg/ml streptomycin under standard culture conditions at a temperature of 37°C and an atmosphere of 5% CO_2. Subculture was routinely performed three times a week. For all experiments, P3 cells were seeded in 24 well plates (6.10^5 cells/well) and grown for 36 h which correspond to the exponential phase. At this time, a series of diluted concentrations of p-PD (5, 2.5 or 1.25 µg/ml) were added to the culture medium. Cells were then incubated for 12, 24, 36 or 60 h, respectively. Each concentration was run in three replicates in a three independent manner.

In other experiments, P3 cells were pretreated with 0.014, 1.4 or 14 µM of $MnCl_2$ or with 1 mM of GSH (Sigma Aldrich, USA) and incubated with 2.5 or 5 µg/ml of p-PD for 12, 24 or 36 h, then the cell viability was determined. The results were reported as the tested groups versus the untreated group. Cells treated with 50 µM of H_2O_2 were used as positive control.

Cell viability

The viability of P3 was assessed from the intactness of the plasma membrane as determined by the trypan blue exclusion test.

Morphologic assessment of apoptosis

Cellular morphology was ascertained by light microscopy following May-Grünwald-Giemsa (MGG) staining of cytocentrifuged cells. Cytospins were prepared using a Shandon cytospin 2. At least, 300cells/slide were counted at 1000 x final magnification.

Evaluation of lipid peroxidation

Lipid peroxidation was estimated by thiobarbituric acid reactive substances (TBARS) reaction with malondialdehyde (MDA) as described by Samokyszyn and Marnett (1990). Briefly, 250 µl of supernatant were added to 10% trichloroacetic (TCA) and 0.375% TBA. The MDA level was expressed as µmol/min/mg protein.

Assay of catalase (CAT) activity

The catalase activity was measured according to Aebi, (1984). The specific activity is given as micromoles of consumed H_2O_2/min/mg protein.

Assay of glutathione reductase (GR) activity

To estimate the GR activity, the method of DI HIO et al. (1983) was used. The enzyme activity was calculated as micromoles of Nicotinamide Adenine Dinucleotide Phosphate (NADPH) oxidized /min/mg protein.

Assay of superoxide dismutase (SOD) activity

The SOD activity was measured according to Paoletti et al. (1986). The enzyme activity was expressed as µmol/min/mg protein.

Determination of protein concentrations:

Protein concentrations were determined by the Bradford (1976) method using bovine serum albumin as a standard.

Statistical analysis

Results were expressed as the mean ± SEM of three to six independent experiments. The statistical evaluation of the data was carried out by applying the student's t-test (paired). Significant difference was taken as $p < 0.05$.

RESULTS

Effects of p-PD on P_3 cells viability

P3 cells, in their exponential phase, were cultured with increasing dose of p-PD (1.25, 2.5 and 5 µg/ml) for 12, 36 and 60 h. Cell viability was evaluated by trypan blue dye exclusion method. p-PD treatment reduced cell viability in a dose dependent manner. After 36 h of treatment, doses of 1.25, 2.5 and 5 µg/ml reduced cell viability to 31, 40 and 51%, respectively (Figure 1). Our results showed also that p-PD reduced cell viability in a time dependent manner. In addition, p-PD inhibited the growth of P3 cells effectively (p=0.001) with a DL_{50} value of 5µg/ml at 36 h (Figure 1). Thus 5 µg/ml of p-PD was selected for some experiments.

p-PD effected cell morphology of P3 cells

To determine whether p-PD induces apoptosis of P3 cells, cells were incubated with 5 µg/ml of p-PD for 12, 24 and 36 h and their morphologies were examined using MGG staining. As shown in Figure 2a, untreated P3 cells retained their normal size and shape. Morphological alterations were appreciated after incubation for 12 h with 5 µg/ml of p-PD. At this time, toxic effect of p-PD was

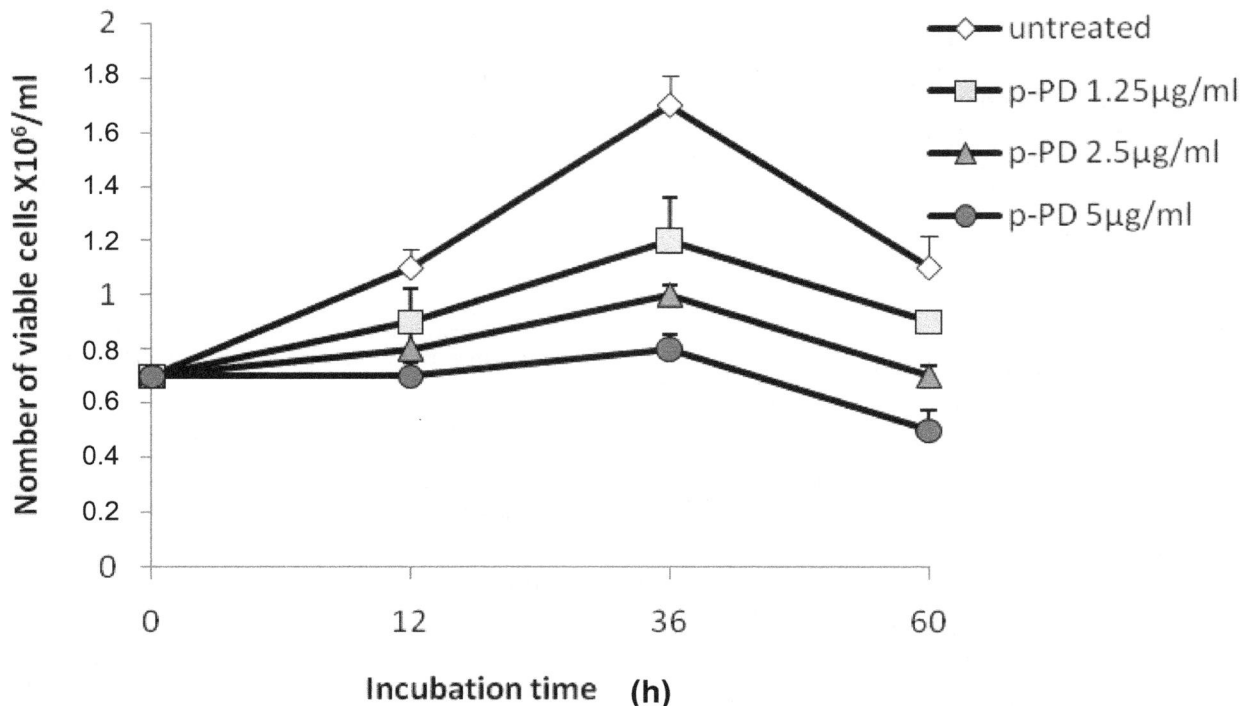

Figure 1. Time course of the viability of P3 cells after treatment with p-PD. Cells in exponential phase were treated with 1.25, 2.5 or 5 µg/ml of p-PD for 12, 36 and 60 h. Results are shown as mean ± SEM from three independent experiments (p<0.05).

characterized by loss of the typical morphology and cell enlargement and by nuclear condensation (Figure 2b).

Effects of p-PD in lipid peroxidation and ROS scavenging enzymes

Since p-PD has been shown to trigger production of ROS, we next assayed for p-PD induced production of ROS in P3 cell. Cells were treated with 1.25, 2.5 and 5 µg/ml p-PD for 12 and 36 h, MDA level and activities of the main antioxidant enzymes (SOD, GR and CAT) were determined (Figures 3 and 4, respectively).

p-PD clearly increase the MDA level dose dependently between 12 and 36 h (p<0.05), reaching 27, 34 and 41 µmol/min/mg protein after treatment for 36 h with 1.25, 2.5 and 5 µg/ml of p-PD, respectively, compared to untreated cells (20 µmol/min/mg protein) (Figure 3).

p-PD significantly reduces SOD activity (p<0.05). This reduction was important after treatment with 5µg/ml for 36 h (p = 0.002) (Figure 4a). Although p-PD induced a decrease in GR activity, a minimal activity of this enzyme was observed when cells were treated with 1.25 µg/ml for 36 h (p = 0.006) (Figure 4b). However, at the same

time, treatment with 5µg/ml increased this activity (p<0.05). The CAT activity increased only when cells were treated with 1.25 and 2.5 µg/ml for 12 h (p=0.006 and p<0.001, respectively). While treatment with 5 µg/ml clearly decreas this activity (p = 0.02). After 36 h no significant effect was observed (Figure 4c).

The antioxidative effect of MnCl₂ on p-PD treated P3 cells viability

To elucidate more on the involvement of oxidative stress in the mechanism of cell death, the effect of MnCl₂ on the p-PD induced apoptosis was examined. P3 cells were pretreated with three concentrations (0.014, 1.14 and 14 µM) of manganese II chloride (MnCl₂) 5 min before treatment with 2.5 or 5 µg/ml of p-PD. Cells then were incubated for 12, 24 and 36 h. H_2O_2 50 µM served as positive control. The three doses of MnCl₂ used in this experiment increased significantly the viability of p-PD treated cells only after 12 h of treatment (p<0.05). They did not affect the H_2O_2 treated cells (Figure 5). After 24 and 36 h of treatment, MnCl₂ had no effect on cell viability (data not shown).

We examined whether MnCl₂ would be more protective

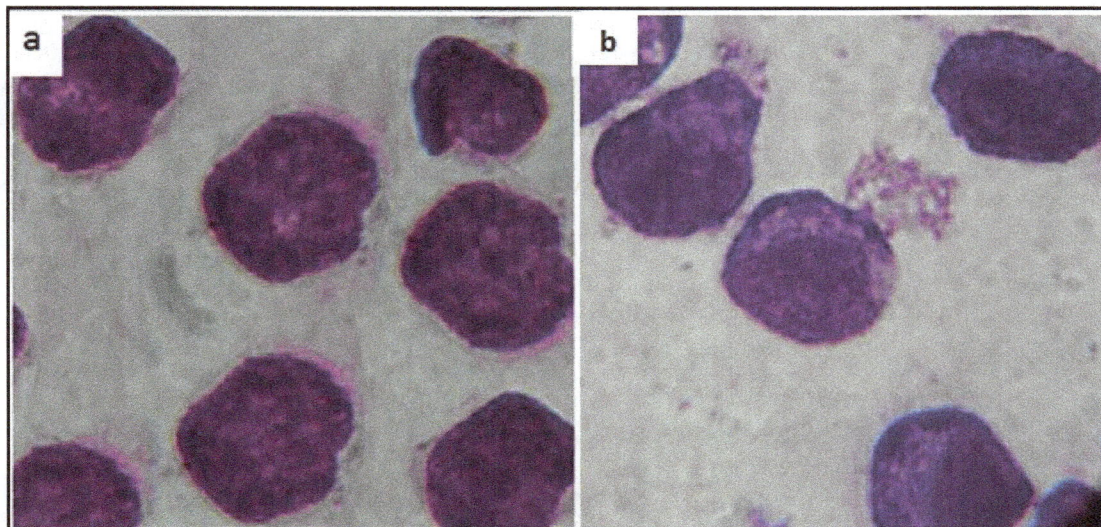

Figure 2. Light microscopy of cultured P3 cells. Cells were incubated for 12 h without (2a) or with 5 µg/ml of p-PD (2b) then stained with MGG and apoptotic cells were evaluated morphologically. At least, 300 cells/slide were counted at 1000 x final magnification.

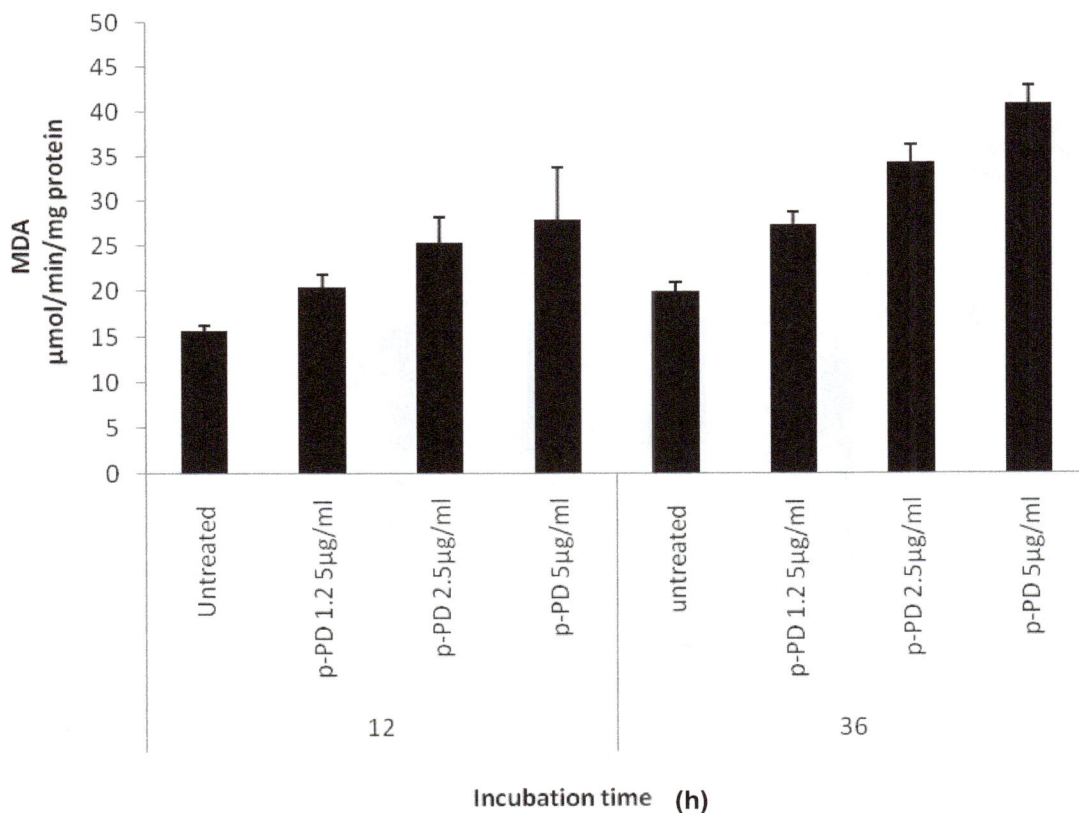

Figure 3. MDA level in P3 cells. Cells were treated with 1.25, 2.5 or 5 µg/ml p-PD for 12 and 36 h. Results are shown as mean ± SEM from three independent experiments ($p < 0.05$).

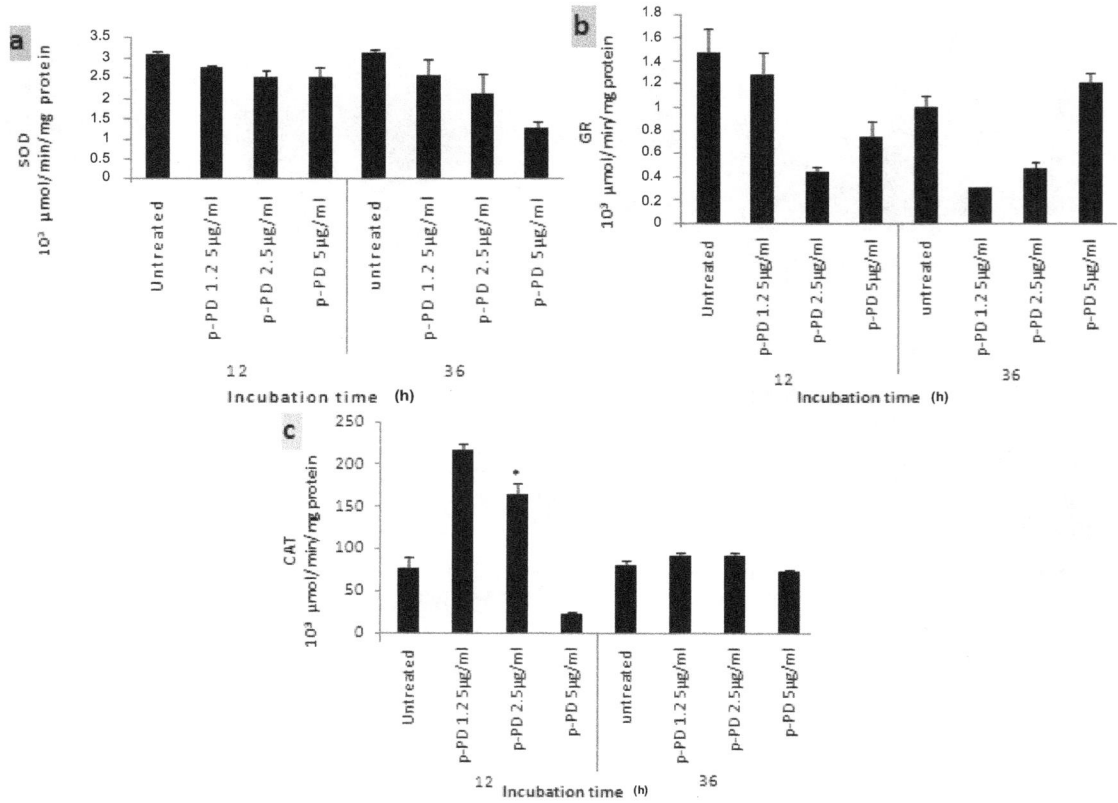

Figure 4. Antioxidant enzymes activities in P3 cells. a) SOD activity. b) GR activity and c) CAT activity. Cells were treated with 1.25, 2.5 or 5 μg/ml p-PD for 12 and 36 h. Results are shown as mean ± SEM from three independent experiments (p<0.05). *p<0.001.

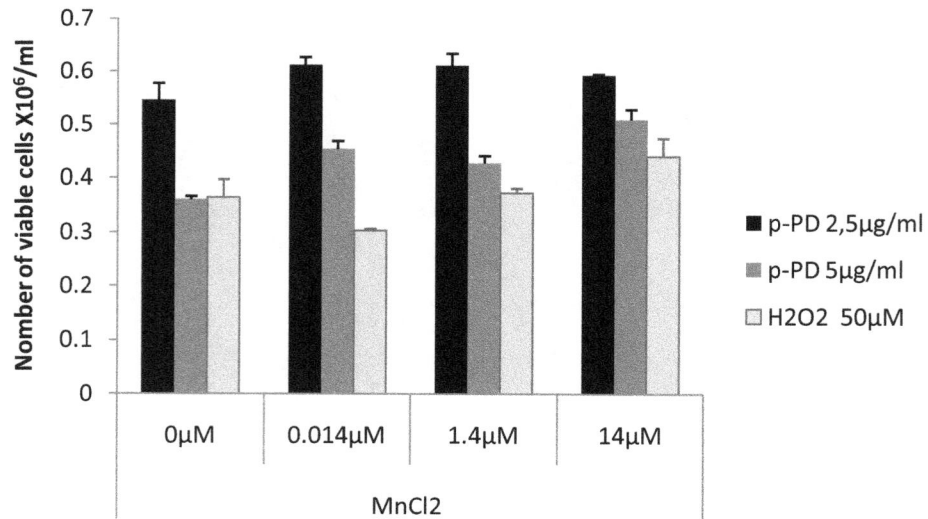

Figure 5. Effects of $MnCl_2$ on P3 cells viability. Cells were pretreated with 0.014, 1.14 or 14 μM $MnCl_2$ 5 min before treatment with 2.5 or 5 μg/ml p-PD. Cells then were incubated for 12 h. H_2O_2 50 μM served as positive control. Results are shown as mean ± SEM from three independent experiments (p<0.05).

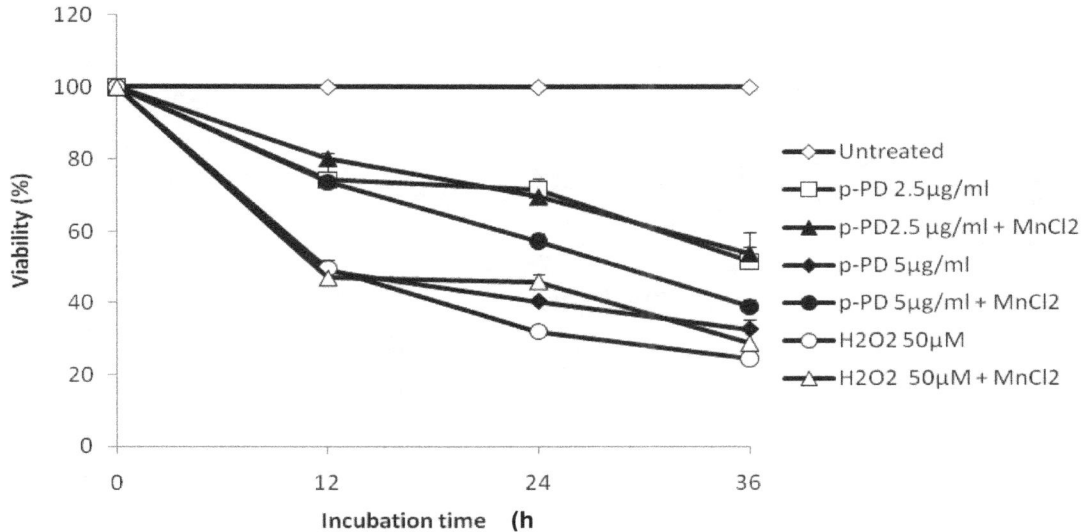

Figure 6. Effects of pretreatment with $MnCl_2$ on the viability of P3 cells. Cells were pre-incubated with 0.014μM $MnCl_2$ for 8h then the same amount was added 5 min before treatment with 2.5 and 5 μg/ml p-PD then incubated for 12, 24 and 36 h. H_2O_2 50 μM served as positive control. Results are shown as mean ± SEM from three independent experiments (p<0.05).

if added hours before treatment with p-PD. Cells were pre-incubated with 0.014μM $MnCl_2$ for 8 h then the same amount was added 5 min before treatment with 2.5 or 5 μg/ml p-PD and incubated for 12, 24 and 36 h. As shown in Figure 6, the $MnCl_2$ added was significantly able to prolong the viability of cells treated with 5 μg/ml of p-PD for 12 to 36 h (p = 0.05, p = 0.002 and p=0.026, respectively). However, $MnCl_2$ had few protective effects on cells treated with 2.5 μg/ml p-PD or 50 μM H_2O_2.

Effects of GSH on p-PD treated P3 cells viability

Reduced glutathione (GSH) and related thiols play important roles in the survival of various cells. In order to investigate the antioxidative effect of GSH on p-PD induced apoptosis we pre-incubated P3 cells for 5 min with 1 mM GSH followed by p-PD treatment (2.5 or 5 μg/ml p-PD). Cells then were incubated for 12, 24 and 36 h. As shown in Figure 7, GSH increased cell viability of treated cells. This increase was significant for cells treated with 5 μg/ml p-PD (p=0.001).

Effects of antioxidants on MDA level of p-PD treated cells

To confirm the involvement of oxidative stress on p-PD induced apoptosis, P3 cells were pretreated for 5 min

with 0.014 μM $MnCl_2$ or 1 mM GSH and treated with 2.5 or 5 μg/ml p-PD then incubated for 12 and 36 h. MDA level was determined. Treatment with $MnCl_2$ or GSH showed a clear protective effect against lipid peroxidation (p= 0.001 and p<0.001, respectively) (Figure 8a and b).

DISCUSSION

Mutagenic/carcinogenic compounds can induce apoptosis in cells. p-PD is a suspected carcinogen that can induce apoptosis (Sontag, 1981; Chen et al., 2006; Chen et al., 2010).

In the present work, p-PD induces cytotoxicity in a dose and time dependent manner and inhibited the growth of P3 cells effectively with DL_{50} value of 5μg/ml at 36 h. A similar effect was observed with SV-40 cells (Huang et al., 2007), human dendritic cells (Coulter et al., 2006) and previously with our team with P3 cells (Benzakour et al., 2011).

After observation of stained P3 cells microscopically, we found that p-PD induces P3 cells apoptosis. This finding was confirmed by DNA fragmentation (data no shown). We have found a similar effect on human neutrophils (Elyoussoufi et al., in press). Studies by Chen et al. (2006) demonstrated that p-PD induced apoptosis of MDCK cells in a dose and time dependent manner.

Also, Huang et al. (2007) revealed that p-PD was able to induce DNA damage dose dependently.

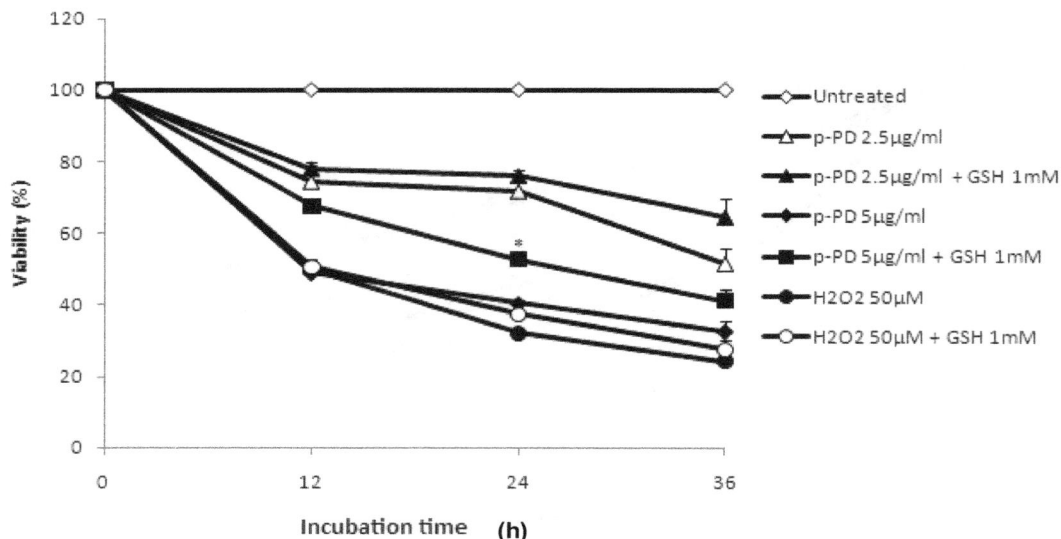

Figure 7. Effects of GSH on the viability of P3 cells. Cells were pre-incubated with 1 mM GSH 5 min before treatment with 2.5 and 5 μg/ml p-PD then incubated for 12, 24 and 36 h. H_2O_2 50 μM served as positive control. Results are shown as mean ± SEM from three independent experiments (p<0.05). *p<0.01.

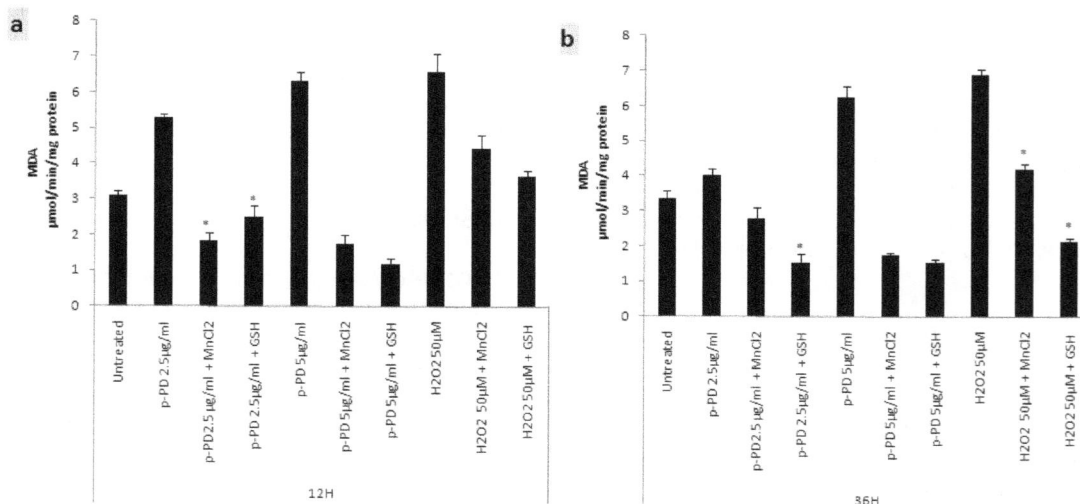

Figure 8. Effect of antioxidants on lipid peroxidation in P3 cells. Cells were pretreated for 5 min with 0.014 μM $MnCl_2$ or 1 mM GSH and treated with 2.5 and 5 μg/ml p-PD then incubated for 12 (a) and 36 h (b). H_2O_2 50 μM served as positive control. Results are shown as mean ± SEM from three independent experiments (p<0.05). *p<0.01.

Oxidation reactions by ROS are regarded as a trigger of the oxidative stress. Several enzymes like SOD, GR and CAT serve as protective antioxidants against oxidative stress (Kuwabara et al., 2008). In this work we detected an enhanced activity of CAT when cells were treated with 1.25 or 2.5 μg/ml p-PD, which shows that defense mechanism is increased to counter the damaging effect of p-PD. However, we noted that with a concentration of 5 μg/ml, p-PD induced a decrease in CAT activity. To explain this decrease two hypotheses can be proposed, either it is only the action of p-PD on the mitochondrial respiratory chain which generates by itself an overproduction of ROS, leading to the inactivation of CAT; or p-PD is able to interact directly with CAT, decreasing its efficiency in detoxifying peroxides and H_2O_2. Our results show also that p-PD reduced SOD and GR activities. It is revealed that activities of ROS scavenging enzymes such SOD and GR are inactivated by ROS (Sampson et al., 2001). According to Chen et al. (2010), p-PD was found to generate ROS in a time dependent manner. However, we found that with a high concentration, p-PD provoked an increase in GR activity. This funding can be explained by the ability of this enzyme to regenerate GSH from GSSG. Moreover, the level of reducing agents, such as GSH, is important in regulating the oxidative intracellular state. This result correlate with those found with human neutrophils (Elyoussoufi et al., in press).

ROS are considered to cause damage to cells by oxidizing lipids in the cell membrane or by attacking DNA directly (Inoue et al., 1994). In the present study, lipid peroxidation was found to be increased by p-PD. It is widely accepted that lipid peroxidation is increased by ROS (Sander et al., 2003). Our finding that p-PD increased lipid peroxidation support the concept that ROS are involved in p-PD induced apoptosis. This finding is in concert with those of Mathur et al. (2005) and Chen et al. (2006) which demonstrate a role of ROS in p-PD induced apoptosis in keratinocytes and in SV-40 cells. It is well known that ROS production serves as an early signal to mediate programmed neuronal death (Greenlund et al., 1995). Furthermore, it has been demonstrated that coordinate regulation of ROS, caspases and p53 facilitates apoptosis induced by p-PD suggesting that production of ROS may be essential for apoptosis (Huang et al., 2007; Chen et al., 2010). The involvement of ROS in p-PD induced apoptosis in P3 cells was validating using $MnCl_2$ and GSH which inhibited ROS generation.

GSH is one of the most important chemical antioxidant agent in mammalian cells (Diaz Vivancos et al., 2010; Jones and Go, 2010). In our work, treatment with GSH enhanced cell viability of p-PD treated P3 cells. In addition, Ozawa et al., 2006 have revealed that addition of GSH to cell culture medium lowered the production of apoptotic cells in blastocysts. Furthermore, many studies have established the importance of GSH redox in cell apoptosis in variety of cell types (Pias et al., 2003; Ekshyyan and Aw, 2005; Circu et al., 2008).

We next hypothesized that manganese has some effect on p-PD treated P3 cells. $MnCl_2$ was then used to confirm the hypothesis; and as expected, $MnCl_2$ showed the inhibitable effect on p-PD induced apoptosis.

Furthermore, Oishi and Machida (1997) have found that $MnCl_2$ inhibited neutrophil apoptosis. The mechanism of Mn^{2+} actions, however, is not clear from these experiments. One possibility is that manganese itself effectively scavenges ROS not only in culture medium, but also in the intracellular space such as nucleus, since Mn^{2+} has SOD-like activity (Archibald and Fridovich, 1982). In fact, ROS scavenging abilities of manganese have been indicated in the brain (Donaldson, 1987) and lactobacilli's spaces (Archibald and Fridovich, 1981).

Whether p-PD itself or oxidative degradation products of p-PD are responsible for the observed results needs further investigations. Coulter et al. (2006) showed inhibition of Bandrowski's base formation in the presence of glutathione in dendritic cells and further the non-enzymatic oxidative degradation of p-PD to different unstable intermediates was prevented by antioxidants like ascorbic acid (Moeller and al., 2008).

In conclusion, our findings show that p-PD induces apoptosis of P3 cells via generation of ROS. The induced apoptosis in P3 cells may be mediated by p-PD, a reactive p-PD intermediate during chemical conversion or by the related ROS itself. To confirm this hypothesis further work is needed to elucidate the cellular mechanism of p-PD induced apoptosis.

ACKNOWLEDGEMENTS

Authors gratefully would like to thank Hassan II University and CNRST for the financial support.

REFERENCES

Aebi H (1984). Catalase in vivo. Methods Enzymol., 105: 121-126.

Archibald FS, Frodovich I (1981). Manganese, superoxide dismutase and oxygen tolerance in some lactic acid bacteria. J Bacteriol., 146: 928-936.

Archibald FS, Frodovich I (1982). The scavenging of superoxide radical by manganous coplexes: in vitro. Arch Biochem Biophys., 214: 452-463.

Benzakour G, Habti N, Oudghiri M, Naya A, Elmaataoui O, Farouqui B, Benchemsi N, Motaouakkil S (2011). p-Phenylenediamine induces apoptosis in murine myeloma cells. Ann Toxicol Anal, 23(1): 41-46.

Bradford M (1976). A rapid and sensitive method for the quantitation of microgram quantities of protein utilizing the principle of protein dye binding. Anal Biochem., 72: 248-254.

Brans R, Dickel H, Bruckner T, Coenraads PJ, Heesen M, Merk HF, Blomeke B (2007). MnSOD polymorphisms in sensitized patients with delayed-type hypersensitivity reactions to the chemical allergen para-phenylene diamine: a case–control study. Toxicology, 212: 148-154.

Chen SC, Chen CH, Chern CL, Hsu LS, Huang YC, Chung KT, Chye SM (2006). p-Phenylenediamine induces p53-mediated apoptosis in Mardin Darby canine Kidney cells. Toxicol. in Vitro, 20: 801-207.

Chen SC, Chen CH, Tioh YL, Zhong PY, Lin YS, Chye SM (2010). Para-phenylenediamine induced DNA damage and apoptosis through oxidative stress and enhanced caspase-8 and -9 activities in Mardin–Darby canine kidney cells. Toxicol in Vitro., 24: 1197-1202.

Circu ML, Rodriguez C, Maloney R, Moyer MP, Aw TY (2008).

Contribution of mitochondrial GSH transport to matrix GSH status and colonic epithelial cell apoptosis. Free Radic. Biol. Med., 44: 768-778.

Corbett JF, Menkart J (1973). Hair colouring. Cutis, 12: 190-193.

Coulter EM, Farrell J, Mathews KL, Maggs JL, Pease CK, Lockley DJ, David A, Basketter B, Park K, Naisbitt DJ (2007). Activation of human dendritic cells by p-phenylenediamine. J Pharmacol Exp Ther., 320: 885–892.

Diaz Vivancos P, Wolff T, Markovic J, Pallardo FV, Foyer CH (2010). A nuclear glutathione cycle within the cell cycle. Biochem J., 431: 169-178.

Di Hio C, Polidoro G, Arduini A, Muccini A, Federici G (1983). Glutathion peroxidase, glutathione reductase, glutathione S-transferase and gamma-glutamyl transpeptidase activities in the human early pregnancy placenta. Biochem Med., 29: 143-148.

Donaldson J (1987). The physiopathologic significance of manganese in brain: its relation to schizophrenia and neurodegenerative disorders. Neurtoxicology, 8: 451-462.

Ekshyyan O, Aw TY (2005). Decreased susceptibility of differentiated PC12 cells to oxidative challenge: relationship to cellular redox and expression of apoptotic protease activator factor-1. Cell Death Differ., 12: 1066-1077.

Elyoussoufi Z, Habti N, Mounaji K, Motaouakkil S, Cadi R. Toxicity of paraphenylendiamine in human neutrophils: apoptosis and oxidative stress. MJB. in press.

Forrest VJ, Kang YH, McClam DE, Robinson DH, Ramaknshnan N (1994). Oxidative stress induced apoptosis prevented by Trolox. Free Radic Biol. Med., 16: 675-684.

Gago-Dominguez M, Castelao JE, Yuan JM, Yu MC, Ross RK (2001). Use of permanent hair dyes and bladder-cancer risk. Int. J. Cancer, 91: 575–579.

Greenlund LJ, Deckwerth TL, Johnson EM Jr (1995). Superoxide dismutase delays neuronal apoptosis role for reactive oxygen species in programmed neuronal death. Neuron., 14: 303-315.

Huang YC, Hung WC, Kang WY, Chen WT, Chai CY (2007). p-Phenylenediamine induced DNA damage in SV-40 immortalized human uroepithelial cells and expression of mutant p53 and COX-2 proteins. Toxicol. Letter, 170: 116-123.

Inoue M, Suzuki R, Koide T, Sakaguchi N, Ogiraha Y, Yabu Y (1994). Antioxidant, gallic acid induces apoptosis in HL-60RG cells. Biochem. Biophys. Res. Commun., 204: 898-904.

Jones DP, Go YM (2010). Redox compartmentalization and cellular stress. Diabetes Obes. Metab., 12(Suppl 2): 116-125.

Kang IJ, Lee MH (2006). Quantification of para-phenylenediamine and heavy metals in henna dye. Contact Dermatitis, 55: 26-29.

Kuwabara M, Asanuma T, Niwa K, Inanami O (2008). Regulation of Cell Survival and Death Signals Induced by Oxidative Stress. J. Clin. Biochem. Nutr., 43(2): 51–57.

Mathur AK, Raizada RB, Srivastava MK, Singh A (2005). Effect of Dermal Exposure to Paraphenylenediamine and Linear Alkylbenzene Sulphonate in Guinea Pigs. Biomed. Environ. Sci., 18: 238-240.

Oishi K, Machida K (1997). Inhibition of neutrophil apoptosis by antioxidants in culture medium. Scand. J. Immunol., 45: 21-27.

Ozawa M, Nagal T, Fahrudin M, Karja NWK, Kaneko H, Noguchi J, Ohnuma K, Kikuchi K (2006). Addition of glutathione or Thioredoxin to culture medium reduces intracellular redox status of porcine IVM/IVF embryos, resulting in improved development to the blastocyst stage. Mole. Reprod. Develop., 73: 998-1007.

Paoletti F, Aldinucci D, Mocali A Carparrini A (1986). A sensitive spectrophotometric method for the determination of superoxide dismutase in tissue extracts. Anal Biochem., 154: 526-541.

Pias EK, Ekshyyan OY, Rhoads CA, Fuseler J, Harrison L, Aw TY (2003). Differential effects of superoxide dismutase isoform expression on hydroperoxide induced apoptosis in PC-12 cells. J. Biol. Chem., 278: 13294-13301.

Picardo M, Zompetta C, Grandinetti M, Ameglio F, Santucci B, Faggioni A, Passi S (1996). Paraphenylene diamine, a contact allergen, induces oxidative stress in normal human keratinocytes in culture. British J. Dermatol., 134(4): 681-685.

Rauscher GH, Shore D, Sandler DP (2004). Hair dye use and risk of adult acute leukemia. Am. J. Epidemiol., 160: 19–25.

Rioux N, Castonguay A (2000). The induction of cyclooxygenase-1 by a tobacco carcinogen in U937 human macrophages is correlated to the activation of NFkappaB. Carcinogenesis, 21: 1745-1751.

Moeller R, Lichter J, Blomeke B (2008). Impact of para-phenylenediamine on cyclooxygenases expression and prostaglandin formation in human immortalized keratinocytes (HaCaT). Toxicology, 249: 167-175.

Samokyszyn VM, Marnett LJ (1990). Inhibition of liver microsomal lipid peroxidation by 13-cis retinoic acid. Free Radic Biol. Med., 8: 491-496.

Sampson JB, Beckman JS (2001). Hydrogen peroxide damages the zinc-binding site of zinc-deficient Cu, Zn superoxide dismutase. Arch. Biochem. Biophys., 392(1): 8-13.

Sander CS, Hamm F, Elsner P, Thiele JJ (2003). Oxidative stress in malignant melanoma and non-melanoma skin cancer. Br. J. Dermatol., 148(5): 913-922.

Sontag JM (1981). Carcinogenicity of substituted-benzenediamines (phenylenediamines) in rats and mice. J. National Cancer Institute, 66: 591–602.

Stanley LA, Skare JA, Doyle E, Powrie R, D'Angelo D, Elcombe CR (2005). Lack of evidence for metabolism of p-phenylenediamine by human hepatic cytochrome P450 enzymes. Toxicology, 210: 147-157.

Steller H (1995). Mechanisms and genes of cellular suicide. Science, 267: 1445-1449.

Thompson CB (1995). Apoptosis in the pathogenesis and treatment of disease. Science, 267: 1456-1462.

Thun MJ, Altekruse SF, Namboodiri MM, Calle EE, Myers DG, Heath Jr CW (1994). Hair dye use and risk of fatal cancers in US women. J. Nat. Cancer Inst., 86: 210-215.

Yu MC, Ross RK, Brady LW (1998). Carcinoma of the Bladder: Innovations in Management. Springer-Verlag, Berlin, pp. 1-13.

20

Occurrence and health implications of high concentrations of Cadmium and Arsenic in drinking water sources in selected towns of Ogun State, South West, Nigeria

Abolanle Azeez A. Kayode[1]*, Joshua Olajiire Babayemi[2], Esther Omugha Abam[1] and Omowumi Titilola Kayode[1]

[1]Department of Chemical Sciences, Biochemistry Unit, Bells University of Technology, Ota, Ogun State, Nigeria.
[2]Department of Chemical Sciences, Industrial Chemistry Unit, Bells University of Technology, Ota, Ogun State, Nigeria.

In this study, we report the quantification of the concentrations of Cd, Pb and As in borehole, well, stream and rain water sources in Ota and some major towns in Ogun State, Nigeria. The pH ranges from 4.8 to 6.98, 5.40 to 7.86 and 6.99 to 8.20 in boreholes, well and rain respectively. Lead was not detected at all in the drinking water sources in all the locations investigated. However, the concentrations of Cadmium and Arsenic were observed to be higher than the maximum allowable limits (MAL) in drinking water by the WHO and the Nigerian Standard for drinking water quality in some of the drinking water sources. The Cadmium levels in boreholes for Ota, Agbara, Ifo, Abeokuta and the Male hostel of BellsTech are greater than the maximum allowable limits while the metal was not detected in the boreholes of CU student hostel and the Female hostel of BellsTech. Similarly, the Arsenic levels in boreholes for Ota, Abeokuta, Agbara and Ifo are greater than the maximum allowable limits whereas the metal was not detected in the boreholes of Male and Female hostels of BellsTech and Covenant University (CU) student hostel. The health implication is that if nothing is done to remove these metals before drinking from these water sources in which the concentrations are significantly higher than the MAL, the consumers of such drinking water are at risk of the health hazards that could be caused by these metals.

Key words: Health, Cadmium, Arsenic, drinking water, Ota, maximum allowable limits.

INTRODUCTION

Water is a basic and indeed an absolute requirement for the survival of human race. The quality of drinking-water is a powerful environmental determinant of health (WHO, 2010) and an adequate supply of good quality safe water is essential for the promotion of public health. Generally, in less developed parts of the world and particularly in the tropic areas, the health hazards caused by polluted water supplies are more numerous than those in the temperate and more developed areas of the world. Water for domestic use should be clear, colorless, odorless, pleasant to drink and reasonably cool and free from impurities harmful to health. It is very well known that human health and survival depend on uncontaminated and clean water for drinking and other domestic uses (Memon, 2002). There is no substitute for water in many of its uses unlike many other raw materials (Sylvester, 2003).

Ground water is the most important source of the domestic, industrial and agricultural water supply in the world (Adeyeye and Abulude, 2004). Although, it is easily accessible from lakes, rivers, streams and springs,

---

*Corresponding author. E-mail: wisdomismine2010@yahoo.co.uk.

Abbreviations: Pb, Lead; Cd, Cadmium; As, arsenic; MAL, maximum allowable limit; BellsTech, Bells University of Technology; CU, Covenant University.

borehole water is of better quality. Rock weathering, atmospheric precipitation, evaporation and crystallization control the chemistry of water. The influence of geology on chemical water quality is widely recognized (Gibbs, 1970, Lester and Birkett, 1999). Developing countries are witnessing changes in ground water which constitute another source of portable water.

The influence of soils in water quality is very complex and can be ascribed to the processes controlling the exchange of chemicals between the soil and water (Hesterberg, 1998). The water chemistry of the ground water will mainly consist of inorganic chemicals and suspended solids as a result of urban run-off (McGregor et al., 2000).

The concern for water resources containing contaminants, such as heavy metals and toxic metalloids, that pose a threat to health, has increased worldwide (Anazawa, 2004). The presence of metals in water results from two independent factors. The first involving the weathering of soils and rocks (Bozkurtoglu et al. 2006; White et al., 2005; Donahue et al., 1983) with its products being transported by air (Moreno et al., 2006; Rubio et al., 2006) and water (Das and Krishnaswami, 2007), and the second involving a variety of anthropogenic activities that have created a societal health risk in rivers that receive a substantial amount of waste from such activities (Espino et al., 2007; Rubio et al., 2004; Rubio et al., 2005).

Arsenic is found widely in the earth's crust in oxidation states of -3, 0, +3 and +5, often as sulfides or metal arsenides or arsenates. In water, it is mostly present as arsenate (+5), but in anaerobic conditions, it is likely to be present as arsenite (+3). It is usually present in natural waters at concentrations of less than 1 to 2 mg/L. However, in waters, particularly ground waters, where there are sulfide mineral deposits and sedimentary deposits deriving from volcanic rocks, the concentrations can be significantly elevated. Apart from occupational exposure, the most important routes of exposure are through food and drinking-water, including beverages that are made from drinking-water. Where the concentration of arsenic in drinking-water is 10 mg/L or greater, this will be the dominant source of intake. In circumstances where soups or similar dishes are a staple part of the diet, the drinking-water contribution through preparation of food will be even greater. Levels in natural waters generally range between 1 and 2 mg/L, although concentrations may be elevated (up to 12 mg/L) in areas containing natural sources (WHO, 2010).

Cadmium is released to the environment in wastewater, and diffuse pollution is caused by contamination from fertilizers and local air pollution. Contamination in drinking-water may also be caused by impurities in the zinc of galvanized pipes and solders and some metal fittings. Food is the main source of daily exposure to cadmium. The daily oral intake is 10 to 35 mg. Smoking is a significant additional source of cadmium exposure (WHO, 2010).

Owing to the decreasing use of lead containing additives in petrol and of lead-containing solder in the food processing industry, concentrations in air and food are declining, and intake from drinking-water constitutes a greater proportion of total intake. Lead is rarely present in tap water as a result of its dissolution from natural sources; rather, its presence is primarily from household plumbing systems containing lead in pipes, solder, fittings or the service connections to homes. The amount of lead dissolved from the plumbing system depends on several factors, including pH, temperature, water hardness and standing time of the water, with soft, acidic water being the most plumbosolvent. Concentrations in drinking-water are generally below 5 mg/L, although much higher concentrations (above 100 mg/L) have been measured where lead fittings are present (WHO, 2010).

Atomic Absorption Spectrophotometry (AAS) is an analytical technique in which the absorption of light of free atoms is measured. In AAS, light of a wavelength characteristic of the element of interest is shone through the atomic vapor. Some of this light is then absorbed by the atoms of that element. The amount of light that is absorbed by these atoms is then measured and used to determine the concentration of that element in the sample (Boss and Fredeen, 1997).

MATERIALS AND METHODS

Sample collection and location

Water samples were randomly collected in Six different areas in Ogun State. These areas include; Abeokuta, Ewekoro, Ifo, Sango-Ota, Ota and Agbara. The samples were collected between the month of July and September, 2010. The drinking water samples were collected in prewashed polyethylene bottles. The samples were obtained directly from the sources and each sample bottle and its cap rinsed three or four times before collection. The composition of the water collected were representative of the water sources. The samples were labeled immediately and transported to the Laboratory (Central Research Laboratory, Bells University of Technology, Ota, Ogun State, Nigeria) where the pH was measured. The electrode of the pH meter with a temperature sensor was rinsed with distilled water and lowered into the beaker containing the water sample. The pH meter was allowed to stabilize and the pH of the sample read. The analysis was carried out at a temperature of 25°C

Sample analysis

The water samples were analyzed for the presence of Cadmium, Arsenic and Lead with wavelengths 228.8, 193.7 and 217.0 nm respectively. AAS (S Series 712354 v1.27) with a deuterium background collector was used in the determination of the trace metals. This analysis was carried out at the Central Laboratory of the University of Agriculture, Abeokuta, Ogun State, Nigeria.

RESULTS

Table 1 shows the pH and concentrations obtained for Pb

Table 1. The pH and concentrations of Lead (Pb) and Arsenic (As) in drinking water samples from boreholes.

Locations	pH	Pb (mgL^{-1})	As (mgL^{-1})
Abeokuta	7.0	ND	0.305
Agbara	4.8	ND	0.478
Ota	5.5	ND	0.237
Ifo	5.7	ND	0.595
FH BellsTech	6.4	ND	ND
MH BellsTech	6.2	ND	ND
CU Hostel	5.9	ND	ND

*ND = Not Detected.

Table 2. The pH and concentrations of heavy metals in drinking water samples from well sources.

Locations	pH	Pb (mgL^{-1})	As (mgL^{-1})
Abeokuta	7.9	ND	0.264
Agbara	5.4	ND	0.638
Ota	7.1	ND	ND
Ifo	5.8	ND	0.350
Ewekoro	7.8	ND	0.560
Brewery Area	5.9	ND	0.290

*ND = Not Detected.

Table 3. The pH and concentrations of heavy metals in drinking water samples from Rain.

Locations	pH	Pb (mgL^{-1})	As (mgL^{-1})
Ota	7.0	ND	ND
Ifo	8.2	ND	0.055

*ND = Not Detected.

and As in drinking water from boreholes in the various locations sampled. The pH values range from 4.80 to 6.98 which suggest that the water from the boreholes in this locations are slightly acidic. Lead was not detected at all in the locations sampled. Similarly, arsenic was not detected in the Male and Female Hostels of BellsTech and CU Student Hostel. We observed that the concentrations of Arsenic are all excessively greater than the WHO standard for MAL in Abeokuta, Agbara, Ota and Ifo. The trend of accumulations of this metal in these locations is as follows: Ifo > Agbara > Abeokuta > Ota. The results for the concentrations of Pb and As in drinking water from wells are depicted in Table 2. The pH values of these sources show that they are mostly slightly acidic. Again Pb was not detected while the As values are higher than the MAL in the locations sampled except Ota where the metal was not detected. The trend of

accumulations of As in these locations is as follows: Agbara > Ewekoro > Ifo > A Brewery area (Sango-Ota) > Abeokuta.

Rainwater was only collected from two locations namely, Ota and Ifo. As was not detected in the rainwater in Ota whereas the concentration observed from rainwater in Ifo is 5.5 times greater than the MAL. Pb was not detected at all in the rainwater of both locations. The results are shown in Table 3. A stream used as a source of drinking water in an area in Abeokuta was analyzed and the pH was found to be weakly alkaline. The As value (0.275 mg/L), is higher than the MAL value but Pb was not detected. Table 4 illustrates this result.

The acceptable limits or MAL for consumption for the metals are as follows: Lead – 0.01 mg/L, Cadmium – 0.003 mg/L, Arsenic 0.01 mg/L. The levels of cadmium in drinking water from boreholes in the sampled locations

Table 4. The pH and concentrations of heavy metals in drinking water samples from stream.

Locations	pH	Pb (mgL^{-1})	As (mgL^{-1})
Abeokuta	7.9	ND	0.275

*ND = Not Detected.

Comparison between Standard and Observed Cadmium Concentrations in Borehole water

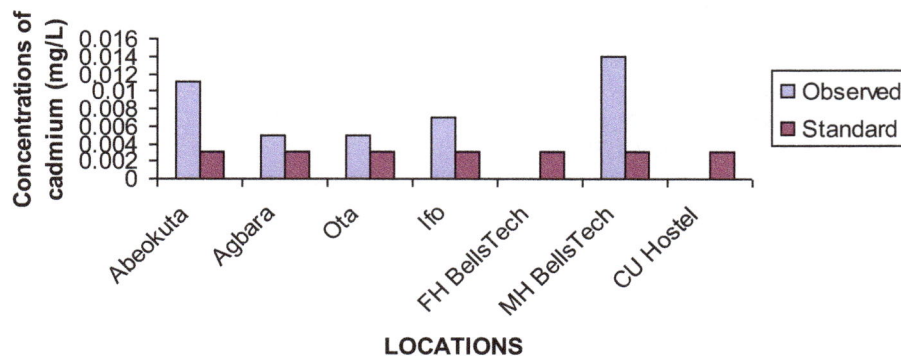

Figure 1. The concentrations of Cadmium compared to the MAL standard in drinking water samples from boreholes and their locations.

Comparison between Standard and Observed Cadmium Concentrations in Well Water

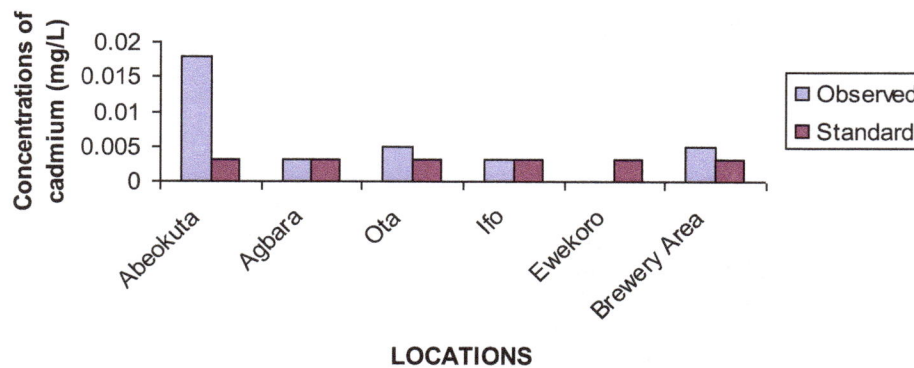

Figure 2. The concentrations of Cadmium compared to the MAL standard in drinking water samples from wells and their locations.

are illustrated in Figure 1. This metal was not detected in the boreholes used by the Female Hostel of BellsTech and the CU Student Hostel. The concentrations of Cadmium are 0.005, 0.007, 0.011 and 0.014 mg/L for Agbara and Ota, Ifo, Abeokuta and the Male Hostel BellsTech respectively and are observed to be greater than the MAL value. Cadmium levels are also observed to be higher in drinking water from Well in Abeokuta, Ota and in a brewery area in Sango-Ota. Cadmium was not detected in the well sampled from Ewekoro. The Cadmium levels from Well in Agbara and Ifo are exactly the same as the MAL value (Figure 2).

The Cadmium levels in the rainwater collected from Ota and Ifo shows a two-fold and three-fold increase

**Comparison Between Standard and Observed
Cadmium Concentrations in Rainwater**

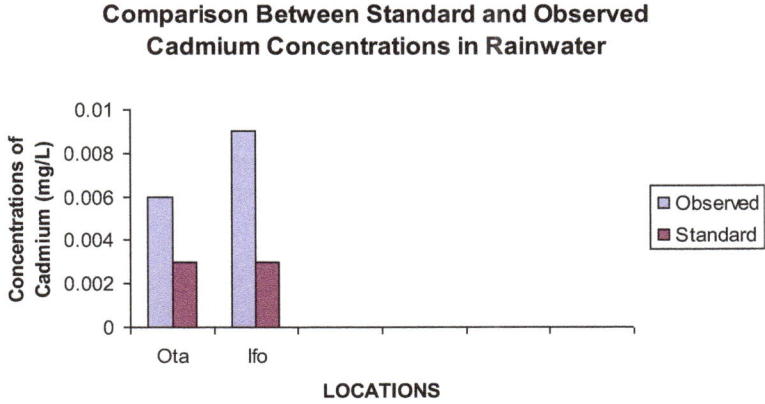

Figure 3. The concentrations of Cadmium compared to the MAL standard in drinking water samples from rainwater and their locations.

**Comparison between Standard and Observed
Cadmium Concentrations from Stream**

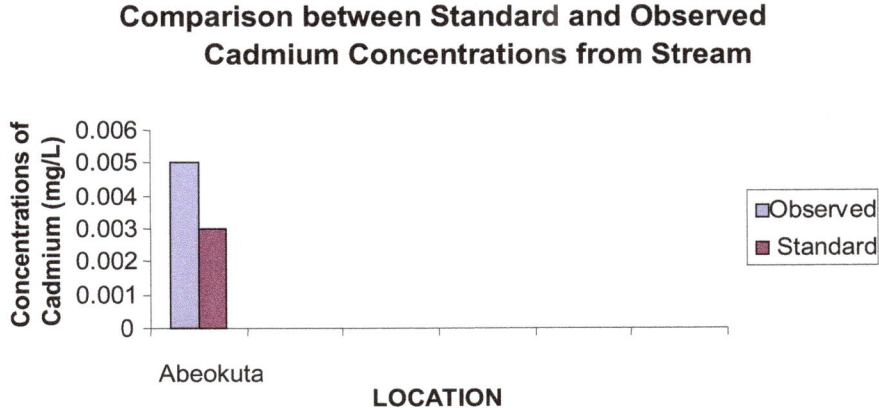

Figure 4. The concentration of Cadmium compared to the MAL standard in drinking water sample from a stream in Abeokuta.

respectively from the MAL value (Figure 3). The Cadmium concentration in the stream used as a source of drinking water in an area in Abeokuta is about twice the MAL value (Figure 4).

DISCUSSION

Borehole, well, stream and rain are common sources of drinking water in Ogun State. Some are treated, while others are consumed untreated. Due to high level of urbanization and industrialization around Ota, Agbara, Abeokuta and some other towns in Ogun State, and indiscriminate waste disposal practices, it is not unlikely that all or some of the drinking water sources are contaminated with some highly dreaded heavy metals in drinking water sources. Domestic sewage, combustion emission, mining operations, metallurgical activities and industrial effluents are among the sources of anthropogenic metal inputs and heavy metals such as

Pb, Cd, Cu, and Zn are released as a result of these processes (Chinni and Yallapragda, 2000). The results of several current analyses of sachet water have implied some levels of microbiological and organic contamination (personal communication); natural sources are now being preferred as safer sources. Hence, these sources have to be assessed for the level of their actual safety. Unfortunately, the results observed from our assessment show that majority of these drinking water sources contain heavy metals such as Arsenic and Cadmium in amounts greater than the maximum allowable limit of the WHO and Nigeria standard for drinking water quality which implies that they may not be safe for drinking. Consequently, there may be accumulation of these metals in the consumers of this drinking water from those sources which may results in cancer (Nigerian Industrial Standard, 2007) and other related diseases.

Lead was not detected in any of the sources in all the locations sampled which suggests that the metal level if present at all is very low and there had not been an accumulation of it in ground water in those areas. It also means that there have been little or no anthropogenic activities resulting in lead accumulation in those areas.

Only the drinking water from boreholes of the CU Student Hostel, and the Female Hostel, BellsTech could be considered to be safe for drinking in our own opinion because the results reveal that the drinking water from these boreholes are free of Lead, Arsenic and Cadmium contamination. However, there is need for monitoring and further assessment of other trace metals that could affect human health in order to fully establish the quality of drinking water from these sources. The drinking water from the borehole of the Male Hostel end of BellsTech contains Cadmium concentration greater than the MAL and should therefore be subjected to further treatment so as to remove or at least reduce the Cadmium level below the MAL value. Also the drinking water from well in a residential building from Ewekoro does not contain detectable levels of Cadmium and Arsenic which suggests that the water is relatively safe for drinking. The lack of detectable levels of these metals is probably due to the location of the well which is about 1.5 km from the famous Ewekoro Cement Factory, Ogun State, Nigeria. However, report has shown that water sources very close to the factory contain high levels of heavy metals (personal communication).

Generally, the drinking water sources from the city of Abeokuta would require adequate water treatment and removal of heavy metals such as Arsenic and Cadmium to improve the quality and make it safe for drinking and other domestic uses. This is very necessary in our own opinion because the results of assessment of the drinking water sampled from borehole, well and stream in Abeokuta gave high concentrations of arsenic and cadmium. Similarly, the drinking water sources from Ota and Agbara would also require treatment and proper monitoring in order to reduce or remove toxic metals. However, the water from wells in Ota and Agbara appears to be safer than the water from boreholes according to the results of these findings.

The quality of water may be described according to their physico-chemical and microbiological characteristics. The quality of ground water is never constant; it is constantly changing in response to daily, seasonal and climatic rhythms. For effective maintenance of water quality through appropriate control measures, continuous monitoring of large number of quality parameters is crucial because the changes in properties of water have far-reaching implications directly to the biota and indirectly to man. Water quality data are thus, essential from the implementation of responsible water quality regulations for characterizing and remediating contaminants and for the protection of the health of humans and ecosystem.

The effect of Cadmium poisoning in humans are very serious. Among them are high blood pressure, kidney damage, destruction of testicular tissue, and destruction of red blood cells. It is believed that much of the physiological action of Cadmium arises from its chemical similarity to Zinc. Specifically, Cadmium may replace Zinc in some enzymes, thereby altering the stereo-structure of the enzymes and impairing its catalytic activity. Diseases symptoms ultimately result (Bhata, 2008). It may be speculated that the high values of Arsenic detected in Ota and the other locations is because the largest number of industries in Ogun State are located in Ota and its environs. Ota is also in the vicinity of the flow of the country's water course into Lagos lagoon hence, the possibility of contamination through surface or aquifer water flow. Moreover, there is a major cement factory located nearby Ota in Ogun State and the direction of water flow is from the cement factory down to Ota town. The findings on high levels of As in drinking water sources reported here, must be an alert to health agencies in the locations concerned because this element has been associated to development of Leukemia (Robinson et al., 2001), abnormalities in children (Steinmaus et al., 2004) cancer (Chen et al., 1988; IARC, 2004) and with various other diseases (Hopenhayn-Rich et al., 1998; NRC, 2001).

Arsenic toxicity strongly depends on the form in which arsenic is present. Inorganic arsenic forms, typical in drinking water, are much more toxic than organic ones that are present in Sea food (Branislav, 2007). The first visible symptoms caused by exposure to low Arsenic concentrations in drinking water are abnormal black-brown skin pigmentation known as *melanosis* and hardening of palms and soles known as *keratosis*, further thickening (*hyperkeratosis)* and can lead on to skin cancer (WHO, 2001). Arsenic may attack internal organs including lungs, kidney, liver and bladder without causing any visible external symptoms, making Arsenic poisoning difficult to recognize. Elevated concentrations in hair, nails, urine and blood can be an indicator of human exposure to Arsenic before visible external symptoms (Rasmussen and Anders, 2002). The disease symptoms caused by chronic Arsenic ingestion are called *arsenicosis* and develop when Arsenic contaminated water is consumed for several years. However, there is no universal definition of the disease caused by Arsenic, and no way of knowing which cases of cancer were caused by drinking Arsenic affected water. A correlation between hypertension and Arsenic in drinking water has also been established in a number of studies. The International Agency for Research on Cancer has concluded that: "There is sufficient evidence in humans that Arsenic in drinking-water causes cancers of the urinary bladder, lung and skin" (IARC 2004).

Conclusion

In the light of the parameters assessed, it can be concluded that the drinking water sources in the selected towns of Ogun State in this study may not have any adverse effect as far as Lead is concerned. On the other hand, the high Arsenic and Cadmium concentrations measured in this study raises a red flag to the health of communities settled in the affected towns. Although, further studies on the assessment of these metals and other heavy metals are ongoing in our research laboratory, we will like to advocate for regular proper treatment and monitoring of these drinking water sources.

REFERENCES

Adeyeye EI, Abulude FO (2004). Analytical Assessments of Some Surface and Ground water resources in Ile-Ife, Nigeria. J. Chem. Soc. Nig., 29: 98 – 103.

Anazawa K, Kaid Y, Shinomura Y, Tomiyasu T, Sakamoto H (2004). Heavy – metal distribution in river waters and sediment around a Firefly Village, Shikoku, Japan. Application of multivariate analysis. Anal. Sci., 20: 79 – 84.

Bhata SC (2008). Environmental Chemistry. Satishkumar Jainfor. 4596/ 1 – A, 11 Darya Ganj, New Delhi – 110002, 5th Edition, India.

Boss BC, Fredeen KJ (1997). Concepts, Instrumentation and Techniques in Inductively Coupled Plasma Optical Emission Spectrometry. The Perkin-Elmer Corporation, 2nd Edition, USA. pp. 16.

Bozkurtoglu E, Vardar M, Suner F, Zambak C (2006). A new numerical approach to weathering and alteration in rock using a pilot area in the Tuzla geothermal area, Turkey. Engineering Geology, 87(1-2): 33-47.

Branislav Petrusevski, Saroj Sharma, Jan C. Schippers (UNESCO – IHE), and Kathleen Shordt (IRC) (2007). Arsenic in Drinking Water. Thematic Overview Paper 17. Reviewed by: Christine van Wijk (IRC). IRC International Water and Sanitation Centre.

Chen CJ, Kuo TL, Wu MM (1988). Arsenic and cancer. Lancet, 1: 414-415.

Chinni S, Yallapragda R (2000). Toxicity of copper, cadmium, zinc and lead to penaeus indicus poslarvae: Effects of individual metals. J. Environ. Biol., 21: 255-258.

Das A, Krishnaswami S (2007). Elemental geochemistry of river sediments from the Deccan Traps, India: Implications to sources of elements and their mobility during basalt-water interaction. Chemical Geology, 242(1-2): 232-254.

Donahue RL, Miller RW, Shickluna JC (1983). Soils, An introduction to soil and plant growth. Prentice may, Inc. Englewood Cliffs, New Jersey, Fifth Edition.

Espino MA, Rubio AH, Navarro CJ (2007). Nitrate Pollution in the Delicias-Meoqui Aquifer of Chihuahua, Mexico. Environmental Health Risk Conference. Republic of Malta June 27-29. Witpress, pp.189-196.

Gibbs RJ (1970). Mechanisms Controlling World water Chemistry. Science, 170: 1088 – 1090.

Hesterberg D (1998). Biogeochemical Cycles and Processes leading to changes in Mobility of Chemicals in Soil. Agric. Ecosyst. Environ., 67: 121- 133.

Hopenhayn-Rich C, Biggs C, Smith MC (1998). Lung and kidney cancer mortality associated with arsenic in drinking water in Cordoba, Argentina. Int. J. Epidemiol, 27:561-569. http://www.who.int/mediacentre/factsheets/fs210/en/

IARC (2004). Some drinking-water disinfectants and contaminants, including arsenic. Lyon, France, International Agency for Research on Cancer. (IARC monographs on the evaluation of carcinogenic risks to humans; vol. 84).

International Agency for Research on Cancer (IARC) (2004). Some drinking-water disinfectants and contaminants including arsenic. IARC Monograr Eval Carcinog Risks Hum., p. 84.

Lester JN, Birkett JW (1999). Microbiology and Chemistry for Environmental Scientists and Engineers. 2nd Edn. E and FN Spon, London and New York.

McGregor DFM, Thompson DA, Simon D (2000). Land – Water Linkages in Rural Watersheds Electronic Workshop. Water Quality and Management in Peri-urban, Iumasi, Ghana. Case study 16. FAO. Of the United Nations. Rome, Italy.

Memon M, Sommor MS, Puno HK (2002). Status of water bodies and their effect on human health in district tharparkar. J. Appl. Sci., 2: 386-389.

Moreno T, Querol X, Castillo S, Alastuey A, Cuevas E, Herrmann L, Mounkaila M, Elvira J, Bibbons, W (2006). Geochemical variations in Aeolian mineral particles from the Sahara-Sahel Dust Corridor. Chemosphere, 65(2): 261-270.

Nigerian Industrial Standard (2007). Nigerian Standard for Drinking Water Quality. NIS 554: p. 16.

NRC. National Research Council (2001). Arsenic in drinking water; 2001 update. Washington, D.C. National Academy Press.

Rasmussen L, Andersen KJ, (2002). Environmental health and human exposure assessment. DHI Water & Environment, Denmark.

Robinson L, Links T, Smith A, Smith M, Guinan M, Todd R, Brown L, Dudding B (2001). Acute Lymphoblastic (Lymphocytic) Leukemia-Review and Recommendations of the expert panel Nevada State Health Division.

Rubio AH, Saucedo TR, Bautista MR, Wood K, Holguin C, Jimenez J (2006). Are crop and range lands being contaminated with cadmium and lead in sediments transported by wind from an adjacent contaminated shallow lake? Geoenvironment and Landscape Evolution. Editors: J.F. Martin-Duque, C.A. Brebbia, D.E. Emmanouloudis, U. Mander. WITPRESS, pp. 135-141.

Rubio AH, Saucedo TR, Lara CR, Wood K, Jiménez, J (2005). Water quality in the Laguna de Bustillos of Chihuahua, México. Water Resources Management III. Editors: M. De Conceicao Cunha, C.A. Brebbia. WITPRESS, pp. 155-160.

Rubio AH, Wood K, Alanis, HE (2004). Water pollution in the Rio Conchos of Northern Mexico. Development and application of computer techniques to Environmental Studies X. Editors; G. Latini, G. Passerini, C.A. Brebbia. WITPRESS, pp.167-176.

Steinmaus C, Lu M, Todd RL, Smith AH (2004). Probability estimates for the unique childhood leukemia cluster in Fallon, Nevada, and risks near other U.S. military Aviation facilities. Environmental Health Perspectives, 112(6): 766-771.

Sylvester A (2003). Quality of Surface Water, River Birim as a Case Study. Department of Chemistry, KNUST, Kumasi, Ghana, p. 34.

White AF, Marjorie SS, Davison VV, Alex EB, Stonestrom DA, Harden JW (2005). Chemical weathering rates of a soil chronosequence on granitic alluvium: III. Hydrochemical evolution and contemporary solute fluxes and rates. Geochimica et Cosmochimica Acta, 69(8): 1975-1996

WHO (2001). Arsenic in drinking water. Fact sheet No. 210, Rev. ed. Available at: 0/en/http://www.who.int/mediacentre/factsheets/fs21.

WHO (2010). Water for Health. Guidelines For Drinking – water Quality. Incorporating First and Second Addendum to Third Edition. Volume 1 Recommendations. 2010. World Health Organization, Geneva, Switzerland.

21

Cadmium toxicity exposure – Induced oxidative stress in postnatal development of wistar rats

P. Dailiah Roopha[1], J. Savarimuthu Michael[2], C. Padmalatha[1] and A. J. A. Ranjit Singh[2]

[1]Department of Advanced Zoology and Biotechnology, Rani Anna Government College for Women, Tirunelveli-627 008, Tamil Nadu, India.
[2]Department of Advanced Zoology and Biotechnology, Sri Paramakalyani College, Alwarkurichi-627 412, Tamil Nadu, India.

Effect of cadmium on ovary increased reactive oxygen species (ROS) was studied in rat. Lipid peroxidation and the reactive oxygen species, hydrogen peroxide were increased in the ovary of Wistar rat administered with cadmium (50 and 200 ppm) in drinking water on 45 and 65 days of post natal development. Elevation of reactive oxygen species and lipid peroxidation is suggestive of tissue damage in the rat ovary due to cadmium poisoning.

Key words: Cadmium, lipid peroxidation, reactive oxygen species, hydrogen peroxide ovary.

INTRODUCTION

Cadmium is used industrially in the following ways as protective plating on steel, stabilizer for poly vinyl chloride (PVC) products, pigments in plastic and glass, electrode material in nickel cadmium batteries and as components of various alloys (Wilson, 1988). Cadmium was found to result in oxidative stress (Hendy et al., 1992; Somashekaraiah et al., 1992) by inducing oxygen free radical production (Balaknina et al., 2005; Demirevska-Kepava et al., 2006).

Reactive oxygen species (ROS) are short lived reactive molecules that can modify cellular components including nucleic acids, proteins and lipids (Etienne and Hall, 2002). Involvement of ROS is implicated in neurodegenerative and other disorders, e.g Alzheimer's disease, Parkinson's disease, multiple sclerosis, Down's syndrome, inflammation, viral infection, autoimmune pathology, and digestive ulcers (Aruoma, 2003; Repetto and Llesuy, 2002; Surh and Ferguson, 2003). Numerous studies support the role of ROS in male infertility (Aitken et al., 2003). Recently reactive oxygen species (ROS) have been shown to have an important role in the normal functioning of reproductive system and in the pathogenesis of infertility in females (Ashok et al., 2004).

The present study was designed to test the effect of cadmium exposure through drinking water that could induce an oxidative damage in ovary of developing female Wistar rats.

MATERIALS AND METHODS

Thiobarbituruc acid, dithio-bis-nitrobenzoic acid, sodium azide, and bovine serum albumin were purchased from Sigma-Aldrich Co, St. Louis, USA. Sulphuric acid, hydrogen peroxide, chloroform and sodium hydroxide were procured from Qualigens Fine Chemicals, Mumbai, India. Pyrogallol and petroleum ether were purchased from E-Merck (India) Limited, Mumbai, India. All other chemicals were purchased from Sisco research Laboratories Private Limited, Mumbai, India.

For toxicity evaluations, cadmium (Cadmium in the form of Cadmium chloride) was selected. After preliminary toxicity evaluation, two test doses 50 and 200 ppm were selected.

EXPERIMENTAL DESIGN

Animals

90 day-old female albino rats of Wistar strain (*Rattus norvegicus*) obtained from National Institute of Nutrition, Hyderabad, weighing 140 ± 10 g was used for the present investigation. Rats were maintained in a temperature controlled animals quarter with 12 h dark: 12 h light schedule and were fed standard rat pellet diet (Broke Bond Lipton India Ltd., India) and drinking water *ad libitum*. The animals were dewormed with albendazole (Bendex - 400,

---
*Corresponding author. E-mail: singhmispkc@gmail.com.

Protec Cipala Ltd., India) (10 mg/kg body weight, orally), before the initiation of the experiment. The females were mated with males at a ratio of 2: 1. Cohabitation began at approximately 16.30 h on each mating day. The following morning, the females were removed from the mating cages and individually, smeared for the presence of sperm in the vaginal lavage. The presence of sperm in the vaginal lavage is indicative of the females that had mated and they were selected for further studies. The pregnant animals were allowed to give birth.

The mother animals with female pups were divided into the following groups: Group I: Control; Group II: 50 ppm; Group III: 200 ppm.

The minimum (50 ppm) and maximum effective doses (200 ppm) were selected (Samuel, 2001), and the mother rats along with female pups were treated with cadmium in the form of cadmium chloride through drinking water from 0 to 65 day pp. Sub group I: 45 days puberty occurred. Sub group II: 65 days full growth of ovary.

At the end of the experimental period, animals were killed by cervical decapitation, and uterus was dissected out and washed with ice-cold saline. A 10% homogenate (100 mg in 1 ml buffer) of washed tissue was prepared in 0.1 M Tris-HCl buffer, pH 7.4 and used for the assay of the following biochemical parameters.

Protein

The protein content was determined by the method of Lowry et al. (1985). The total protein content was expressed as g/100 g tissue.

Lipid peroxidation

Tissue lipid peroxidation was measured by the method of Devasagayam and Tarachand (1987). The malondialdehyde content of the samples was expressed as nmoles of MDA formed/mg protein.

Reactive oxygen species

Hydrogen peroxide

Hydrogen peroxide production was assessed by the spectrophotometric method of Holland and Storey (1981). The malondialdehyde content of the sample was expressed as μ moles of MDA formed/mg protein.

Statistical analysis

All data were presented as means ± standard error of the mean (SEM). Statistical significance was calculated using student's 't' test to test the significance of individual variations. Where n_1 and n_2 are the number of observations in the two classes being compared (Ostle, 1966). The value of probability was obtained from the degree of freedom by using standard table value, given by Fisher and Yates (1948). If the calculated value was more than the table value, it was significant at the probability level. The level of significance was assessed at $P < 0.05$.

RESULTS

Lipid peroxidation (LPO)

LPO levels in the ovary of cadmium exposed rat and control group were shown in Figure 1. A gradual dose

dependent significant increase was observed in treated groups when compared to control. The peak level was observed in the 65 days of postnatal development (PND) rat ovary.

Reactive oxygen species

Hydrogen peroxide

Figure 2 shows the hydrogen peroxide levels in the ovary of control and the exposed groups. The production of ROS (hydrogen peroxide) significantly increased with age and dosage of cadmium when compared to control.

DISCUSSION

Lipid peroxidation is one of the main manifestations of oxidative damage, which plays an important role in the toxicity of many xenobiotics (Stohs and Bagchi, 1995; Anane and Creppy, 2001). Lipid peroxyl radicals react with other lipids, proteins and nucleic acids, propagating thereby the transfer of electrons and bringing about the oxidation of substrates. Cell membranes, which are structurally made up of large amounts of poly unsaturated fatty acids are highly susceptible to oxidative attack and consequently changes in membrane fluidity, permeability and cellular metabolic functions result (Bandyopaddhyay et al., 1999). Moreover, the end products of lipid peroxidation like melondialdehyde (MDA) can also cause tissue injury by interacting with bio macromolecules (Freeman and Crapo, 1982; Valentine et al., 1998; Mylonas and Kouretas, 1999). Cadmium, arsenic and mercury toxicity all involve similar pathways of cellular damage, inhibition of mitochondrial enzymes, and suppression of protein synthesis and production of free radicals (Fowler, 1978). Ognianovic et al. (2003) reported that Cd stimulated reactive oxygen species (ROS) thus causing oxidative damage in various tissues. Cadmium causes a significant increase of LP concentration in liver and kidney of rats, since it causes lipid peroxidation in numerous tissues both in vivo and in vitro (Kostic et al., 1993; Sarkar et al., 1998; Ognijanovic et al., 2003; El-Demerdash et al., 2004). The present study revealed an oxidative stress in rats exposed to 50 and 200 ppm cadmium through drinking water during postnatal development of rats. The dose dependent increase in lipid peroxidation (LPO) was observed in both the treatment groups (Figure 1).

Cadmium may be inducing oxidative stress by producing hydroxyl radicals (O'Brien and Salasinski, 1998), superoxide anions, nitric oxide and hydrogen peroxide (Kozzumi et al., 1996, Waisberg et al., 2003). Watanabe et al (2003), showed generation of non-radical hydrogen peroxide which by itself became a significant source of free radicals via the Fenton chemistry. Hydrogen peroxide at micromolar levels also appears

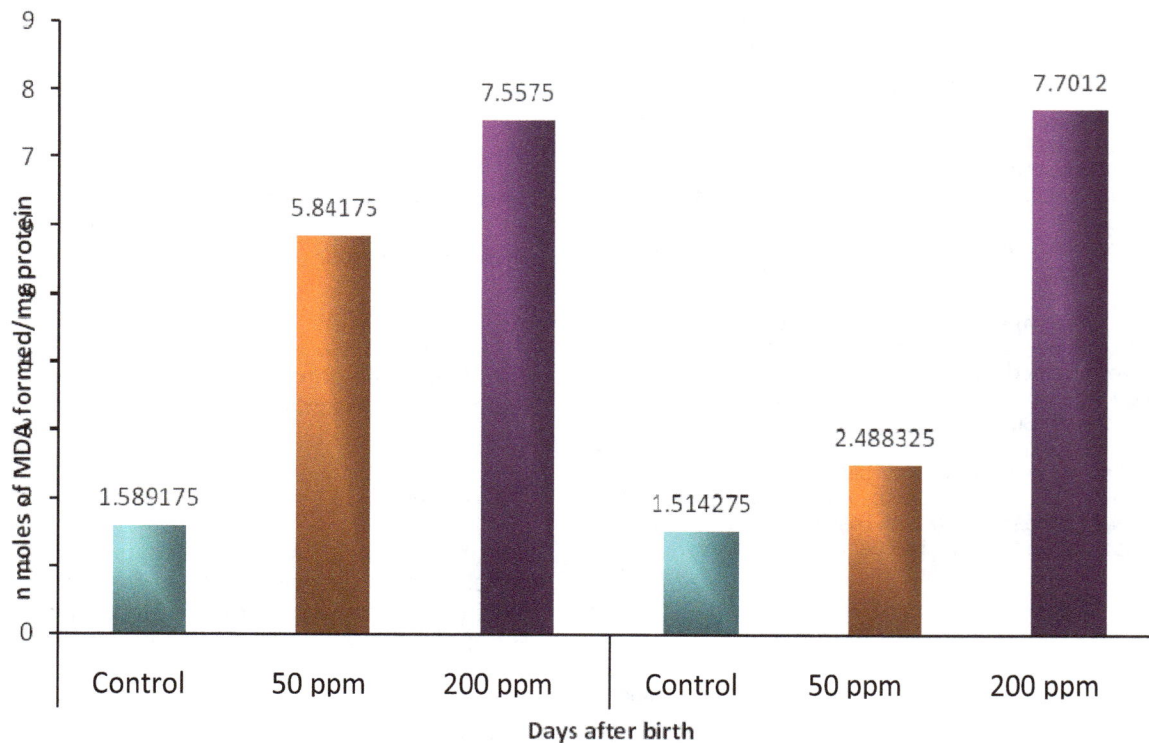

Figure 1. Effect of Cd exposure on ovarian lipid peroxidation in developing female rats.Each bar represents the mean and SEM (n=4). Statistical significance of difference among groups at p < 0.05; Control versus experiment: 50 versus 200 ppm.

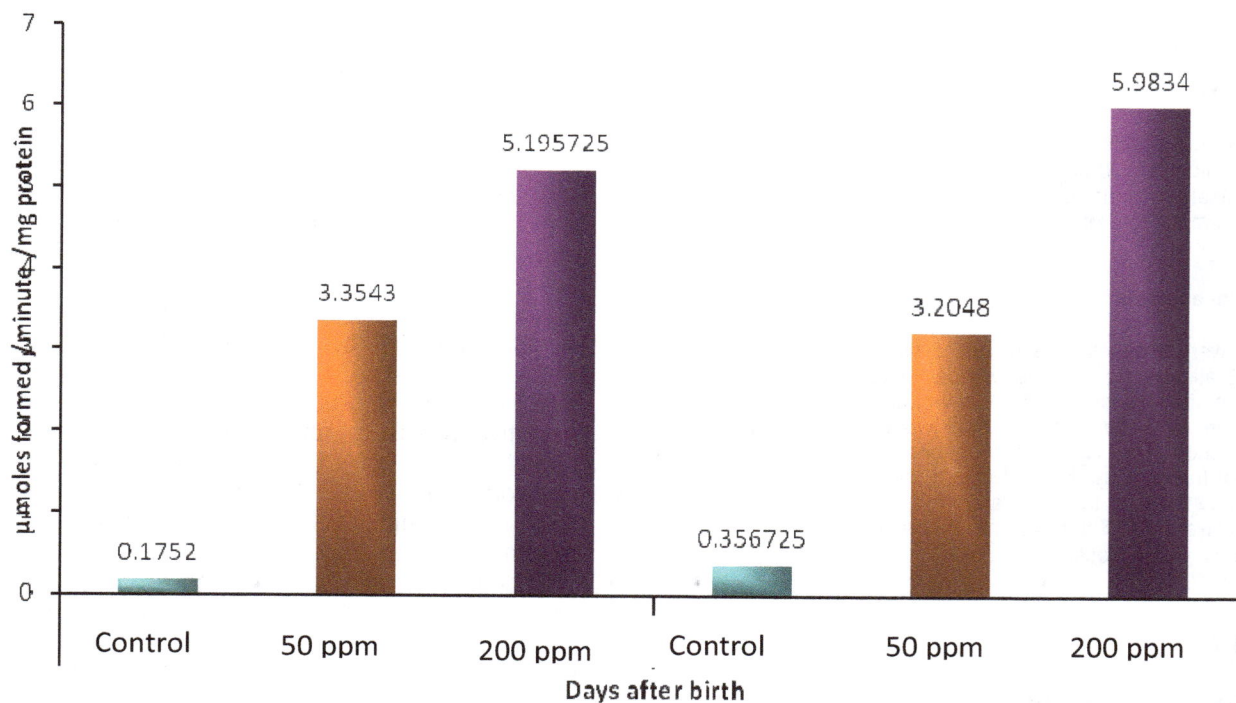

Figure 2. Effect of Cd exposure on hydrogen peroxide production in ovaries of developing female rats. Each bar represents the mean and SEM (n=4) Statistical significance of difference among groups at p < 0.05; Control versus experiment; 50 versus 200 ppm.

poorly reactive, but higher levels of H_2O_2 can attack certain cellular targets (Chance et al., 1979; Meler et al., 1990). The increase in hydrogen peroxide observed in the ovary of 65 day of PND was higher than 45 day PND rat ovary (Figure 2). This may be due to the formation of radicals by metal dependent reactions.

The reaction stating the formation of ROS, via the sequential univalent process of O_2 reduction forming several intermediates should be supplied.

The observed uterus LPO and H_2O_2 could have resulted in an elevated oxidative stress to the developing organ like ovary thereby altering ovary development and processing. Though low concentration of ROS is essential for normal reproductive process in the female reproductive tract, high in appropriate production of ROS is known to cause opposite effects (Man'kovs'ka and Serebrovs'ka, 1998). Therefore, the increase in ovarian LPO observed in the present study could be due to the concomitant increase in the generation of free radicals like hydrogen peroxide in the ovary of cadmium exposed rats.

REFERENCES

Aitken RJ, Baker MA, Sawyer D (2003). Oxidative stress in the male germ line and its role in the actiology of male infertility and genetic disease. Reprod. Biomed. Online, 7: 65-70.

Anane R, Creppy EE (2001).Lipid peroxidation as pathway of aluminum cytotoxicity in human skin fibroblast cultures: Prevention by superoxide dismutase and catalase and vitamins E and C. Hum. Exp. Toxicol., 20: 477-481.

Aruoma OI (2002). Methodological considetations for characterizing potential antioxidant actions of bioactive components in food plants. Mut. Res., 9-20: 523-524.

Ashok A, Shyam SR, Allamaneni SSR (2004). Role of free radicals in female reproductive diseases and assisted reproduction. Reprod. Biomed. Online, 9(3): 338-347.

Bandyopaddhyay U, Das U, Banerjee RK (1999). Reactive oxygen species; Oxidative damage and pathogenesis Curr. Sci., 77(5): 658-666.

Balaknina T, Kosobryukhov A, Ivanov A, Kreslauskii V (2005). The effect of cadmium on CO_2 exchange, variable fluorescence of chlorophyll and the level of antioxidant enzymes in pea leaves. Russian J. Plant Physiol., 52: 15–20.

Chance B, Sies H, Boveris A (1979). Hydroperoxide mechanism in mammalian organs. Physiol. Rev., 59: 527-605.

Demirevska KK, Simova–Stoilova L, Stoyamova Z, Feller U (2006). Cadmium stress in barlely. Growth, leaf pigment and protein composition and detoxification of reactive oxygen species. J. Plant Nutr., 29: 451–468.

Devasagayam TPA, Tarachand V (1987). Decreased lipid peroxidation in the rat kidney during gestation. Biochem. Biophys. Res. Commun., 145: 134-138.

El-Demerdash FM, Youself MI, Kedwany FS, Baghidadi HH (2004). Cadmium induced changes in lipid peroxidation blood hematology, biochemical parameters and semen quality of male rats: Protective role of vitamin E and beta-carotene. Food Chem. Toxicol., 42: 1563-1571.

Fischer RA, Yates T (1948). In: Statistical tables for biological, agricultural and medical research, Oliver and Boyd, London, p. 62.

Fowler BA (1978) . General subcellular effects of lead, mercury, cadmium, and arsenic. Environ. Health Perspect 22:37-41.

Freeman BA, Crapo JD (1982). Free radicals and tissue injury. Lab. Invest., 47: 412-416.

Hendy GAF, Baker AJM, Evart CF (1992). Cadmium tolerance and toxicity, oxygen radical processes and molecular damage in cadmium tolerant and cadmium – sensitive clones of Holcuse lanatees. Acta. Bot. Neerl., 41: 271–281.

Holland MK, Storey BT (1981). Oxygen metabolism of mammalian spermatozoa. Generation of hydrogen peroxide by rabbit spermatozoa. Biochem., 198: 273-280.

Kozzumi T, Shirakura G, Kumagi H, Tataumoto H, Suzuki KT (1996). Mechanisms of cadmium-induced cytotoxicity in rat hepatocytes: Cadmium-induced active oxygen-related permeability changes of the plasma membrane. Toxicol., 114: 124-134.

Kostic MM, Ognijanovic B, Dimitrijevic S, Zikic RV, Stajn A, Rosic GL (1993). Cadmium induced changes of antioxidant and metabolic status in red blood cells of rats: In vivo effects. Eur. J. Haematol., 51: 86-92.

Lowry OH, Rosebrough, NJ, Farr AL, Randall RJ (1951). Protein measurement with the folin phenol reagent. J. Biol. Chem., 193: 265-275.

Man'kovs'ka IM, Serebrov'ka ZO (1988). The role of oxygen radicals in physiology and pathology of human sperm. Fiziol. Zh., 44: 118-125.

Meier B, Radeke H, Selle S, Habermehil G, Resch K, Sies H (1990). Human fibroblasts release reactive oxygen species in response to treatment with synovival fluids from patients suffering from arthritis. Free Radic. ROS Commun., 189: 149-160.

O'Brien P, Salasinski HJ (1998). Evidence that the reactions of cadmium in the presence of metallothionein can produce hydroxyl radicals. Arch. Toxicol., 72: 690-700.

Ognianovic BS, Pavlovic SD, Maleyic SD, Zikic RV, Stanjn AS, Radijicic RM, Saicic ZS, Petrovic VM (2003). Physiol. Res., 52: 563-570.

Ostle B (1996). Statistics in Research, Oxford and IBH Publications, New Delhi, p. 82.

Repetto MG, Lilescy SF (2002). Antioxidant properties of natural compounds used in popular medicine for gastric ulcers. Braz. J. Med. Biol. ROS, 35(5): 523- 534.

Samuel BJ (2001). Reproductive toxicity of chromium on ovarian development in female wistar rats - An Endocrine and a Biochemical Approach, (Ph.D., Thesis) Chapters I and III.

Sarkar S, Yadov P, Bhatnagar D (1998). Lipid peroxidative damage on cadmium exposure and alterations in antioxidant system in rat erythrocytes: A study with relation to time. Biol. Metals, 11: 153-157.

Stohs SJ, Bagchi D (1995).Oxidative mechanisms in the toxicity of metal ions. Free Radic. Biol. Med., 18: 321-326.

Somashekaraiah BV, Padmaja K, Prasad APK (1992). Phytotoxicity of cadmium ion in germinating seedling of mung beam (Phaseoles vulgaris): Involvement of lipid peroxides in chlorophyll degradation. Phys. Plant, 85: 85–89.

Valentine JS, Wertz DL, Lyons TJ, Liou LL, Gato JJ, Gralla ED (1998). The dark side of dioxygen biochemistry. Curr. Opin. Chem. Biol., 2: 253-247.

Waisberg M, Joseph P, Hale B, Beyersmann D (2003). Molecular and cellular mechanisms of cadmium carcinogenesis: A review. Toxicol., 192: 95-117.

Watanabe M, Henmi K, Ogwa K, Suzuki T (2003). Cadmium-dependent generation of reactive oxygen species and mitochondrial DNA breaks in photosynthetic and non-photosynthetic strains of Euglena gracitis. Comp. Biochem. Physiol. C. Toxicol. Pharmacol., 134: 227-234.

Wilson DN (1988). Cadmium – Marked trends and influences. In: Cadmium 87th proceedings of the 6th International Cadmium conference, London, Cadmium Association, pp. 9–16.

Surh YZ, Fergeuson LR (2003). Dietary and medicinal antimutagens and anticorcinogens molecular mechanisms and chemopreventive potential – Highlight of a symposium.

22

Effect of lead nitrate on survival rate and chronic exposure on growth performance of grass carp (*Ctenopharyngodon idella*)

Esmail Gharedaashi, Hamed Nekoubin*, Alireza Asgharimoghadam, Mohammad Reza Imanpour and Vahid Taghizade

Department of Fishery, Gorgan University of Agricultural Sciences and Natural Resources, Gorgan, Iran.

The acute toxicity of lead nitrate to grass carp (*Ctenopharyngodon idella*) juveniles was assessed in a static renewal bioassay for 96 h. In addition, an experiment was conducted to determine the growth performance during 60-day sublethal ($Pb(NO_3)_2$) exposure. The results indicated that median lethal concentration (LC_{50}) of lead nitrate to Grass carp for 96 h of exposure was 246.455 µ. The chronic exposure to sublethal concentration of lead nitrate to the studied fish showed a significant decrease in final body weight in comparison to control group. The lead nitrate also had significantly decreased effect on body weight in comparison to the control. Also, the food conversion ratio (FCR) was significantly increased in comparison to control ($P < 0.05$). The lead nitrate also caused a significant decrease in the survival rate ($P < 0.05$).

Key words: Lethal concentration (LC_{50}), lead nitrate, growth, grass carp, *Ctenopharyngodon idella*.

INTRODUCTION

Heavy metals have long been recognized as serious pollutants of the aquatic environment. The accumulation of metals in the aquatic environment has direct effect on man and aquatic ecosystem. While the metals were required for metabolic activities in organisms lies in the narrow range between their essentiality and toxicity (Fatoki et al., 2002). Heavy metal contamination usually causes depletion in food utilization in fish and such disturbance may result in reduced fish metabolic rate and hence cause reduction in their growth (Javed, 2005a). Growth is a sensitive and reliable endpoint in chronic toxicological investigations (De Boeck et al., 1997).

The present work was design to investigate acute toxicity and toxic effect of lead on the growth performance of Grass carp under chronic sublethal concentrations to evaluate its potential to growth in contaminated water. Metal concentrations in aquatic organisms appear to be of several magnitudes, higher than concentrations present in the ecosystem (Laws,2000) and this is attributed to bioaccumulation whereby metal ions are taken up from the environment by the organism and accumulated in various organs and tissues. Metals also become increasingly concentrated at higher trophic levels, possibly due to food-chain magnification (Wyn et al., 2007). Metals are non-biodegradable and considered as major environmental pollutants causing cytotoxic, mutagenic and carcinogenic effects in animals (More et al., 2003).

Lead occurs in environment in a wide range of physical and chemical forms that influence the behavior of fish adversely at concentration higher than normal. Most of the lead in the environment is in the inorganic form and exists in several oxidized states (Jackson et al., 2005). Pb is the most stable ionic species present in the environment and is thought to be the form in which the maximum bioaccumulation of Pb occurs in aquatic organisms. However, the toxicity of Pb depends upon many factors including fish age, pH and hardness of the water (Nussey et al., 2000). Also lead is not necessary for the biological functions of animals even at low concentrations. It is being

---

*Corresponding author. E-mail: nekoubin.hs@gmail.com.

discharged to aquatic systems mainly from petroleum, chemistry, dyes and mining industries, which have toxic effects and can cause mortality to aquatic animals (Sorensen, 1991; Heath, 1995).

Chronic lead poisoning has similar toxic effects in fish as in mammals, these include hematological and neural disorders and tetanic spasms together with some morphological changes such as darkening in caudal fin, deformation of vertebrate, anomalies in pigment formation and covering of the gills by a mucus layer (Tulasi et al., 1992; Shah, 2006b).

MATERIALS AND METHODS

Uniform juveniles of grass carp were obtained from the Institute of Pond Fish Culture in Gorgan (Agh Ghala), Iran. They were weighted (initial weight 4.3 ± 0.5 g). The total length was also measured accurately (8.2 ± 0.44 cm). Fish were fed with aquatic plant food (*Lemna* species) at least twice a day before the experiments, and the fish were not fed during the experiments. Nutritional compositions of experimental diets (*Lemna* spp.) are given in Table 1. Proximate composition of diets was carried out using the Association of Analytical Chemists AOAC (2000) methods. Protein was determined by measuring nitrogen ($N \times 6.25$) using the Kjeldahl method; crude fat was determined using petroleum ether (40 to 60 bp) extraction method with Soxhlet apparatus and ash by combustion at 550°C. Physicochemical parameters of water, temperature, pH, total hardness, dissolved O_2, total NH_3, Na, K and CO_2 of the treated and control media were monitored on daily basis by following the methods of American Public Health Association (APHA) (1998). However, water temperature (24 ± 1°C), pH (7 to 7.5) and hardness (275 ± 2.58 mg L^{-1}) was kept constant throughout the study.

LC_{50} determination

Firstly, to investigate acute toxicity of lead, all aquaria (60 L) capacity were filled with 50 L of dechlorinated tap water. A total of 24 aquaria that each stocked with 10 fish were used in LC_{50} experiments. Stock solutions of lead nitrate were prepared by dissolving analytical grade lead nitrate $[Pb(No_3)]_2$ (from Merck) in double distilled water. 30 fishes were used per concentration of Pb. Ninety-six hours acute bioassays were performed following in general Organization for Economic Co-operation and Development (OECD) guidelines for fish acute bioassays (guideline OECD203, 92/69/EC, method C1) (OECD, 1993). For determination of the LC_{50}/96 h (lethal concentration) values, following a range finding test, seven Pb (100, 200, 240, 260, 280, 300 and 320 mg/l) concentrations were chosen for Grass carp. For each metal-treated and control, three replicates were conducted.

Metal solutions were prepared by dilution of a stock solution with dechlorinated tap water. A control with dechlorinated tap water only was also used. The number of dead fish was counted every 12 h and removed immediately from the aquaria. The mortality rate was determined at the end of 24, 48, 72 and 96 h. During the toxicity test, the fishes were not fed. Acute toxicity test was conducted in accordance with standard methods (OECD, 1993). In this study, the acute toxic effect of lead on the Grass carp was determined by the use of Finney's Probity Analysis LC_{50} Determination Method (Finney, 1971). Confidential limits (Upper and Lower) were calculated and also used SPSS18 for LC_{50} value of lead with the help of probity µ analysis.

Table 1. Diet composition and proximate chemical analysis (%).

Ingredient	Percentage (%)
Protein	28
Lipid	11.4
Fiber	2.7
Ash	6

Growth performance

Thereafter, to investigate toxic effect of lead nitrate on the growth performance of Grass carp under chronic sublethal concentrations, an experiment was conducted in a completely randomized design with 3 treatments (tow concentration of lead and a control), and three replicates per treatment for a total of six fiberglass tanks (each with a capacity of 200 L), 60 fishes were used per concentration of lead. Separate groups of 60 fish each served as control for lead. 5 and 10% of LC_{50}/96 h concentration for lead nitrate (12.32 and 24.64 mg L^{-1}) was used as sublethal level for Grass carp. In control experiment set up, water with no metal was added. Throughout the experimental period of 60 days, fish were fed to satiation daily [(4% body weight (3 times a day)]. The treated fish were kept in the fiberglass tank containing sub lethal concentration of lead and grown for 60 days.

The fish were weighed individually at the beginning and at the end of the experiment. In the termination of experiment, total larvae from each tank were sampled and the final weight and length of body were measured. Growth parameters of fish were calculated based on the data of biometry of Grass carp larvae. One-way ANOVA and Duncan's multiple range tests were used to analyze the significance of the difference among the means of treatments by using the SPSS program.

RESULTS

LC_{50}/96 h of lead for grass carp

Acute toxicity of lead showed that mortality is directly proportional to the concentration of the lead nitrate while the percentage of mortality is virtually absent in control (Table 2) showing the relation between the lead concentration and the mortality rate for 96 h of Grass carp. Results according to SPSS18 analysis showed that the median lethal concentration (LC_{50}) of lead nitrate to Grass carp for 96 h of exposure is 246.455 µ (Table 3).

Growth performance

The results clearly showed that the lead nitrate had harmful effects on the growth parameters on Grass carp. The feeding and growth parameters of Grass Carp are presented in (Table 4). The chronic sublethal lead nitrate exposure to the fish exerted that larvae had significantly decreased final body weight in T1 and T2 when compared to control ($P < 0.05$). The lead nitrate also had significant effects on specific growth rate (SGR) and body weight increased in comparison to the control. The food

Table 2. Cumulative mortality of grass carp during sub-lethal exposure to lead nitrate (n = 30, each concentration).

Concentration (mg L^{-1})	N	Mortality rate (%) on 96 h
0	30	0
100	30	0
200	30	0
240	30	16.6
260	30	43.3
280	30	63.3
300	30	83.3
320	30	100

Table 3. Lethal concentrations (LC_{50}) of lead nitrate depending on time (24 to 96 h) for grass carp.

Point	Concentration (mg L^{-1})	95% confidence limits
LC_1	179.821	157.052-194.214
LC_5	199.341	182.339-210392
LC_{10}	209.747	195.665-219.170
LC_{15}	216.768	204.548-225.201
LC_{50}	246.455	239.668-253.143
LC_{85}	176.141	267.914-287.957
LC_{90}	283.162	273.972-296.814
LC_{95}	293.568	282.774-310.116
LC_{99}	313.089	298.976-335.380

Table 4. Growth parameters and survival rate of Grass carp in experimental treatments (Trial 1 to 2) and control.

Growth Index	Treatment		
	Control (Free of metal)	T1 (12.32 mg L^{-1} lead nitrate)	T2 (24.64 mg L^{-1} lead nitrate)
IW[1]	4.30±0.01	4.32±0.02	4.30±0.01
FW[2]	6.82±0.01[a]	6.28±0.02[b]	5.96±0.32[b]
WG[3]	2.52±0.02[a]	1.96±0.04[b]	1.66±0.33[b]
SGR[4]	0.77±0.01[a]	0.62±0.01[b]	0.54±0.09[b]
FCR[5]	22.94±0.47[a]	26.00±0.75[a]	34.68±2.06[b]
SR[6]	96.18±1.47[a]	89.07±1.47[a]	69.12±2.60[b]

Groups with different alphabetic superscripts at the same row differ significantly at P<0.05. IW = Initial weight (g), FW = final body weight (g), WG = body weight increased (g), SGR = specific growth rate for weight (% BW day^{-1}), FCR = feed conversion ratio (%), SR = survival rate (%).

conversation ratio (FCR) was significantly increased in T2 when in comparison with the control and T1 (P < 0.05). Between the two different concentrations of lead nitrate to Grass carp, the greatest effect appeared to be obtained in treatments T2 (concentration 24.64 mg L^{-1} of lead nitrate). This is particularly false for food conversation ratio (FCR) where the lower was obtained in the experimental control treatment, and survival rate where the highest was obtained in the experimental control treatment not significantly by T1 (12.32 mg L^{-1}) (P > 0.05). Of course, final body weight, body weight increased, specific growth rate and body weight gain in T1 (12.32 mg L^{-1}) were not significant by T2 (24.64 mg L^{-1}) (Table 4).

DISCUSSION

Heavy metal pollution in water is, in large part, due to agricultural run-off, industrial waste and mining activities. Mining is by far the biggest contributor to metal pollution. Mine drainage water, effluent from the tailing ponds and drainage water from soil heaps continue to extrude unwanted metals into the aquatic environment (Rani and Sivaraj, 2010).

The present study was initiated to find the susceptibility of the Grass carp to potentially hazardous lead nitrate on the survival and growth performance. The results showed that median lethal concentration (LC_{50}) of lead nitrate to Grass carp for 96 h of exposure is 246.455 ppm. The median lethal concentration 96 h (LC_{50}) value of lead in other aquatic organisms was reported as 300 μ for lead as in *Tench tinca* (Shah and Altindag, 2005a), which were higher than present study. The toxicity reported by other studies differs from this study probably due to different species used, age, size of the organism, test methods and water quality such as water hardness, as this can affect toxicity (Hodson et al., 1982; McCahon and Pascoe, 1988). Toxicity of metals may vary depending on their permeability and detoxification mechanisms (Darmono and Denton, 1990).

Toxic effect of lead on the growth performance of grass carp under chronic sublethal concentration showed that the lead nitrate had harmful effects on the growth para-meters on Grass carp. These results are in accordance with the findings of Javed et al. (1993b) that reported the fish (*C. mrigala*) stressed with sublethal concentration of lead showed significantly lower weight increment (42.20 ± 35.52 g) than control fish (55.55 ± 29.47 g).

Also, these results are in accordance with the findings of Hayat et al. (2007) who exposed the fingerlings of three major carps: *C. catla, Labeo rohita* and *Cirrhina mrigala* to sublethal concentrations of manganese for 30 days. During this exposure period, all the fish species showed negative growth. Also, these results are in accor-dance with the findings of Javed et al. (1993b) that observed low feed conversion ratios in major carps (*C. catla, L. rohita* and *C. mrigala*) due to exposure of these fish to water-borne zinc (Javed, 2005a).

The results of these studies may provide guidance to select acute toxicity to be considered in field bio-monitoring efforts designed to detect the bioavailability of lead nitrate and early warning indicators of this heavy metal toxicity in Grass carp.

REFERENCES

AOAC (2000). Official Methods of Analysis. Association of Official Analytical Chemist. ESA.

APHA (1998). Standard Methods for the Examination of Water and Wastewater, 20th edition. American Public Health Association, New York.

Darmono D, Denton GRW (1990). The pathology of cadmium and nickel toxicity in the banana shrimp *Penaeus merguiensis* de Man. Asian Fish Sci. 3(3):287-297.

De Boeck G, Vlaeminck A, Blust R (1997). Effects of sublethal copper exposure on copper accumulation, food consumption, growth, energy stores, and nucleic acid content in common carp. Arch. Environ. Contam. Toxicol. 33:415-422.

Fatoki OS, Lujiz N, Ogunfowokan AO (2002). Trace metals pollution in Umtatma River. Water S. Afr. 28: 83–89.

Finney D (1971). "Probit Analysis." Cambridge University Press. Cambridge, UK.

Hayat S, Javed M, Razzaq S (2007). Growth performance of metal stressed major carps viz. *Catla catla, Labeo rohita* and *Cirrhina mrigala* reared under semi-intensive culture system. Pak. Vet. J. 27:8-12.

Heath AG (1995). Water pollution and fish physiology. CRC press, Boca Raton. pp. 141-170.

Hodson PV, Dixon DG, Spry DJ, whittle DM, Sprague JB (1982). Effect of growth rate and size of fish on rate of intoxication by waterborne lead. Canadian J. Fish. Aquat. Sci. 39(9):1243-1251.

Jackson RN, Baird D, Els S (2005). The effect of the heavy metals, lead (Pb^{2+}) and zinc (Zn^{2+}) on the brood and larval development of the burrowing crustacean, *Callianassa kraussi*. Water S. Afr. 31(1):107-116.

Javed M (2005a). Heavy metal contamination of freshwater fish and bed sediments in the River Ravi stretch and related tributaries. Pak. J. Biol. Sci. 8:1337-1341.

Javed M, Hassan M, Javed K (1993b). Length weight relationship and condition factor of *Catla catla, Labeo rohita* and *Cirrhina mrigala* reared under polyculture condition of pond fertilization and feed supplementation. Pak. J. Agric. Sci. 30(2):167-172.

Laws E (2000). Aquatic Pollution: An introductory text. John Wiley & Sons. New York, USA. pp. 309-430.

McCahon C, Pascoe D (1988). Use of *Gammarus pulex* (L.) in safety evaluation tests: culture and selection of a sensitive life stage. Ecotoxicol. Environ. Saf. 15(3):245-252.

More TG, Rajput RA, Bandela NN (2003). Impact of heavy metals on DNA content in the whole body of freshwater bivalve, *Lamelleiden marginalis*. Environ. Sci. Pollut. Res. 22:605-616.

Nussey G, Vuren VJHA, Preez HH (2000). Bioaccumulation of chromium, manganese, nickel and lead in the tissues of the moggel, (*Labeo umbratus*) from Witbank Dam. Mpumalanga. Water S. Afr. 26(2):264-284.

OECD (1993). Guidelines for testing of chemicals. OECD, Paris.

Rani AMJ, Sivaraj A (2010). Adverse effects of chromium on amino acid levels in freshwater fish *Clarias batrachus* (Linn.). Toxicol. Environ. Chem. 92(10):1879-1888.

Shah SL (2006b). Hematological parameters in tench *Tinca tinca* after short term exposure to lead. J. Appl. Toxicol. 26:223-228.

Shah SL, Altindag A (2005a). Effects of heavy metal accumulation on the 96-h LC_{50} values in tench *Tinca tinca* L., 1758. Turk. J. Vet. Anim. Sci. 29:139-144.

Sorensen EM (1991). Metal poisoning in fish. CRC Press, Boca Raton. pp. 175–234.

Tulasi SJ, Reddy PUM, Romano-Rao JV (1992). Accumulation of lead and effects on total lipids on lipid derivatives in the freshwater fish *Anabas testudineus* (Bloch), Ecotoxicol. Environ. Saf. 23: 33-38.

Wyn B, Sweetman JN, Leavitt PR (2007). Historical metal concentrations in lacustrine food webs revealed using fossil ephippia from Daphnia. Ecol. Appl. 17(3):754-764.

23

Acute and subacute toxicity of aqueous extract of leaves mixture of *Aloe buettneri* (Liliaceae), *Dicliptera verticillata* (Acanthaceae), *Hibiscus macranthus* (Malvaceae) and *Justicia insularis* (Acanthaceae) on Swiss mice and albinos Wistar female rats

Pone K. B.[1,2]*, Telefo, P. B.[2], Gouado, I.[3] and Tchouanguep, F. M.[2]

[1]Central Institute of Medicinal and Aromatic Plants, P. O. CIMAP, Kukrail Picnic Spot Road, 226015-Lucknow, Uttar Pradesh, India.
[2]Department of Biochemistry, Faculty of Science, University of Dschang. P. O. Box 67 Dschang, Cameroon.
[3]Department of Biochemistry, Faculty of Science, University of Douala, P. O. Box 24157, Douala, Cameroon.

Aloe buettneri (Liliaceae), *Justicia insularis* (Acanthaceae), *Hibiscus macranthus* (Malvaceae) and *Dicliptera verticillata* (Acanthaceae) (ADHJ) are medicinal plants generally found in tropical and subtropical areas. The leaf mixture of these plants is used in the Western Region of Cameroon to increase fertility, regularize the menstrual cycle and to treat dysmenorrhoea or cases of infertility in women. In order to evaluate the toxicity of the leaf mixture extract of these plants, the values of their LD_{50} and LD_{100} were determined in Swiss mice and the subacute toxicity studied in albinos Wistar female rats. The herbal drug induced changes in the physiological (body and vital organ weights), toxicological [alanine aminotransferase (A.L.T), aspartate aminotransferase (A.S.T), creatinine] and biochemical (total proteins) parameters. The LD_{50} and LD_{100} were 27 and 32 g/kg in male mice, respectively, whereas in female mice the values were 18 and 24 g/kg, respectively. The leaf mixture of the plants has significantly increased the reproductive organ weights of treated female rats. The serum, uterine and ovarian proteins as well as the creatinine levels was increased significantly, while the hepatic protein level was decreased. The rate of AST remained unchanged whereas that of ALT increased when animals were treated at the dose of 100 mg/kg. These results suggests on one hand that aqueous extract is not short-term poisonous but presents unfavourable effects in the long run (60 days) and on the other hand, the aqueous extract have a direct action on the reproductive organs and cause disturbances on cellular metabolism.

Key words: Toxicity, mice, female rats, toxicological and biochemical parameters.

INTRODUCTION

Medicinal plants are those which contain active substances on living organisms and are used as precursor in drug synthesis (Abayomi, 1984). Some of these plants contain secondary metabolites which can affect human and mammal reproduction (Butenandt and Jacobi, 1933). The leaf mixture of *Aloe buettneri, Dicliptera verticillata, Justicia insularis* and *Hibiscus macranthus* are vegetal species used in the Western region of Cameroon to increase fertility, regularize the menstrual cycle and treat dysmenorrhea or some cases of infertility in women. Burkill (1982) and Chopra (1933) have shown that the

---

*Corresponding author. E-mail: pon2812@yahoo.fr.

decoction of *J. insularis* alleviates the childbirth pain during the last month of pregnancy. Telefo et al., 1998 have reported that, different doses of aqueous extracts from the leaves of *Aloe buettneri*, *Justicia insularis*, *Hibiscus macranthus* and *Dicliptera verticillata*, when given daily to 22 day old rats for 5, 10, 15, 20 and 25 days by gastric intubation increases the serum oestradiol level attesting the presence of some oestrogenic compounds in the plant extracts. Furthermore, the aqueous extract of the leaf mixture of *Aloe buettneri*, *Dicliptera verticillata*, *Justicia insularis* and *Hibiscus macranthus* given by oral route to the immature female rats at the doses of 13, 49 and 94 mg/kg/day for 15 days induced a significant increase in ovarian and uteri weight as well as serum and ovarian oestradiol (Telefo et al., 2001). Guemo (2002) have shown that the decoction of the leaves of *A. buettneri*, *D. verticillata*, *J. insularis* induce the uterus contraction, whereas *H. macranthus* releases it.

Despite the various pharmacological data on the leaf mixture extract of these plants, no data on their toxicity is available in the literature. Therefore, the aim of the present study is to identify the adverse effects of the leaf mixture of these plants on Swiss mice and Wistar albino female rats. To attain our objective we have evaluated specifically: the values of LD_{50} and LD_{100} of the aqueous extract (AE); the effect of the aqueous extract on the body and reproductive organ weights (liver, kidney, ovary and uterus) and the effect of aqueous extract on toxicological (alanine aminotransferase [ALT], aspartate aminotransferase [AST], creatinine) and biochemical parameters (total proteins).

MATERIALS AND METHODS

Leaf samples of *A. buettneri*, *D. verticillata* and *J. insularis* were collected in March and May 2005 in Dschang region, while the leaf sample of *H. macranthus* was collected in Batoufam area (West-Cameroon) in May and June 2005. The aerial part of each specimen were air-dried in shadow at room temperature and reduced into fine powder using an electric grinder (Moulinex).

Extract preparation

The powders were mixed in the proportion of 25% *A. buettneri*, *D. verticillata*, *H. macranthus* and *J. insularis* (ADHJ). Then, the crude aqueous extract (AE) of leaves was prepared by adding 1000 ml of distilled water to 100 g of mixture leaf powder and kept at 100°C for 30 min. The extract was filtered and then concentrated in rotary evaporator at temperature lower than 45°C. The yield of the dry residue was 26.56% w/w.

Experimental animals

Adult male and female Swiss mice (weighing 20-30g) and 21 day old female Wistar rats (weighing 22-35g) were used. They were bred in the animal house of the Faculty of Science at the University of Dschang-Cameroon under approximately 20°C. They were receiving food and water *ad libitum*.

Toxicity testing

Acute toxicity

The mice were treated orally with doses ranging from 2 to 32 g/kg of the crude extract. The animals were observed for the clinical signs of toxicity (locomotion, sensibility to pinch, agressivity, tail and faeces aspects) during 24 h. Furthermore, the LD_{50} values were calculated according to Berhens and Karber method (1983).

Sub acute toxicity

The extract was administered to rats by force-feeding at the rate of 0; 12.5; 50 and 100 mg/kg doses for 40 (group 1) and 60 (group 2) days. All the animals were observed for appearance of toxicity signs or behavioural alterations during the experimental period. At the end of the treatment, the animals were anaesthetized with chloroform vapours and the blood was collected through cardiac puncture using sterile syringes. The animals were sacrificed and the uterus, ovary, spleen, liver, suprarenal gland and kidneys were collected, cleaned and weighed.

Preparation of the homogenates

After weighing the ovaries and uteri, they were grinded separately in mortars containing fresh phosphate buffer (1/150 M; pH 7.0) to obtain 1% homogenate. The liver homogenate was prepared by the same procedure but at 10%. The homogenates were centrifuged for 20 min at 4°C and 2000 rpm. The supernatants were separated from the pellets and stored at -20°C until use.

Serum preparation

After the sacrifice of the animals by decapitation, the blood was collected in the test tubes and allowed to rest for 2 h at 2°C. Further, all the tubes were centrifuged for 20 min at 4°C and 1500 rpm. The supernatants (serum) were collected using the micro-pipette and kept at -20°C.

Biochemical analysis

Serum and hepatic total proteins were measured by Biuret method described by Gornall et al. (1949), while uteri and ovaries total proteins were assessed by Bradford method (1976). Measurements of alanine aminotransferase (ALT), aspartate aminotransferase (AST) and creatinine levels were done with their respective kits.

Statistical analysis

The differences among the experimental and control groups were determined using the ANOVA test and comparison of average performed with Paraphenylenediamine (PPD)-Fischer test. Level of significance was set at $p < 0.05$.

RESULTS

Acute toxicity of AE of (ADHJ)

Diarrhoea, decrease of the activity and difficulties in the locomotion were observed within the first 6 h of treatment in the groups receiving the extract at doses varying

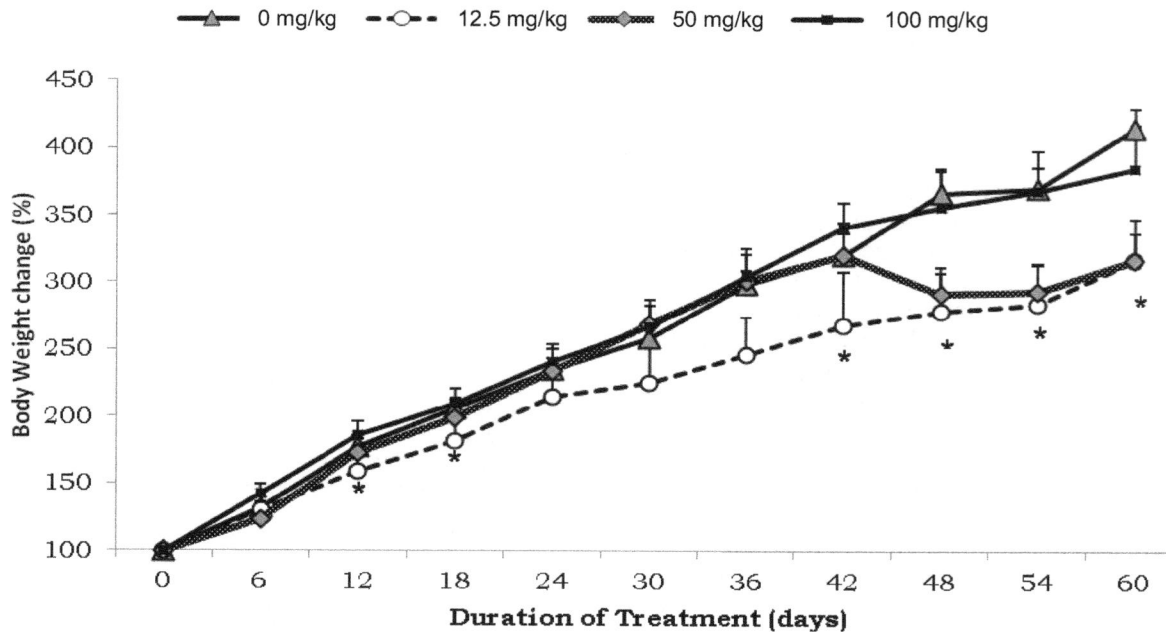

Figure 1. Effect of AE of ADHJ on body weight of rats after 40 and 60 days treatment.

Table 1. Effect on liver weight, hepatic and serum proteins after oral administration of AE of ADHJ during 40 and 60 days in Wistar female rats.

Dose (mg/kg)	Treatment duration					
	40 days			60 days		
	Liver weight (g/ 100 g of bw)	Serum protein (mg/ml)	Hepatic protein (mg/g of bw)	Liver weight (g/ 100 g of bw)	Serum protein (mg/ml)	Hepatic protein (mg/g of bw)
0	5.43 ± 0.21	132.03 ± 2.32	12.27 ± 0.57	4.05 ± 0.16	160.67 ± 4.90	15.17 ± 0.48
12.5	5.97 ± 0.30	177.84 ± 4.61***	14.01 ± 0.52*	4.09 ± 0.24	173.92 ± 11.09*	11.86 ± 0.84*
50	4.80 ± 0.18	171.28 ± 1.12***	13.39 ± 0.46*	3.73 ± 0.06	178.65 ± 10,43*	12.92 ± 0.69*
100	5.86 ± 0.16	180.73 ± 2.23***	19.67 ± 0.09**	3.70 ± 0.09	216.89 ± 23.78**	11.98 ± 1.34*

The values represented Mean ± ESM of liver weight, hepatic and serum proteins levels in each group (n=6). *$p<0.05$; **$P<0.01$; ***$P<0.001$ (PPD-Fischer test). Bw = Body weight.

between (16 to 32 g/kg). No abnormalities were observed in treated groups at varying doses (between 2 to 4 g/kg) compared to the vehicle control. After 48 h, the death of all the animals was recorded at the doses of 24 and 32 g/kg, respectively for female and male mice. The LD_{50} values were 18 and 26.8 g/kg for male and female mice, respectively.

Subacute toxicity of AE of (ADHJ)

Effect of AE of ADHJ on body weight

The body weights were linearly proportional to the period of treatment and the values were increased at the end of the treatment in comparison to the initial values. This is materialized by the appearance of the curves from 0 to

60 days, but as compare to the vehicle control the weights of the rats treated with 12.5 mg/kg had decreased significantly (Figure 1).

Effect of AE of ADHJ on hepatotoxic biochemical parameters

No variation was found in relative liver weight (Table 1). The hepatic protein levels decreased in the groups treated during 60 days, while the values were increased in the groups treated during 40 days. The administration of AE of ADHJ did not change the AST level. However, at the rate of 12.5; 50 and 100 mg/kg, the ALT levels were decreased compared to the vehicle control (Figure 2) in rats treated during 40 days. The ALT level was also increased when the animals were treated during 60 days

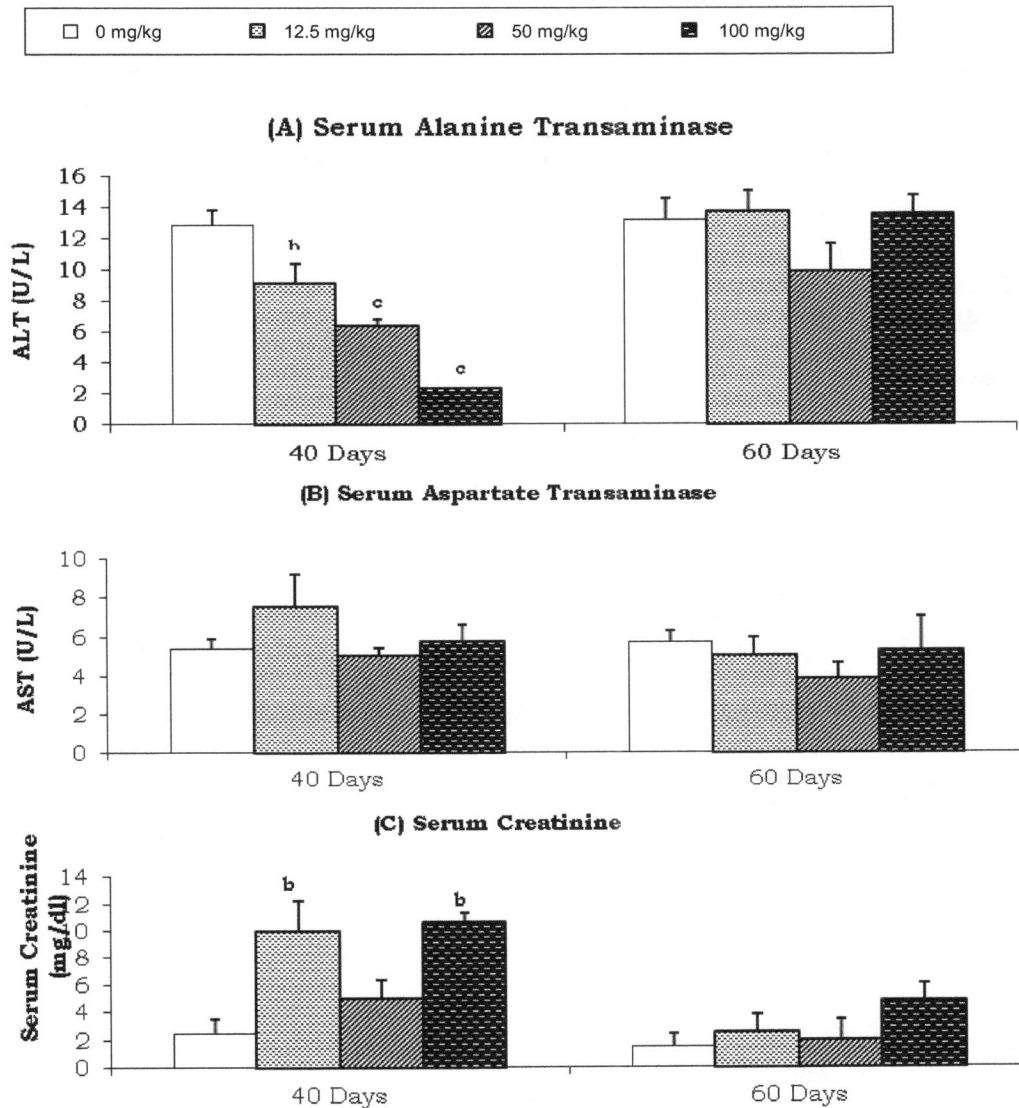

Figure 2. ALT (A), AST (B) and creatinine (C) levels in the serum of treated rats during 40 and 60 days. The values represented Mean ± ESM of liver weight, hepatic and serum proteins levels in each group (n=6). *p<0.05; **p<0.01; ***p<0.001 (PPD-Fischer test).

at 100 mg/kg (Figure 2).

Nephrotoxicity of AE of ADHJ

The administration of the plant mixture extract has induced a significant increase of the kidney weights in the rats treated during 40 days whereas in those treated during 60 days the values remain unchanged, all compared to the vehicle control (Table 2). The serum creatinine level significantly increased in all the groups treated during 40 days with the leaf mixture AE (Figure 2).

Effect of AE of ADHJ on uterus and ovary of treated female rats

The administration of AE of ADHJ did not change ovarian weight but increased uterine weight, corpus luteum number, ovarian and uterus protein levels (Table 3).

DISCUSSION

Acute toxicity of AE of ADHJ

Generally, the oral administration of AE of ADHJ to male

Table 2. Effect of AE of ADHJ on relative organ weights of kidneys, spleen and suprarenal glands on Wistar female rats.

Dose (mg/kg)	Treatment duration					
	40 days			60 days		
	Kidney weight (g/ 100 g of bw)	Spleen weight (g/ 100 g of bw)	Weight of surrenal gland (g/ 100 g of bw)	Kidney weight (g/ 100 g of bw)	Spleen weight (g/ 100 g of bw)	Weight of surrenal gland (g/ 100 g of bw)
0	0.83 ± 0.05	0.66 ± 0.05	0.031 ± 0.003	0.96 ± 0.03	0.46 ± 0,09	0.0439 ± 0.0017
12.5	0.99 ± 0. 06*	0.61 ± 0.09*	0.033 ± 0.003	0.94 ± 0.04	0.42 ± 0.09	0.0378 ± 0.002
50	1.15 ± 0.07**	0.47 ± 0.02*	0.035 ± 0.002	0.93 ± 0.02	0.48 ± 0.018	0.035 ± 0.0032**
100	1.01 ± 0.02**	0.50 ± 0.05*	0.032 ± 0.001	0.99 ± 0.03	0.48 ± 0.02	0.0307 ± 0.015***

The values represented Mean± ESM of relative organ weights of kidneys, spleen and suprarenal glands in each group (n=6). *p<0.05; **P<0.01; ***P<0.001 (PPD-Fischer test).

Table 3. Effect of AE of ADHJ on corpus luteum number, ovarian and uterus protein levels and weights.

Treatment duration (days)	Dose (mg/kg)	Ovarian weight (mg/100 g bw)	Ovarian protein (µg/mg)	Uteri weight (mg/100 g bw)	Uteri protein (µg/mg)	Corpus luteum number
40	0	36.8 ± 3,10	20.51 ± 1.48	65.10 ± 6.00	23.46 ± 1.34	0.00 ± 0.00
	12.5	28.00 ± 3.70	32.06 ± 3.02**	69.00 ± 12.00	27.06 ± 1.29*	2.67 ± 2.67*
	50	45.10 ± 2.30	31.18 ± 1.46**	105.50 ± 25.30	25.00 ± 0.46*	2.33 ± 2.33*
	100	38.00 ± 3.10	31.03 ± 2.97**	99.40 ± 32.00	30.57 ± 1.50***	10.00 ± 0.00***
60	0	41.80 ± 1.90	14.64 ± 1.23	65.00 ± 6.60	14.64 ± 10.23	2.00 ± 1.26
	12.5	53.60 ± 4.30	25.00 ± 0.59***	88.90 ± 6.55	25.00 ± 0.59	17.00 ± 4.60**
	50	42.40 ± 4.00	29.33 ± 1.17***	82.70 ± 9.10	29.33 ± 1.17	14.17 ± 6.77**
	100	44.90 ± 2.80	29.69 ± 2.13***	78.30 ± 7.90	29.69 ± 2.13	11.17 ± 2.44**

The values represented Mean ± ESM of corpus luteum number, ovarian and uterus protein levels and weights in each group (n=6). *p<0.05; **P<0.01; ***P<0.001 (PPD-Fischer test).

and female mice at the doses of 16 – 32 g/kg induced difficulties in locomotion and agressivity. These signs may have resulted from the attack of the nervous system (Bep and Bever, 1986) owing to the high intake of the herbal drug.

Similar signs have also been recorded in the study of toxicological evaluation of the aqueous extract of *Allium sativum* bulbs on laboratory mice and rats (Gatsing et al., 2005). The low reaction of the mice after pinching the tail and decreased sensibility in touch may result from the low level of prostaglandins induced by the treatment with the extract. Prostaglandins are hormones that regulate the perception of pain (Lehninger, 1982). Prostaglandins are not stored, and their release is dependent on biosynthesis. Evidently, various medications that prevent the perception of pain inhibit the conversion of arachidonic acid by inhibiting the release of prostaglandins synthetase or by interfering in some other way with the synthesis of prostaglandins (Eisenhauer et al., 1998). In spite of the above side effects, the LD_{50} values higher than 5g/kg permit to classify the AE of ADHJ among the non toxic substances according to the Hodge and Steiner criteria (Delongeas et al., 1983; Schorderet, 1992).

Subacute toxicity of AE of ADHJ

The AE of ADHJ induced various degrees of activities on some biochemical parameters of toxicological study. The oral administration of AE of ADHJ during 40 days increased body and organs (suprarenal glands, uterus ovary and kidneys) weights. The increase in reproductive organ weight gain was certainly due to the fixation of estrogenic compounds present in the AE on uterus and ovary receptors (Jensen and Desombre, 1972; Katzenellenbogen et al., 1979). Similar results were obtained by Ettebong et al. (2011) during the study of the contraceptive, estrogenic and anti-estrogenic potentials of methanolic root extract of *Carpolobia lutea* in rodents where in very low doses the plant extract has increased the uterine wet weight. Meanwhile, the standard error of the uteri weights were high and might be due to the fact that the uterus mass was affected by womb fluid collection. The decrease of the body weights in rats treated with the plant mixture at the dose of 12.5 mg/kg might be due to the less food and water intake as reported by Joseph et al., 1989. The significant increase of ALT serum levels in rats treated with the AE of ADHJ

(100 mg/kg) compare to that of the control group suggested that the extract might be harmful to the liver as described by Kakeno et al. (1997) and Emerson et al. (1993). The ALT in blood increases when the hepatic cellular permeability varies or when necrosis and cellular injury occur. The significant ($p < 0.01$) increase of creatinine level in rats treated with AE during 40 days might be due to their release in urine during the excretion. Moreover, the creatinine value was higher than 0.04 g/L; consequently a nephrotoxicity may be suggested with reference to the work done by James and Kathleen (1992).

Conclusion

Despite the fact that the AE of ADHJ is not short term poisonous, the increase of ALT and creatinine serum levels in treated rats at the rate of 100 mg/kg during 40 days and 12,5, 50, 100 mg/kg during 60 days is a proof that this extract may exhibits some adverse effects in the long run. Thus, considering the important and widespread traditional use of the plants in combination, for the sake of the safety of populations, the histopathological analyses need to be performed in order to study the manifestations on the liver and kidneys through microscopic examinations.

REFERENCES

Abayomi S (1984). Medicinal plants and Traditional Medicine in Africa. John Wiley and Sons Ltd. p.247

Bep O, Bever (1986). Medicinal plants in tropical West Africa. Cambridge University Press, New York p.375.

Berhens B, Karber G (1983). "Mathematics for naturalists and agriculturalists". PWN, Warsaw pp.218-219.

Bradford M (1976). A rapid and sensitive method for the quantification of microgram quantities of protein utilizing the principle of protein-dye-binding. Anal. Biochem. 72:248-254.

Burkill HM (1982). The useful plants of west tropical Africa.2nd Ed. Vol.1, Royal Botanic Garden K.E.W. pp.120-145.

Butenandt A, Jacobi H (1933). Female sexual hormone preparation from a plant, tokokinin and its identification with the α-follicular hormone. Zeitschrift für physiologische Chemie 218:204-222.

Chopra UN (1933). Indigenous drugs of India. Their medical and economic aspects. The Art Press, Calcutta, India p.550

Delongeas JL, Burnel D, Netter P, Grignon M, Mor J, Royer RJ (1983). Toxicité et pharmacocinétique de l'oxychlorure de zincinium chez la souris et chez le rat. J. Pharmacol. 14(4):437-447.

Eisenhauer L, Nichols WL, Spencer TR, Bergan WF (1998). Clinical Pharmacology and Nursing Management. Philadelphia, New York. Lippincott pp.779-809.

Emerson SP, Sharada AC, Devi UP (1993). Toxic effects of crude nut extract of plumbago nosea (Raktachitraka) on mice and rats. J. Ethnopharmacol. 38:79-84.

Ettebong EO, Nwafor PA, Ekpo M, Ajibesin KK (2011). Contraceptive, estrogenic and anti-estrogenic potentials of methanolic root extract of Carpolobia lutea in rodents. Pak. J. Pharm. Sci. 24(4):445-449.

Gatsing D, Reseline A, Jules RJ, Garray IH, Jaryum KH, Nestor T, Chouanguep FM, Godwin A (2005). Toxicological evaluation of the aqueous extract of Allium sativum bulbs on laboratory mice and rats. Cameroon J. Exp. Biol. 01(1):39-45.

Gornal AG, Bardwil G, David MM (1949). Determination of serum proteins by mean of Biuret reactions. Biochemistry 177:751-756.

Guemo TC (2002). Effets des extraits aqueux de Dicliptera verticillata GJH Amshoff, (Acanthacées), Hibiscus macranthus Hochst ex A. Rich (Malvacées) Justicia insularis T. Anders (Acanthacées), A. buettneri A. Berger (Liliacées) et de leur mélange sur le muscle utérin de la ratte en oestrus. Mémoire de maîtrise. Département de Biochimie. Université de Dschang.

James TP, Kathleen D (1992). Mosby's diagnostic and laboratory test reference. Most by year book, St. Louis, USA. p.843.

Jensen EV, Desombre ER (1972). Mechanism of action of the female sex hormones. Annu. Rev. Biochem. 41:203-230.

Joseph PK, Rao KR, Sundaresh CS (1989). Toxic effects of garlic extracts and garlic oil in rats. Indian J. Exp. Biol. 27:977-979.

Kakeno JJ, Harvey JW, Bruss ML (1997). Clinical Biochemistry of Domestic Animals, 5th éd. Academic Press, San Diego p.932.

Katzenellenbogen BS, Bhakoo HS, Ferguson ER, Lan NC, Tatee J, Tsia TLS, Katzenellenbogen JA (1979). Estrogens and antiestrogens action in reproductive tissues and tumours. Recent Progress in Hormone Research 35:259-292.

Lehninger LA (1982). Principles of Biochemistry. New York, Worth Publishers. Inc. p.1011.

Schorderet M (1992). Pharmacologie des concepts fondamentaux aux applications thérapeutiques. Eds Frisson Roche et Iatkine. Paris Grenoble p.920.

Telefo PB (1998). Contribution à l'étude des plantes médecinales du Cameroun: Influence de l'extrait aqueux du mélange de feuilles d'Aloe buettneri A. Berger (Liliacées), Dicliptera verticillata GJH Amshoff (Acanthacées), Hibiscus macranthus Hochst ex A-rich (Malvacées), Justicia Insularis T. Anders (Acanthacées), sur certains paramètres biochimiques et physiologiques de la reproduction chez la rate. Thèse de Doctorat 3e cycle en biochimie. Université de Yaoundé pp.1-154.

Telefo PB, Moundipa PF, Tchouanguep FM (2001). Influence de l'extrait aqueux de feuilles d'Aloe buettneri, Dicliptera verticillata, Hibiscus macranthus, Justicia insularis sur la fertilité et quelques paramètres biochimiques de la reproduction chez la rate. Revue de l'Academie des Sciences du Cameroun 1(30):7: 144-150.

24

Heavy metal contamination risk through consumption of traditional food plants growing around Bindura town, Zimbabwe

P. Dzomba*, S. Nyoni and N. Mudavanhu

Chemistry Department, Faculty of Science Education, P Bag 1020 Bindura, Zimbabwe.

Mining activities are a leading cause of heavy metal contamination of soils and food plants leading to health risk fears. The concentration of heavy metals in soils, *Bidens pilosa* and *Fadogia ancylantha* in the vicinity of Bindura town were investigated. Results show that heavy metal concentration of soils and plants found around Bindura town was higher than that of the control area where there were no mining activities suggesting heavy metal contamination. The higher values of heavy metals may place consumers of these plants at health risk with time via bioaccumulation. It is suggested therefore that measures should be taken to lower plant and soil heavy metal accumulation.

Key words: Heavy metals, food plants, health risk, metal accumulation.

INTRODUCTION

Mining activities can contribute to heavy metal pollution of the environment (Navarro et al., 2008; Singh et al., 2005). Progressive accumulation of heavy metals in soils surrounded by mines, result in increased heavy metal uptake by food plants. This is worrisome because of potential health risk to the people leaving in the surrounding areas (Pruvot et al., 2006). Elements like Pb, Cd, Cr and Ni, are said to be non biodegradable thus, persist everywhere in the environment and have the ability to be deposited in various body organs which poses a great threat to the human health (Chen et al., 2005). Several researches have shown that food plants, growing in heavy metal contaminated soils have higher concentrations of heavy metals than those grown in uncontaminated soil (Akan et al., 2010). It has been reported that serious health problems may develop as a result of excessive accumulation of heavy metals such as Cd, and Pb in the human body (Zhuang et al., 2008). Despite Zn and Cu being essential elements in the diet, excessive concentration in food plants is of great concern because they are toxic to humans and animals (Kabata-Pendias and Mukherjee, 2007). Pb and Cd metals are believed to be potential carcinogens and are implicated in the ontology of many diseases, especially cardiovascular diseases, kidney, nervous system, blood as well as bone ailments (Jarup, 2003). Recent studies reported that soil and vegetables polluted with Pb and Cd in Copsa Mica and Baia Mare, Romania, significantly contributed to decreased human life expectancy within the affected areas, reducing average age at death by 9 to 10 years (Zhuang et al., 2008). Another study reported that children living around a smelter in France (Pruvot et al., 2006) and Brazil (Bosso and Enzweiler, 2008) had high concentration of Pb in their blood. Türkdoğan et al. (2002) believed that the increased prevalence of upper gastrointestinal cancer rates observed in the Van region of Turkey was related to the high concentration of heavy metals in the soil, fruit and vegetables. Dietary consumption is the major route of exposure to heavy metals for most people, apart from inhalation (McBride, 2003; Khan et al., 2008).Therefore periodic monitoring of heavy metal concentrations in food plants is very important for assessing their risk to human health. Bindura town is surrounded by several mines which can be potential sources of heavy metals. Therefore, the aim

---

*Corresponding author. E-mail: pdzomba@gmail.com or pdzomba@buse.ac.zw.

Table 1. Certified reference material concentrations (mg kg^{-1}).

Metal	CRM 1			CRM 2		
	Certified value	Measured value	Recovery (%)	Certified value	Measured value	Recovery (%)
Cu	120 ± 0.6	119 ± 0.5	99	2.5 ± 0.03	2.5 ± 0.01	100
Zn	260 ± 0.5	258 ± 0.4	99	13.1 ± 0.5	12.6 ± 0.5	96
Cd	1.2 ± 0.01	1.1 ± 0.05	92	0.10 ± 0.00	0.11 ± 0.00	110
Pb	73 ± 0.5	73 ± 0.5	100	0.11 ± 0.01	0.12 ± 0.00	109

% Recovery = mean measured value/mean certified value × 100%. Values are expressed as mean ± SD, n = 10.

of the present study was to quantify the concentration of heavy metals in soils, *Fadogia ancylantha* and *Bidens pilosa,* herbs growing in the natural environment. *F. ancylantha* and *B. pilosa* leaves are fused to make a kind of tea believed to have medicinal properties. *B. pilosa* leaves are also used as relish. The two plants are highly consumed by people surrounding Bindura town.

MATERIALS AND METHODS

All reagents were of analytical grade. HCL, HNO$_3$ and H$_2$O$_2$ were obtained from Merck, New Jersey, USA. Calibration and certified reference materials were obtained from Aldrich, Steinheim, Germany. Double deionized water was used for all analysis. Analysis of heavy metal content was performed using AAS, Varian Spectra AA 600 flame Atomic Absorption spectrophotometer.

Sample collection and preparation

Plant samples were collected from grass lands around Bindura town in April, June and July 2011 and these were compared to samples collected from St Albert's in Chiweshe where there is no mining activities (control). Soil samples were collected from five points of 50 m apart each randomly selected. In each, soil samples were collected at a depth of 0 to 10 cm. In all cases, samples were collected in clean polythene bags and transported straight to the laboratory. Soil samples were then dried at room temperature, crushed and passed through 2 mm sieve. The samples were then put in clean plastic bags and sealed. Dry plant samples (air dried at room temperature) were crushed in a mortar and the resulting powder digested by weighing 0.5 g of dried ground and sieved (1 mm stainless-steel mesh) into an acid-washed porcelain crucible and placed in a muffle furnace for four hours at 500℃. The crucibles were removed from the furnace and cooled. 10 ml of 6 M HCl were added covered and heated on a steam bath for 15 min. Another 1 ml of HNO$_3$ was added and evaporated to dryness by continuous heating for one hour to dehydrate silica and completely digest organic compounds. Five milliliters of 6 M HCl and 10 ml of Milli-Q water were then added and the mixture heated on a steam bath to complete dissolution. The mixture was cooled and filtered through a Whatman no. 1 filter paper into 50 ml volumetric flasks and made up to the mark with Milli-Q water. Two grammes of the soil samples were weighed into acid-washed glass beakers and digested by the addition of 20 cm^3 of aqua regia (mixture of HCl and HNO$_3$, ratio 3:1) and 10 cm^3 of 30% H$_2$O$_2$. The H$_2$O$_2$ was added drop wise to avoid any possible overflow leading to loss of material. The beakers were covered by means of a watch glass, and heated over a hot plate at 90℃ for two hours. The beaker walls and watch glass were washed with Milli-Q water and the samples were filtered

and the filtrate collected. The volumes were adjusted to 100 cm^3 with Milli-Q water. Blanks were made in the same way but without samples. All samples and blanks were stored in plastic containers.

Physico-chemical assay

The pH was measured using a 1:2 soil: water ratio (Mclean, 1982).

Elemental analysis of samples

Determination of Cu, Zn, Fe, Cr, Cd, As, Ni and Pb in soil and vegetable samples were made directly on each of the final solution under atomic absorption spectroscopy (AAS).

Quality assurance and quality control

Suitable quality control and assurance protocols were carried out to ensure reliability of the results. Milli-Q water was used throughout the study. Glassware was properly cleaned, and the reagents used were of analytical grade. Reagents blank determinations were used to correct the instrument readings. All analyses were performed ten times. Certified reference materials (CRM) were used for validation of the analytical procedure. The results of measurements of CRMs are summarized in Table 1.

Statistical analysis

The data was presented as mean ± standard deviation (SD) of ten determinations. Bio-accumulation factor (BAF) values for each metal were subjected to Student's t test to compare values for Bindura samples to that of control site and that of the two plant species. ANOVA was applied to compare BAF values between sites followed by multiple comparisons using the least significant difference (LSD) test. Differences were considered statistically significant at $p < 0.05$. All statistical analyses were performed using SPSS 16 software (SPSS Inc., Chicago, IL, USA).

Bio-accumulation factor (BAF)

Bio-accumulation factor (BAF) is an index measuring the ability of a plant to accumulate a particular metal with respect to its concentration in the soil substrate (Ghosh and Singh, 2005). This was calculated as follows:

$$BAF = C_{plant}/C_{soil}$$

where C_{plant} and C_{soil} represent the heavy metal concentration in the useful part of plants and soils, respectively.

Table 2. Heavy metal concentration (mg kg^{-1}) of soil collected around Bindura town.

Site	pH	Mean total concentration							
		Cu	Zn	Fe	Cr	Cd	As	Ni	Pb
1	5.4[a]*	420[d]	254[d]	360[c]	218[d]	0.12[a]*	0.52[a]*	354[d]	14[d]*
2	5.5[b]*	243[b]	256[a]	380[d]*	220[a]	0.10[a]*	0.55[a]*	330[d]	16[b]
3	6.9[b]	331[a]	230[d]	380[b]*	215[a]	0.20[a]	0.48[b]	360[c]	13[d]*
4	5.5[a]*	345[d]	266[d]	366[a]	225[d]	0.09[a]	0.51[a]	352[c]	15[d]
5	6.8[a]	336[d]	250[d]	350[d]	230[d]	0.11[a]*	0.56[b]	365[d]	22[b]

n = 10; a = ± 0.00, b = ± 0.04, c = ± 0.33, and d = 0.18. ANOVA analysis was applied to compare results for site 1 to 5, Student t test was applied to compare results for each metal between control and Bindura town; *, = do not differ at p = 0.05; (LSD) test.

Table 3. Heavy metal concentration (mg kg^{-1}) of soil collected from Chiweshe (Non mining area).

Site	pH	Mean total concentration							
		Cu	Zn	Fe	Cr	Cd	As	Ni	Pb
1	6.8[a]*	200[a]	210[a]	360[c]	200[c]	0.01[a]	0.03[a]*	10[a]	8[a]*
2	6.6[a]*	210[c]	220[d]	370[d]	180[c]	0.06[a]	0.06[a]	15[a]	10[a]
3	6.4[b]	230[d]	211[c]	378[d]	205[d]	0.02[a]*	0.04[a]*	20[a]	6[a]
4	6.8[a]*	215[b]	213[c]*	358[d]	212[d]	0.03[a]	0.04[a]*	30[b]	12[b]
5	6.6[b]*	220[d]	214[c]*	330[b]	208[d]	0.02[a]*	0.03[a]*	28[a]	8[a]*

n = 10; a = ± 0.00, b = ± 0.01, c = ± 0.13, and d = 0.09. ANOVA analysis was applied to compare results for site 1 to 5; Student t test was applied to compare results for each metal between control and Bindura town; *, do not differ at p = 0.05; (LSD) test.

Table 4. Heavy metal concentration (mg kg^{-1}).of mostly consumed indigenous food/medicinal plants from Bindura town.

Metal	Plant species (Mean total concentration of metal)	
	F. ancylantha	B. pilosa
Cu	1.11 ± 0.02	0.60 ± 0.01
Zn	13.20 ± 0.31	25 ± 0.05
Fe	0.82 ± 0.01	0.91 ± 0.02
Cr	0.08 ± 0.00	0.09 ± 0.00
Cd	0.05 ± 0.00	0.06 ± 0.00
As	0.02 ± 0.00	0.01 ± 0.00
Ni	0.68 ± 0.01	0.88 ± 0.01
Pb	0.95 ± 0.04	0.86 ± 0.02

Values are expressed as mean ± SD, n = 10 (fresh weight basis). Student t test to compare results of the two plants and results of control area and Bindura town for the same plant.

RESULTS AND DISCUSSION

The results of heavy metal contamination in plant leaves soils are shown in Tables 2 to 5. The pH values of soils presented can be classified as acidic to neutral. The concentration of heavy metals in soils and plant species obtained from areas around Bindura town differs significantly to that of samples obtained from the control area (p = 0.05). The concentration of heavy metals in Bindura soils were highest for copper followed by iron with cadmium and arsenic being the lowest. Among the plants, the mean concentrations were highest for zinc in both plants. The mean concentrations for control soil samples were highest for iron, copper and zinc. Arsenic and cadmium were once more the lowest. The difference in mean heavy metal concentration for the control and the experimental area indicates the presence of some soil contamination.

Heavy metal contamination is more pronounced in areas around mines due to physical contamination by dust and through translocation (Itanna, 2002; Muchuweti et al., 2006). Soil to plant transfer is the major component

Table 5. Heavy metal concentration (mg kg^{-1}) of mostly consumed indigenous food/medicinal plants from Chiweshe (Non mining area)

Metal	Plant species (Mean total concentration of metal)	
	F. ancylantha	B. pilosa
Cu	0.09 ± 0.00	0.05 ± 0.05
Zn	10.12 ± 0.05	16.10 ± 0.03
Fe	0.08 ± 0.00	0.06 ± 0.00
Cr	0.02 ± 0.00	0.02 ± 0.00
Cd	0.00 ± 0.00	0.01 ± 0.00
As	0.00 ± 0.00	0.00 ± 0.00
Ni	0.01 ± 0.00	0.00 ± 0.00
Pb	0.01 ± 0.04	0.00 ± 0.00

Values are expressed as mean $\pm$ SD, n = 10 (fresh weight basis); Student t test to compare results of the two plants and results of control area and Bindura town for the same plant.

of human exposure to heavy metals through the food chain. The present results show that BAF values e. g. for zinc 0.00264 – 0.005045 differed significantly for the five chosen areas (ANOVA, p = 0.05). The difference can be rationalized by the fact that soil nutrient management differs due to soil properties Liu et al. (2005). Comparing the two studied species, BAF values for the metals Cu – Pb differed significantly. The result of the present study was compared with previous studies (Khan et al 2008; Cui et al., 2004). BAF values obtained in these studies differed significantly. For a given metal, the transfer value varies greatly with plant species (Cui et al., 2004).

Conclusions

The present study reveals that soil samples collected from areas around Bindura town consist of higher heavy metal concentration which might cause health risk to the local inhabitants through contamination of indigenous and medicinal plants. The present studied plants show higher heavy metal content as compared to control samples. It is therefore recommended that measures should be taken to arrest heavy metal contamination risk.

ACKNOWLEDGEMENT

The researchers would like to thank Bindura Research Center and Kutsaga Research Unit for their kind help.

REFERENCES

Akan JC, Abdulrahaman FI, Sodipo OA, Lange AG (2010). Physicochemical parameters in soil and vegetable samples from Gongulon Agricultural site, Maiduguri, Borno state, Nigeria. J. Am. Sci., 6: 12.

Bosso ST, Enzweller J (2008). Bioaccessible lead in soils, slag and mine waste from abandoned mining district in Brazil. Environ. Geochem. Health, 30: 219-229.

Chen Y, Wang C, Wang Z (2005). Residues and source identification of persistent organic pollutants in farmland soils irrigated by effluents from biological treatment plants. Environ. Intern., 31: 777-783.

Cui YJ, Zhu YG, Zhai RH, Chen DY, Huang ZH, Qiu Y (2004). Transfer of metals from soil to vegetables in an area near a smelter in Nanning, China. Environ. Int., 30: 785-91.

Itanna F (2002). Metals in leafy vegetables grown in Addis Ababa and toxicological implications. Ethiop. J. Health Dev., 6: 295-302.

Jarup L (2003). Hazards of heavy metal contamination. Brit. Med. Bull., 68: 167-182.

Kabata-Pendias A, Mukherjee AB (2007). Trace elements from soil to human. NewYork: Springer-Verlag.

Khan S, Cao Q, Zheng YM, Huang YZ, Zhu YG (2008). Health risks of heavy metals in contaminated soils and food crops irrigated with waste water in Beijing, China. Environ. Pollut., 152: 686-692.

Liu HY, Probst A, Liao BH (2005). Metal contamination of soils and crops affected by the Chenzhou lead zinc mine spill (Hunan, China). Sci. Total Environ., 339: 153-166.

McBride MB (2003). Toxic metals in sewage sludge-amended soils: has promotion of beneficial use discounted the risks? Adv. Environ. Res., (8): 5-19.

Mclean EO (1982). Soil pH and lime requirement. In: A.L. Page, R.H. Miller, and D.R. Keeney (eds.), Methods of Soil Analysis. Part 2: Chemical and Microbiological Properties. 2nd ed. American Society of Agronomy. Madison. WI, pp. 199-224.

Muchuweti M, Birkett JW, Chinyanga E, Zvauya R (2006). Scrimshaw MD, Lester JN. Heavy metal content of vegetables irrigated with mixtures of wastewater and sewage sludge in Zimbabwe: implications for human health. Agric. Ecosyst. Environ., 112: 41-48.

Navarro MC, Perez-Sirvent C, Martinez-Sanchez MJ, Vidal J, Tovar PJ, Bech J (2008). Abandoned mine sites as a source of contamination by heavy metals: A case study in a semi-arid zone. J. Geochem. Explor., 96: 183-193.

Pruvot C, Douay F, Herve F, Waterlot C (2006). Heavy metals in soil, crops and grass as a source of human exposure in the former mining areas. J Soils Sediments, 6: 215-220.

Singh AN, Zeng DH, Chen FS (2005). Heavy metal concentrations in redeveloping soil of mine spoil under plantations of certain native woody species in dry tropical environment, India. J Environ. Sci., 1: 168-174.

Türkdoğan MK, Kilicel F, Kara K, Tuncer I, Uygan I (2002). Heavy metals in soil, vegetables and fruits in the endemic upper gastrointestinal cancer region of Turkey. Environ. Toxicol. Pharmacol., 13: 175-179.

Zhuang P, McBride MB, Xia H, Li H, Li Z (2008). Heavy metal contamination in soils and food crops around Dabaoshan mine in Guangdong, China: implication for human health. Environ. Geochem. Health, 31: 707-715.

25

Prevention of renal toxicity from lead exposure by oral administration of *Lycopersicon esculentum*

Salawu Emmanuel O.[1]*, Adeleke Adeolu A.[1], Oyewo Oyebowale O.[2], Ashamu Ebenezer A.[2], Ishola Olufunto O.[1], Afolabi Ayobami O.[1] and Adesanya Taiwo A.[3]

[1]Department of Physiology, Ladoke Agential University, Ogbomoso, Nigeria.
[2]Department of Anatomy, Ladoke Akintola University, Ogbomoso, Nigeria.
[3]Department of Biochemistry, Ladoke Akintola University, Ogbomoso, Nigeria.

For decades, lead (Pb) has being known for its adverse effects on various body organs and systems. In the present study, the ability of Pb to lower renal clearance (RC), as an index of renal function, was investigated and tomato (*Lycopersicon esculentum*: source of antioxidants) paste (TP) was administered orally to prevent the Pb's adverse effects. 54 Sprague Dawley rats, randomly divided into 3 groups (A, B and C) n = 18, were used for this study. Group A animals served as the control and were drinking distilled water. Group B and Group C animals were drinking 1% lead(II)acetate (LA). Group C Animals were, in addition to drinking LA, treated with 1.5 ml of TP/day. All treatments were for 8 weeks. Mann–Whitney U-test was used to analyse the results obtained. The results of this study showed that Pb caused a significant reduction in the weight gain, 24 h urine volume, RC, plasma and tissue superoxide dismutase (SOD) and catalase (CAT) activities, but a significant increase in plasma and tissue malondialdehyde (MDA) concentration. Administration of TP, however, prevented these Pb's adverse effects. These findings lead to the conclusion that oral administration of TP prevents Pb's adverse effects on the kidney mainly by preventing oxidation.

Key words: Renal clearance, tomato, lead, *Lycopersicon esculentum*, heavy metals, oxidative stress.

INTRODUCTION

Lead, a dangerous heavy metal, is harmful even in small amounts. Nevertheless, humans get exposed to Pb through their environment and diet (Gidlow, 2004). The manifestations of Pb poisoning in humans are nonspecific. They may include: loss of appetite, weight loss, anaemia (Khalil-Manesh et al., 1994; Waldron, 1966), sluggishness, memory loss (Hopkins, 1970), nephropathy, infertility (Patocka and Cerný, 2003) etc. However, oxidation accompanies lead toxicity (Hande et al, 2004), and its treatment include elimination of exposure, chelation therapy and often diet modification to ensure adequate essential metal (calcium and iron) intake (Markowitz, 2003).

Tomato, on the contrary, is a source of antioxidants

(Lisa, 2002; Jeanie, 2007) and is made up by components (e.g. Lycopene, Glutation, Vitamin C, Vitamin A, Potassium and Calcium) very appropriate for detoxification, illnesses prevention (Nguyen and Schwartz, 1999), attaining growth (John and Marc, 2000), helping the immunologic system (Sandhu et al., 2000), maintaining blood in good state (Khalil-Manesh et al., 1994) etc. Lycopene has similar properties to the betacarotenes of the carrots and has anti-cancerous properties (John and Marc, 2000). Glutation has been shown to have antioxidant properties that help to eliminate free radicals. It is also very important in the elimination of the body toxins, especially heavy metals that produce deterioration of the organism by its accumulation. In addition glutation has the ability to lower blood pressure, favour the good state of our liver and prevent eczema (The world of plants, 2008). The vitamin A present in tomato helps the body to attain cellular growth (John and Marc, 2000), maintain the bones and the teeth in good state, help the immunologic system

*Corresponding author. E-mail: seocatholic@gmail.com or seocatholic@yahoo.com.

in combating infections (Sandhu et al., 2000) and to maintain sight in good state.

In addition the potassium and calcium components of tomato ensure availability of essential metals that are can compete with and displace lead, thus reducing its toxicity. These metals even play beneficial roles in bone formation, regulation of the corporal liquids, nerves, heart (Haddy et al., 2006) and the muscles (Ford and Podolski, 1970; Endo et al., 1970). It has also been documented that oral administration of tomatoes may increase blood parameters such as Haematocrit, RBC, WBC etc (Khalil-Manesh et al., 1994).

Furthermore, the detoxifying components and the health-protective antioxidants of tomato are more available for absorption and more potent in cooked tomato than in raw uncooked one (Gärtner et al., 1997; John and Marc, 2000; Thompson et al., 2006).

This research, therefore, centres on whether oral administration of cooked tomatoes prevents Pb induced renal toxicity or not, using renal clearance and tissue oxidation as major indicators.

MATERIALS AND METHODS

54 adult male Sprague Dawley rats (180 - 220 g) were used for this study. They were inbred at the animal house section of the department of physiology, Ladoke Akintola university of technology, Ogbomoso. The animals were acclimatized over a period of 2 weeks.

Preparation of tomato paste (TP)

TP was prepared by grinding tomatoes and heating it in water a bath for 45 min at 80°C.

Grouping of animals and treatment

The rats were randomly grouped into three (A, B and C), n = 18. Animals in group A served as the control group and were drinking distilled water. Animals in group B and group C were drinking 1% lead (II) acetate (LA) (Marchlewicz et al., 1993). Group C animals were, in addition to drinking LA, treated with 1.5 ml of TP/day. All treatments were for 8 weeks.

Animal sacrifice and collection of samples

24 h after the last treatment, each animal was transferred to metabolic cage equipped with accessory for collecting urine. The 24 h urine sample was collected and its volume recorded for each animal. Each rat was weighed, then and sacrificed by cervical dislocation. Blood samples were collected by cardiac puncture. Blood collected from each rat was divided into 2: one half in plain bottle and the other half in EDTA bottle. Plasma and serum were obtained from blood samples by spinning at 3000 rpm for 20 min.

Collection of data and statistical analysis

Each kidney was homogenized. Kidney homoginate was used in

determining kidney SOD activity, kidney CAT activity and kidney MDA concentration. The Weight Increase and 24-hour Urine volume (ml) were recorded. To obtain creatinine and urea clearance, urine and serum creatinine concentration were determined using Alkaline Picrate Method described by Jaffe (1886); urine and serum Urea concentration were determined using Diacetylmonoxime Method described by Ceriotti and Spandro (1963). Renal clearance was then calculated using the formula "Clearance of Y = (Urine concentration of Y * 24hr Urine volume) /Plasma concentration of Y" as documented by Guyton and Hall (2001). Plasma and tissue superoxide dismutase (SOD) activity were determined using the method described by Fridovich (1986). Plasma and tissue catalase (CAT) activity were determined using the method described by Sinha (1972). Plasma and Tissue Malondialdehyde (MDA) Concentrations were determined using the procedure described by Varshney and Kale (1990).

The data obtained are presented as mean ± SD. The "Control Group" and the "Test Groups" were compared using the Mann–Whitney U-test. The significance level was set to a P-value < 0.05.

RESULTS

The following results were obtained and are presented as mean ± SEM. Level of significance is taken at "P value < 0.05" (*) and/or "P value < 0.01" (**).

Weight increase (g)

Comparing their final and initial weight showed that there was significant weight gain (P value < 0.05) is in all the groups over the 8 weeks of the research. There was, however, no significant difference (P value > 0.05) in weight gain of group C and control, while group B had a significantly smaller weight gain.

Kidney weight (g)

The kidney weight of group B was significantly lower (P value < 0.05) than that of the control, while there was no significant difference in the kidney weight of group C and that of the control.

24 h urine volume (24 HrUV) (ml)

The 24 HrUV for group B was significantly (p value < 0.05) lower than that of the control (Group A). While 24 HrUV of group C D showed no significant difference (P value > 0.05) from that of the control.

Creatinine clearance

A significant (p value < 0.05) decrease was noticed in the renal creatinine clearance of group B when compared to control; while group C showed no significant (p value > 0.05) difference from the control.

Urea clearance

Group B had renal urea clearance (RUC) that is signifi-

Table 1. Weight increase across the three groups during the 8 weeks of research.

	Group A	Group B	Group C
Weight before sacrifice (g)	219.3 ± 0.987	198.9 ± 0.543	218.5 ± 0.324
Initial weight (g)	183.2 ± 1.021	182.1 ± 0.342	184.5 ± 0.443
Weight increase (g)	36.1 ± 0.334	16.8 ± 0.987*	34.0 ± 0.943
P value (when compared with control)		0.0439	0.1163

* "P value < 0.05"

Table 2. Comparison of kidney weight across the groups.

	Group A	Group B	Group C
Kidney weight (g)	0.6028 ± 0.055	0.5235 ± 0.033*	0.6421 ± 0.057
P value (when compared with control)		0.03064	0.2269

cantly (p value < 0.05) lower than that of the control. While RUC of group C was not significantly (p value > 0.05) lover than that of the control.

Plasma superoxide dismutase (SOD) activity

Group B showed a highly significant (P value < 0.01) decrease in plasma SOD activity. Group C was, however, not significantly (P value > 0.05) different from the control in terms of plasma SOD activity.

Plasma catalase (CAT) activity

Group B showed a highly significant (P value < 0.01) decrease in plasma CAT activity. However, group C showed no significant (P value > 0.05) difference in CAT activity from the control.

Plasma malondialdehyde (MDA) concentration

Group B showed significant (P value < 0.05) increase in plasma MDA concentration; while group B showed no significant (P value > 0.05) difference from control.

Tissue superoxide dismutase (SOD) activity

Group B showed a highly significant (p value < 0.01) decrease in plasma SOD activity, there was, however, no significant (p value > 0.05) difference between the Tissue SOD Activity of Group C and that of the Control.

Tissue catalase (CAT) activity

Group B, showed a significant (p value < 0.05) decrease in tissue CAT activity. However, group C showed no significant (p value > 0.05) difference from control.

Tissue malondialdehyde (MDA) concentration

Group B showed a significant (p value < 0.01) increase in

tissue MDA concentration. While MDA concentration in group C was found to be significantly lower when compared with group C.

DISCUSSION

The results of this study shows that exposure to Pb for 8 weeks significantly (p value < 0.05) reduces weight gain (Table 1), this is in support of the findings of Suzan et al. (1999) and can be linked to the less efficient metabolic processes associated with Pb toxicity (Struzyńska et al., 1997). Administration of 1.5 ml TP/day, however, annuls this Pb's adverse effect on weight gain. This may be partly due to the fatty acid composition of TP (Cantarelli et al., 1993) and more importantly due to presence of health-protective antioxidants such as lycopene, vitamin C and vitamin A in TP (Jeanie, 2007) despite its relatively low caloric value (21 Kcal/100 g) and low protein content (0.85% by weight) (The world of plants, 2008). These can also explain the significant (p value < 0.05) decrease in kidney weight (Table 2) noticed in animals exposed to Pb (Group B) and the no significant (p value > 0.05) decrease in kidney weight noticed in animals administered TP alongside Pb exposure (Group C) since organ weights are normally fractions of the body weight (within specific range).

There was no significant (p value > 0.05) decrease in the 24 HrUV (Table 3) of animals treated with TP even though they were as well exposed to Pb. On the contrary, the lead only group (Group B) showed significant (p value < 0.05) reduction in 24 HrUV. This is because Pb (like most other heavy metals) interferes with glomerular filtration rate (GFR) and tubular processes, tubular reabsorption and/or tubular secretion, (Oberley et al., 1995; Machiko et al., 1978) which are the major determinants of urine volume. The administered tomato would therefore be responsible for the prevention of these lowering effects of Pb on 24 HrUV by preventing the lowering of GFR.

Table 3. Comparison of 24 HrUV (ml) across the groups.

	Group A	Group B	Group C
24 HrUV (ml)	3.12 ± 0.112	2.30 ± 0.108*	2.94 ± 0.068
P value (when compared with control)		0.0139	0.1298

* "p value < 0.05"

Table 4. Comparison of creatinine clearance across the 3 groups

	Group A	Group B	Group C
Creatinine clearance	3.455 ± 0.121	2.831 ± 0.148*	3.303 ± 0.097
P value (when compared with control)		0.0341	0.2291

* "p value < 0.05"

Table 5. Comparison of urea clearance across the 3 groups.

	Group A	Group B	Group C
Urea clearance	0.3718 ± 0.062	0.2578 ± 0.042*	0.3209 ± 0.094
P value (when compared with control)		0.0222	0.3252

* "p value < 0.05"

Table 6. Plasma SOD activity across the Groups

	Group A	Group B	Group C
Plasma SOD activity	1.766 ± 0.052	1.123 ± 0.061**	1.701 ± 0.092
P value (when compared with control)		0.0098	0.4113

** "P value < 0.01"

In a similar way renal creatinine clearance (RCC) and renal urea clearance (RUC) of animals treated with tomato alongside Pb were not significantly different (p value > 0.05) from those of the control (Table 4 and Table 5 respectively). Meanwhile, animals treated with Pb only showed significant (p value < 0.05) decrease in RCC and RUC. This supports the findings of Machiko et al. (1978) which say that heavy metal toxicity brings about reduction in renal clearance among other renal dysfunctions. But TP significantly (p value < 0.05) reduced the Pb's adverse effects on RC, such that there was no significant difference in RC of control and that of Pb + TP group.

There was no significant (p value > 0.05) difference in SOD activity of both the plasma and tissue (Tables 6 and 9 respectively) of the control and that of the animals treated with tomato alongside Pb. On the contrary, there was a highly significant (p value < 0.01) decrease in plasma and tissue SOD activity in animals treated with Pb only compared to the control. This finding is in agreement with Ping-Chi and Yueliang (2002) and is at the same time in support of *Lycopersicon esculentum* (tomato) as an antioxidant.

Furthermore, there was a significant decrease in both plasma CAT Activity (p value < 0.01) and tissue CAT Activity (p value < 0.05) of animals treated with Pb only relative to control (Tables 7 and 10 respectively). There was, however no significant (p value > 0.05) difference between the control and the animals treated with tomato alongside Pb in this respect. This further establishes that it was TP that reduced the oxidative stress that Pb could cause.

These significant decreases in the activities of both plasma and tissue SOD and CAT activity, resulting from Pb exposure, would have markedly reduced the level of anti-oxidation defenses in the body. This is in support of the fact that oxidation through free radical (e.g. Reactive Oxygen Species, ROS) accompanies Pb toxicity (Hande et al, 2004).

Finally, there was no significant (p value > 0.05) difference in both plasma and tissue MDA concentration (Tables 8 and 11 respectively) of control and those of the animals treated with tomato alongside Pb. While animals treated with Pb only showed a significant increase in both plasma (p value < 0.05) and tissue (p value < 0.01) MDA concentration. This confirms that it was TP, source of antioxidants (Lisa, 2002; Jeanie, 2007), that reduced the

Table 7. Plasma CAT activity across the groups.

	Group A	Group B	Group C
Plasma CAT activity	0.4101 ± 0.082	0.2173 ± 0.032**	0.3997 ± 0.095
P value (when compared with control)		0.0093	0.5192

** "P value < 0.01"

Table 8. Plasma MDA concentration across the groups.

	Group A	Group B	Group C
Plasma MDA concentration (μg/g protein)	1400.3 ± 23.01	1813.6 ± 11.18*	1419.5 ± 22.07
P value (when compared with control)		0.0424	0.5370

* "P value < 0.05"

Table 9. Tissue SOD activity across the groups.

	Group A	Group B	Group C
Tissue SOD activity	1.534 ± 0.076	1.203 ± 0.074**	1.497 ± 0.085
P value (when compared with control)		0.0092	0.3646

** "p value < 0.01"

Table 10. Tissue CAT activity across the groups.

	Group A	Group B	Group C
Tissue CAT activity	0.3357 ± 0.052	0.2256 ± 0.056*	0.3739 ± 0.046
P value (when compared with control)		0.0214	0.1754

* "p value < 0.05"

Table 11. Tissue MDA concentration across the 3 groups.

	Group A	Group B	Group C
Tissue MDA concentration(μg/g protein)	1367.9 ± 1.94	1954.1 ± 3.79**	1305.1 ± 2.76
P value (when compared with control)		0.0069	0.2633

** "p value < 0.01"

oxidative stress that Pb exposure could have caused in the animals.

It can, thus, be concluded that exposure to Pb lowers renal clearance due to Pb's ability to cause oxidative stress by interfering with the activities of SOD and that of CAT and thereby given freedom to free radicals (e.g. ROS) to cause oxidation which manifests as increase in the concentration of MDA (in the case of lipid peroxidation). Oral administration of *L. esculentum* (in the form of tomato paste, TP), however, prevented these Pb induced reduction in renal clearance. This would be mainly due to the anti-oxidant characteristics of the constituents (e.g. lycopene) of *L. esculentum* (administered as TP).

REFERENCES

Cantarelli PR, Regitano-d'Arce MAB, Palma ER (1993). Physico-chemical characteristics and fatty acid composition of tomato seed oils from processing wastes. Sci. Agric. (Piracicaba, Braz.). 50(1).

Ceriotti G, Spandro L (1963). A spectrophotometric. method for determination of urea. Clin. Chim. Acta. 8: 295-299.

Endo M, Tanaka M, Ogawa Y (1970). Calcium induced release of calcium from the sarcoplasmic reticulum of skinned skeletal muscle fibres. Nature. 228: 34-36.

Ford L, Podolski R (1970). Regenerative calcium release within muscle cells. Science 167: 58-59.

Fridovich I (1986). Superoxide dismutases. Adv Enzymol Relat Areas Mol Biol. 58: 61-97.

Gärtner C, Stahl W, Sies H (1997). Lycopene is more bioavailable from tomato paste than from fresh tomatoes. ASCN Annual Meeting. 66

(1): 199-222.

Gidlow DA (2004). Lead toxicity. Occupational Med. 54: 76-81.

Guyton AC and Hall JE (2001). Textbook of medical physiology. 10th ed. Elsevier India. New Delhi. pp. 309-310.

Haddy FJ, Vanhoutte PM, Feletou M (2006). Role of potassium in regulating blood flow and blood pressure. Am J Physiol Regul Integr Comp Physiol. 290: R546-R552

Hande GO, Handan US, Hilal Ö (2004). Correlation between clinical indicators of lead poisoning and oxidative stress parameters in controls and lead-exposed workers. Toxicology 195(2-3): 147-154.

Hopkins A (1970). Experimental lead poisoning in the baboon. Brit. J. Ind. Med. 27(2): 130-140. PMCID: PMC1009086.

Jaffe M (1886). Ueber den Niederschlag, Welchen Pikrinsaeure in normalem Harn erzeught und ueber eine neue Reaktion des Kreatinins, Z. Physiol. Chem. 10: 391-400.

Jeanie LD (2007). Tomato and Broccoli_ 2 Antioxidant Power Blasts. Available from: http://www.webmd.com/food-recipes/features/tasty-tomato-antioxidant-power-blast?src=RSS_PUBLIC.

John S, Marc LM (2000). Lycopene in Tomatoes: Chemical and Physical Properties Affected by Food Processing. Crit Rev Food Sci. Nutr. 40(1): 1-42.

Khalil-Manesh F, Tartaglia-Erler J, Gonick HC (1994). Experimental model of lead nephropathy. Correlation between renal functional changes and hematological indices of lead toxicity. J. Trace Elem. Electrolytes Health Dis. 8(1): 13-9.

Lisa T (2002). The Top 10 Antioxidant Foods. Better Nutrition. Available from: http://www.crumcreek.com/library/ antioxidant.html.

Machiko T, Kosuke N, Yoshinori I (1978). Effect of cadmium administration on growth, excretion, and tissue accumulation of cadmium and histological alterations in calcium-sufficient and -deficient rats: An equalized feeding study. Toxicol. Appl. Pharmacol. 45(2): 591-598.

Marchlewicz M, Protasowicki M, Rózewicka L, Piasecka M, Laszczyńska M (1993). Effect of long-term exposure to lead on testis and epididymis in rats. Folia Histochem. Cytobiol. 31(2): 55-62.

Markowitz M (2003). Lead poisoning: A disease for the next millennium. Current Problems Pediatrics. 30(3): 62-70.

Nguyen ML, Schwartz SJ (1999). Lycopene: chemical and biological properties. Food composition: Diet and diet-related diseases. 53(2): 38-45.

Oberley TD, Friedman AL, Moser R, Siegel FL (1995). Effects of Lead Administration on Developing Rat Kidney: II. Functional, Morphologic, and Immunohistochemical Studies. Toxicol. Appl. Pharmacol. 131(1): 94-107.

Patocka J, Cerný K (2003). Inorganic lead toxicology. Acta Medica (Hradec Kralove) 46(2): 65-72.

Ping-Chi H, Yueliang LG (2002) Antioxidant nutrients and lead toxicity. Toxicol. 180(1): 33-44.

Sandhu JS, Krasnyanski SF, Domier LL, Korban SS, Osadjan MD, Buetow DE (2000). Oral immunization of mice with transgenic tomato fruit expressing respiratory syncytial virus-F protein induces a systemic immune response. Transgenic Res. 9(2):127-135.

Sinha AK (1972). Colorimetric assay of catalase. Anal. Biochem. 47: 389-394.

Struzyńska L, Dabrowska-Bouta B, Rafałowska U (1997). Acute lead toxicity and energy metabolism in rat brain synaptosomes. Acta Neurobiol. Exp. (Wars) 57(4): 275-81.

Suzan AW, Ghayasuddin A (1999). Effects of Lead on the Male Reproductive System in Mice. J. Toxicol. Environ. Health. Part A, 56(7): 513- 521.

The world of plants (2008). Properties of tomatoes. Available from: http://www.botanical-online.com/tomatesangles.htm.

Thompson KA, Marshall MR, Sims CA, Wei CI, Sargent SA, Scott JW (2006). Cultivar, Maturity, and Heat Treatment on Lycopene Content in Tomatoes. J. Food Sci. 65(5): 791-795.

Varshney R, Kale RK (1990). Effects of Calmodulin Antagonists on Radiation-induced Lipid Peroxidation in Microsomes. Int. J. Radiat. Biol. 58(5): 733-743.

Waldron HA (1966). The Anaemia of Lead Poisoning: A Review. Brit. J. Ind. Med. 23(2):83-100.

26

Mutagenecity testing of phamarceutical effluents on *Allium cepa* root tip meristems

Abu, Ngozi E.* and Mba, K. C.

Department of Botany, University of Nigeria, Nsukka, Nigeria.

Using macro analysis of root growth parameters and micro assay of the root tip meristems through the estimation of the mitotic indices and aberrant cells of *Allium cepa* root meristems, a study was conducted on three Pharmaceutical effluents discharged into agricultural lands. The chemical analysis of the effluents showed the presence of some potential mutagenic heavy metals (Pb, Cu and Zn) and Cyanide. The experimental design was a factorial in completely randomized design (CRD). The results obtained showed significant reduction in the number of roots and the length of roots grown on the effluents and their dilutions. Highly reduced cell reproduction was observed in all the effluents. The mitotic index (MI) ranged from 49.2 - 2.95, 52.6 - 1.77 and 48.8 - 2.63 in Jutrim, Flu-J and Ampiclox effluents, respctively. Both concentration of the effluents and treatment time had significant effect on the mitotic index. The MI as a percentage of the control at undiluted effluent concentrations and at 24 h exposure time were below 22% of the control value ranging from 6.02, 3.6 and 5.4% in Jutrim, Flu-J and Ampiclox effluents, respectively. Aberrant cells oberved ranged from a mild C-mitotic effect to anaphase bridge and even an induction of multiple nuclei in a cell. The effects of the cytotoxic and genotoxic substances in these wastewaters discharged into the environment on biolife are discussed in this report.

Key words: Mutagenicity, pharmaceutical effluents, mitotic index, *Allium cepa*.

INTRODUCTION

Plant comprise a large portion of our biosphere and constitute a vital link in the food chain. Due to the highly conserved structure of the genetic material, it is possible to use a broad variety of species in genetoxicity tests. Several higher plant bioassays for screening and monitoring environmental mutagens have been esterblished (Maluszynaska and Juchimiuk, 2005). The detection of genetoxicity of hazardous chemicals has been studied for many years using higher plants as biological systems (Rank et al., 2002). Plant assays are useful for testing of complex environmental samples such as wastewater (Grover and Kaur 1999), wastewater sludge (Rank and Nielson, 1998), contaminated river water (Ivanova et al., 2005) and soils (Chang et al., 1997; Kovalchuk et al., 1998; Cotelle et al., 1999). Mutagenecity testing is utilizing one of the oldest methods, namely the observation of chromosomal aberrations.The common onion, *Allium cepa* L. makes a convinient test system for estimating harmful effects of chemicals on biological materials (Fiskesjo,1985; Fiskesjo, 1993; Rank and Nielson, 1998). Because of its excellent chromosome conditions, the *Allium* genetic material has been widely exploited for such purposes since A. Levan first introdced it as a test system in 1938. Early in the thirties, Levan (1938) showed that colchicine could cause spindle disturbances and polyploidy in Allium root meristem cells and later in 1945, he demostrated that various inorganic salt solutions induced different kinds of chromosome aberrations in *A. cepa* root cells. Odeigah et al. (1997) reported that plant roots are very useful in this testing because the root tips are often the first to be exposed to chemicals in the soil and water. Since then, many new mutagenecity assays using microrganisms, mammalian cells and other biological systems have been developed, but plant tests are still used routinely for genotoxicity testing all over the world (Grant, 1999). Fiskesjo and Levan (1993) reported that *Allium* test has been found to have a high correlation with other test system (MIT-217 cell test with mice, rats or

---
*Corresponding author. E-mail: abu_ngo@yahoo.com.

Table 1. Treatment combinations of effluents (A) and concentrations (B).

Effluents	Concentrations(%)				
	0	25	50	75	100
A1 (Jutrim)	A1 0	A1 25	A1 50	A1 75	A1 100
A2 (Flu-J)	A2 0	A2 25	A2 50	A2 75	A2 100
A3 (Ampiclox)	A3 0	A3 25	A3 50	A3 75	A3 100

humans *in vivo*) and could be used as an alternative to laboratory animal in toxicological research.

Pharmaceutical effluents are wastes generated by pharmaceutical industries during the process of drug manufacturing. The steps involved in the compounding of drugs generate air emission, liquid wastes and solid wastes. Some pharmaceutical effluents are known to contain high concentration of toxic materials with varying pH values (Osaigbovo and Orhue, 2006). Ivanova et al. (2002) reported that heavy metals are amongst the most toxic and environmentally dangerous pollutants. It has been reported that there is relationship between the heavy metal amounts in natural and industrial environment and an increased frequency of chromosome mutations and the cancerous processes in organisms (Bruning and Chronz, 1999). The increase of pollution by the release of genotoxic chemicals and the increase of radiation levels have affected the ecosystem and the health of organisms, including humans (Houk, 1992).

It has been keenly observed that industrialization is increasing in this nation. This is of great benefit to the country, however there are no wastewater treatment system and industries discharge the untreated waste into agricultural land and water bodies. These wastewaters are complex mixtures, often times with high concentrations of potencially mutagenic heavy metals and cyanides. The published work on the effects of industrial wastewater are few in our environment, however, the results of previous investigations made by various authors pointed out that industrial wastewaters in Nigeria are environmental risks and ecological threat by their being mitodepressive in addition to inducing diverse kinds of chromosomal aberration, cytokinetic problems and nuclear disolutions (Odeigah et al., 1997; Abu and Ezeugwu 2008; Abu and Ogbonna 2009). The objective of this work is therefore, to assess the mutagenic potential of three pharmaceutical wastewaters on *A. cepa* root meristem as a pointer to their possible delecterious effects on the ecosystem.

MATERIALS AND METHODS

The three effluents were collected from a pharmaceutical industry in Enugu, Nigeria. The samples were collected at the discharge end of the sewer system during the production of Jutrim (cotrimoxazole), Flu-J (chlorphiramine, paracetamol and Ascorbic acid) and Ampiclox (ampicillin and cloxiacillin), respactively. Samples were collected and stored in opague plastic gallons in the refrigerator.

The pH of the samples were taken at the point of collection. The samples were analysed for total disolved solid (TDS), chlorine, sulphate, cyanide and some heavy metals which include zinc, lead and copper. Approximately, equal sized bulbs of the common onion (*A. cepa*) were bought from the open market. The wastewater samples were diluted with distilled water to produce three dilutions - 25, 50 and 75%, the undiluted wastewater was 100% while distilled water was used as the control (0%).

Macro assessment of the root

The root initiation and growth assessment was done using the procedure of modified *Allium* test (Rank and Nielson, 1993). The bases of the bulbs were gently scrapped to expose the root primordia. The bulbs were planted direclty on the effluents and their dilution without an initial rooting over water. The experimental set up had five replicates.

Micro assessment of the root tips

The experimental set up for the micro assay is shown in Tables 1 and 2. The experimental design was a factorial in completely randomized design (CRD). The set up (3 x 5 x 2 factorial in CRD) was replicated three times. The bulbs were germinated over water before being transferred to each of the treatment combination. The root tips were harvested and fixed in aceto-alcohol (1 part of glacial acetic acid to 3 parts of absolute alcohol). The root tips were hydrolysed in IN HCL at 60°C for 5 min. The squashing was over aceto - orcein stain on a slide. The slides were examined, data were collected on number of dividing cells out of 1000 cells. The Mitotic index was estimated as number of dividing cells over 1000 cells expressed in percentage and photomicrographs were taken. The chromosomes were transfered to a computer system using the motic camera fixed on the ordinary light microscope.

RESULTS

Chemical composition of the effluents

The data from the chemical analysis of the effluents are presented on Table 3 with the required environmental standard according to Federal Environmental Protection Agency (FEPA) of Nigeria (FEPA 1991). The pH values from the effluent analysis are 3.0, 2.9 and 4.1 for Jutrim, Flu-J and Ampiclox respectively while the standard pH values ranged from 6 - 9. Thereby puttting all the wasters as too acidic on environment based on the recommended value. The total dissolved solids (TDS) in the effluents ranged from 12,000 - 42,000 mgl^{-1} while the recommended value is 2000 mgl^{-1}. The chlorine and

Table 2. Treatment combinations of effluent, concentration(%) and time (Factor C).

Factors A and B combinations	Treatment C6	Durations (hours) C12
A1 0	A1 0 C6	A1 0 C1 2
A1 25	A1 25 C6	A1 25 C1 2
A1 50	A1 50 C6	A1 50 C12
A1 75	A1 75 C6	A175 C12
A1 100	A1 100 C6	A1 100 C12
A2 0	A2 0 C6	A2 0 C12
A2 25	A2 25 c6	A2 25 C12
A250	A2 50 C6	A2 50 C12
A2 75	A275 C6	A2 75 C12
A2 100	A2 100 C6	A2 100 C12
A3 0	A3 0 C6	A3 0 C12
A3 25	A3 25 C6	A3 25 C12
A3 50	A3 50 C6	A3 50 C12
A375	A375 C6	A3 75 C12
A3 100	A3 100 C6	A3 100 C12

Table 3. Chemical analysis of the wastewater and the Nigeria environmental standard.

Effluents and std.	pH	TDS mgl^{-1}	Cu mgl^{-1}	Zn mgl^{-1}	Pb mgl^{-1}	Cl mgl^{-1}	SO4 mgl^{-1}	Cn mgl^{-1}
Jutrim	2.9	42,000	20.0	17.3	2.2	0.02	696.2	0.09
Flu-J	2.8	14,000	8.5	117.6	0.05	0.03	56	0.04
Ampiclox	4.1	12,00	6.0	147.6	1.23	0.01	79	0.07
Nig. Std*	6 - 9	2,000	<1	-	<1	600	500	-

*Recommeded environmental standard in Nigeria (FEPA, 1991).

sulphate amounts in the wastewaters are within the recommended range except that the sulphate was slightly higher in Jutrim effluent. Cyanide was present in all the effluent ranging from 0.04 - 0.09. The potentially mutagenic heavy metals, (Cu, Zn, Pb) recommended to be less than 1.0 mgl^{-1} in the environment varied from 6.0 - 20.0 mgl^{-1}, 17.4 - 147.6 mgl^{-1} and 0.05 - 2.2 mgl^{-1}, respectively.

Macro assessement of root growth

Root growth parameters was used to assess the general toxicity of the effluents. The results showed a strong root growth retardation by the three effluents. There was a drastic decline in the number of roots from the control to the undiluted (100%) effluent concentration, ranging from 109 - 5.3, 93 - 20.4 and 122 - 8.3 in Jutrim, Flu-J and Ampiclox effluents,respectively (Figures 1 and 2). The root length as a percentage of the control values showed similar decline as the concentration of each effluent increased.

Microscopic analysis

The microscopic analysis showed that the effluents are mitodepressive. The main effects of the effluents on MI are shown in Figure 3. The MI values in the effluents varied from 49.3% in the control to 8.3% in Jutrim. All the effluents were significantly lower than the control value but Jutrim had the most adverse effect. The main effects of concentrations on MI were highly significant shown in Figure 4. The MI estimated from the data decreased as the concentrations of the effluents increased except for a few variations. The effects of effluents, concentrations and time with the interaction F- LSD (P = 0.01) as the error bar are shown in Figure 5. All the effluents had more adverse effects at undiluted concentration excepting Ampiclox which did not vary from it's effect at 75% over 24 h treatment durations, as seen in Figure 5. The MI ranged from 49.2 - 2.95%, 52.6 - 1.77% and 48.8 - 2.63% in Jutrim, Flu-J and Ampiclox effluents, respectively, across concentrations and treatment durations. Jutrim effluent at both 12 and 24 h of treatment had significant effect on MI. Higher reduction in cell reproduction were observed as concentrations and

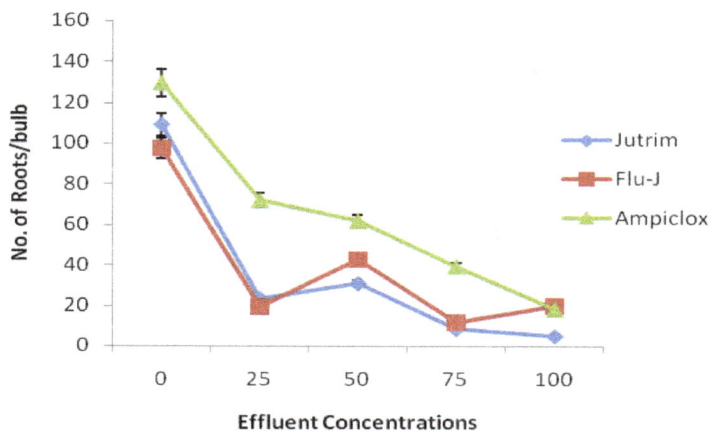

Figure 1. Number of roots per bulb in the three effluents and their dilutions.

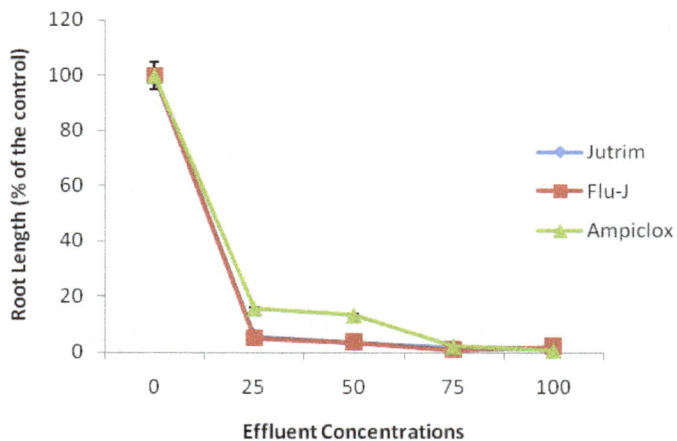

Figure 2. Root lenght as a percentage of the control in the three effluents and their dilutions.

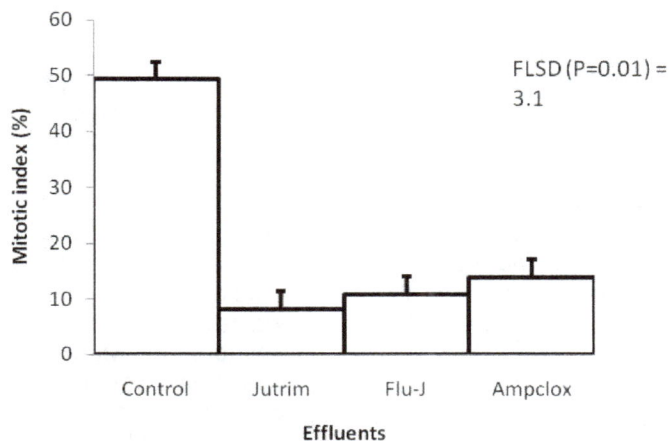

Figure 3. The main effect of control and effluents on the mitotic index.

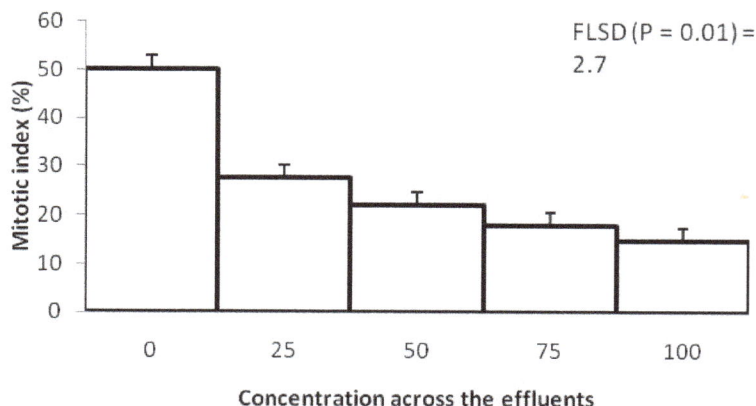

Figure 4. The effect of concentration across the effluent and treament time on the mitotic index.

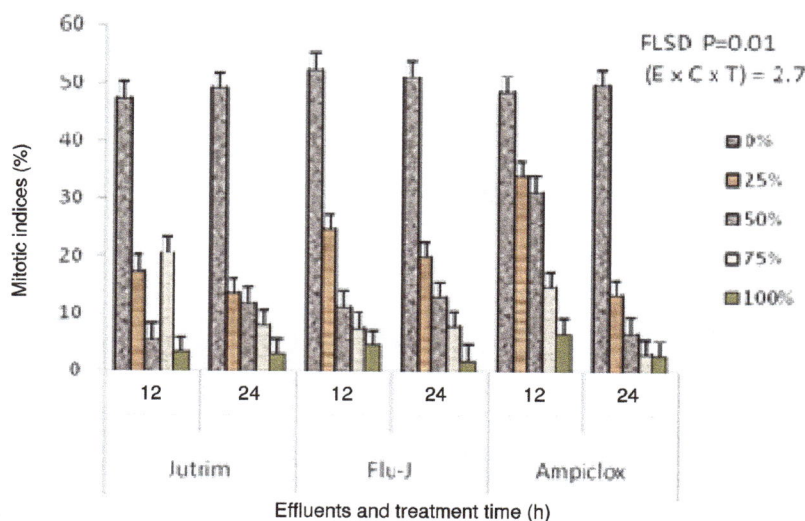

Figure 5. The effect of effluent and treatment time at different concentrations on the mitotic index.

exposure time increased in the different effluents. Chromosomal aberrations observed in this work include C-mitotic effects, degenerating chromatin materials, Anaphase bridge, precocious chromosomes at anaphase and telophase stages. Cytokinetic failure and degenerating nucleus were induced by Jutrim effluent while multinucleate cells were induced by both Jutrim and Ampiclox effluents. The normal mitotic stages are shown on Plate 1 for a quick comparison with the abnormal ones on Plates 2 and 3.

DISCUSSIONS

The presence of heavy metals and cyanides in wastewater are implicated in provoking mutagenic effects in contact organisms (Ivanova et al., 2002 and 2005; El-Shahaby et.al., 2005). The heavy metal amounts in these three wastewaters are more than the recomended safe levels in the environment. Cyanides are not tolerated to be in the environmental even at low levels (FEPA 1991) but all the wastewaters had cyanide present in them. The results of the chemical analysis seems to imply that these wastewaters are too toxic to be discharged into the environment without a pretreatment system. However, it has been reported that the complexity of industrial wastewater makes it impossible to carry out a hazard assessment based on chemical analysis only (Fiskesjo, 1985; El-Shahaby et al., 2005).

In search of test systems which can be combined with

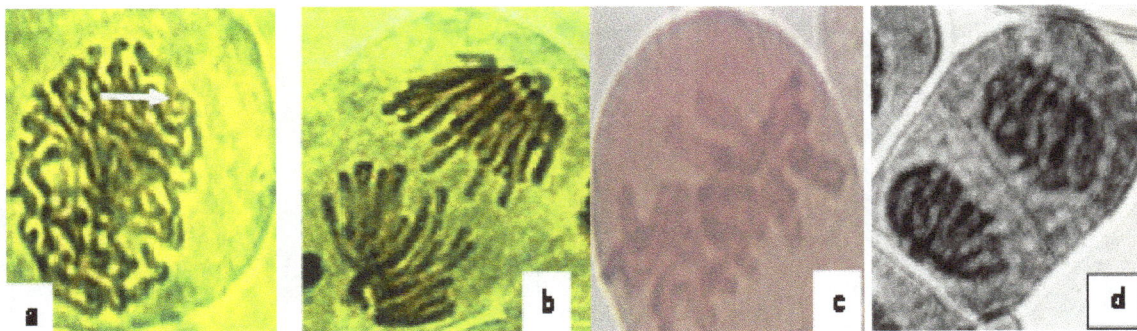

Plates 1a-d. Normal mitotic stages in *Allium cepa* root meristems (a) prophase, (b) metaphase,(c) anaphase and (d) telophase with signs of cell plate formation (arrowed).

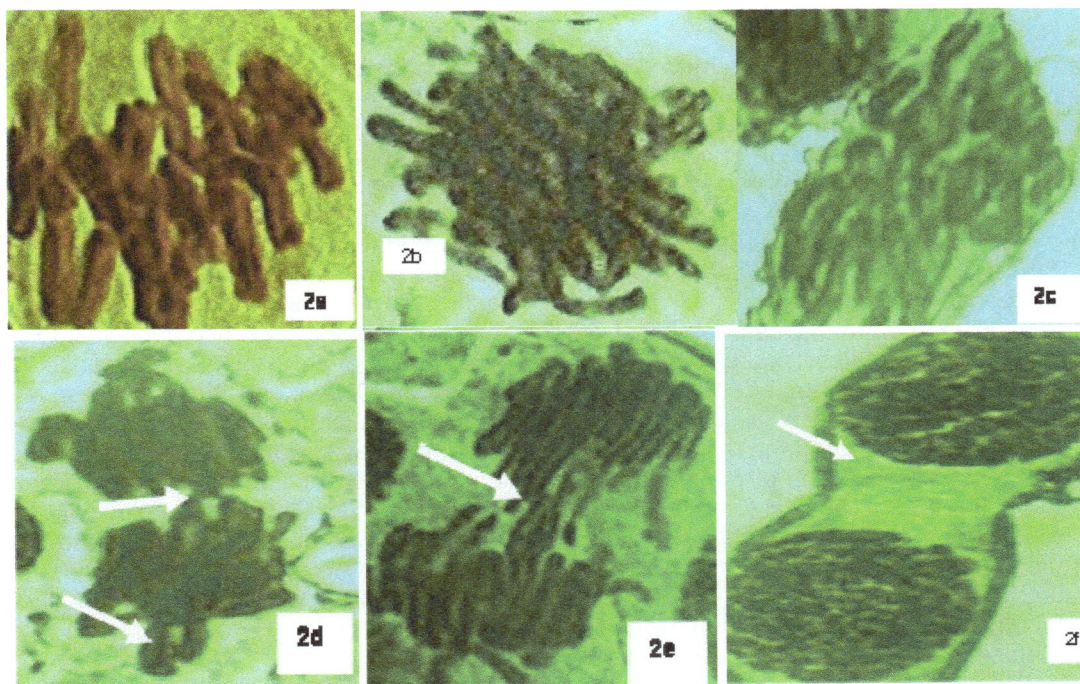

Plates 2a-e. Abnormal dividing cells (a - c) Diverse forms of C — mitotic effect, (d) Anaphase bridge with vagrant chromosomes(arrowed) and signs of chromatin degeneration and (e) Anaphase with multiple bridges, (f) Cytokinetic failure.

chemical analysis to provide data as a scientific basis for regulating the discharge of potential hazardous substances into the environment and suitable for performance of toxicity evaluation, the Allium test (Fiskesjo, 1985) was proposed. In the macro assay, Jutrim and Flu-J suppressed root growth more than Ampiclox. The effect of these effluents are dose dependent as there are more adverse effects at high concentrations. The root length as a percentage of the mean control value was also adversely affected as concentrations of each effluent increased. Root length as

low as 0.66% of the control was obtained in undilutated Jutrim effluent concentration. The general toxicity gives a quick assessment of the toxicity of these wastewaters on plants at the point of discharge. Odeigah et al. (1997) reported that plant roots are very useful in environmental monitoring of wastewaters because they are often the first to be exposed to chemicals in the soil and water.

The high acidity coupled with high amount of heavy metals and cyanide in these wastewaters may have affected the root initiation and growth via their action on cell reproduction. Fiskesjo (1985) had also reported that

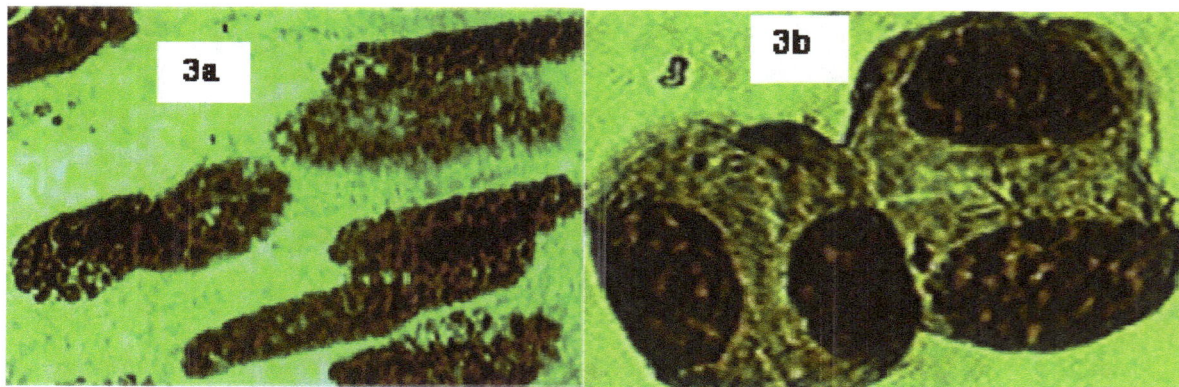

Plates 3a and b. Abnormal nuclei (a) Nucleus with nuclear wall lesions, (b) Binucleate cells.

root growth retardation and root wilting are as a result of the supression of cell division and chromosomal aberrations. The effluents, concentrations and exposure time were mitodepressive. The presence of heavy metals in complex mixtures like industrial wastewater has been implicated as the cause of cytotoxicity and genotoxicity effects in both plant and animal test systems (Grover and Kaur, 1999). Panda and Sahu (1985) and Antonsie - Wiez (1990) had reported that a mitotic index decrease below 50 and 22% of the control causes sub-lethal and lethal effects, respectively, on test organisms. Copper has been reported to have the effect of rapidly decreasing the mitotic index (Jiang et al., 1996). In the present study, Jutrim wastewater had the highest amount of copper and the least MI value thereby agreeing with the findings of Jiang et al. (1996). Moreover all the effluents in undiluted form (100%) at both 12 and 24 h treatment durations had less than 22% of the control MI value. The values are 6.56 - 6.0%, 8.23 - 3.37% and 13.26 - 5.4% in Jutrim, Flu J and Ampiclox at 12 and 24 h respectively. These values are therefore, suggesting that these wastewaters are lethal in undiluted forms. This implies that when these effluents are discharged into agricultural land for at least 12 h without dilution from rain or other sources, they would cause lethal effects on contact plants, however, if discharged into water bodies, the effects would be more at the point of discharge than down stream.

The lethal effects of these effluents may be attributed to the high levels of heavy metals and cyanide in them. The heavy metal amounts are far more than the recommended safe levels in the environment by Federal Republic of Nigeria Environmental Protection Agency (FEPA, 1991). Several authors have reported high correlation between the heavy metals amounts in natural and industrial environment to the increased freguency of chromosome mutations and the cancerous processes in organisms (Bruning and Chronz, 1999). C-mitotic effect is a common sign of spindle inhibition. Dash et al. (1988)

reported that lead has diverse effects on the cell, some of which are enzyme inhibition, chromosome aberration and mutation and its clastogenic effects leading to spindle impairment and malfunction. Cytokinetic failure leading to multinucleate cells is a sever deleterious effect that can lead to cancerous cells in tissues while nuclear disintegeration could lead to cell death. The nuelear leisions suggest indication of nuclear poison. Mercykutty and Stephen (1980) and Akaneme and Iyioke (2008) reported that the presence of nuclear leisions and nuclear dissolution offer cytological evidence for the inhibitory action on DNA biosynthesis. The severe retardation of root growth, lethal effects based on very low MI, aberrant cells, nuclear leisions and disintegration are signs of the mutagenecity of these effluents. These could cause irreversible health and ecological damages. These untreated wastewater are therefore considered to be lethal based on the above results and are environmental risk. A sound sewerage system in industries is therefore essential.

REFERENCES

Abu NE, Ezeugwu SC (2008). Risk evaluation of industrial wastewater on plants using onion (Allium cepa) chromosome aberration assay. Agro – Science 7(3): 242-248.

Abu NE, Ogbonna OH (2009). Cytotoxicity and genotoxicity screening of industrial wastewater: A case study of Cosmetics and Soap industrial effluents, Biores, 7(2): 534-539.

Akaneme FI, Iyioke IV (2008). Mutagenic potentials of the sterilizing fluid – Purital on root tip mitosis of Allium cepa. Bio – Research, 6(1): 293-297.

Antonsie- Wiez D (1990). Analysis of cell cycle in the root meristem of Allium cepa under the influence of Leda Krin, folia Histochem. Cytobiol., 26: 79-96.

Bruning T, Chronz C (1999). Occurance of Urinary Tract Tumour's in Miners Highly Exposed to Dimtroluene, YOEM, 3: 144-149.

Chang LW, Meier JR, Smith T (1997). Application of plant and earthworm bioassays to evaluate remediation of a lead-contaminated soil. Arch. Environ. Contam. Toxicol., 32: 166-171.

Cotelle S, Masfaraud JF, Fe'rard JF (1999). Assessment of the genotoxicity of contaminated soil with the Allium / Vicia-micronucleus and the Tradescantia micronucleus assays. Mutation Res., 426: 161-171.

Dash S, Panda KK, Pada BB (1998). Biomonitoring of low levels of mercurial derivatives in water and soil by Allium micronucleus assay. Mutation Res., 203: 11-21.

El-Shahaby OA, Abdel HM, Soliman MI, Mashaly IA (2005). Genotoxicity screening of industrial wastewater using the Allium cepa chromosome aberration assay. Pak. J. Biol. Sci., 6(1): 23-28.

Fedral Environmental Protection Agency FEPA (1991). Guidelines and standards for Environmental control in Nigeria, Lagos. In : Umeh, I.C. and Uchegbu, S.N. (1997) (Eds.) Principles and procedures of environmental impact assessment (EIA). Amazing Grace Printing and Publishing Company, Lagos, p. 167.

Fiskesjo G (1985). Allium test on river water from Braan and Sexan before and after closing of a chemical factory, Ambio, 14: 99-103.

Fiskesjo G (1993). The Allium Test in wastewater monitoring. Environ. Toxicol. Water Quality, 8(3): 291-298.

Fiskesjo G, Levan A (1993). Evaluation of the first ten MEIC chemicals in the Allium test. ATLA, 21: 139-149.

Grant WF (1999). Higher plant assays for the detection of chromosomal aberrations and gene mutations - a brief historical background on their use for screening and monitoring environmental chemicals. Mutation Res., 426: 107-112.

Grover IS, Kaur S (1999). Genotoxicity and wastewater samples form sewage and industrial effluent detected by the Allium root anaphase aberration and micronucleus assays. Mut. Res., 426: 183-188.

Houk WS (1992). The genotoxicity of industrial wastes and effluents. Mutat Res., 277: 91-138.

Ivanova E, Staikova T, Velcheva I (2002). Mutagenic effect of water polluted with heavy metals and cyanides on Pisum sativum plant in vivo. J. Balkan Ecol., 3: 307-310.

Ivanova E, Staikova TA, Velcheva I (2005). Cytogenetic testing of heavy metal and cyanide contaminated river waters in a mining region of South Bulgaria. J. Cell Mole. Biol., 4: 99-106.

Jiang W, Liu D, Li H (1996). Effect of coper on root growth, cell division and and nucleus of Helianthus annus L. Sci. Total Environ., 265: 59-65.

Kovalchuk O, Kovalchuk I, Arkipov A, Telyuk P, Hohn B, Kovalchuk L (1998). The Allium cepa chromosome aberration test reliably measures genotoxicity of soils of inhabited areas in Ukraine contaminated by the Chernobyl accident. Mutation Res., 415: 47-57.

Levan A (1938). The effect of colchicine on root mitosis in Allium. Hereditas, 24: 471-486.

Levan A (1938). Cytological reactions induced by inorganic salt solutions. Nature, 156(2973): 751-752.

Maluszynaska J, Juchimiuk J (2005). Plant genotoxicity: A molecular and cytogenetic approach in plant bioassays. Plant Genotoxicity, 56: 177-184.

Mercykutty VC, Stephen J (1980). Adriamycin induced genetic toxicity as demonstrated by the Allium test. Cytologia, 45: 769-777.

Odeigah PGC, Nurudeen O, Amund OO (1997). Genotoxicity of oil field wastewater in Nigeria. Hereditas, 126: 161-167.

Osaigbovo A, Orhue E (2006). Influence of pharmaceutical effluents on some soil chemical properties and early growth of maize (Zea mays). Afr. J. Biotechnol., 5(12): 1612-1617.

Panda BB, Sahu UK (1985). Induction of abnormal spindle function and cytokinesis inhibition in mitotic cells in Allium cepa by organophosphorus insecticides fensulfothion. Cytobios, 42: 147-155.

Rank J, Nielson MH (1993). A modified Allium test as a tool in the screening of genotoxicity of complex mixtures. Herditas, 118: 49-53.

Rank J, Nielson MH (1998). Genotoxicity testing of wastewater sludge using the Allium cepa anaphase-telophase chromosome aberration assay. Mutation Res., 418: 113-119.

Rank J, Lopez LC, Nielsen MH, Moretton J (2002). Genotoxicity of maleic hydrazide, acridine and DEHP in Allium cepa. Hereditas, 136: 13-18.

27

Occupational health hazards of fabric bag filter workers' exposure to coal fly ash

Jacobus Engelbrecht[1]*, Phanuel Tau[1] and Charles Hongoro[1,2]

[1]Department of Environmental Health, Faculty of Science, Tshwane University of Technology, Private Bag X680, Pretoria, 0001, South Africa.
[2]Health Systems Research Unit, South African Medical Research Council, South Africa.

The objective of the study was to assess employees' exposure to coal fly ash dust during the replacement of fabric bag filters in bag houses at a power station. Personal and environmental sampling were conducted on a random sample of workers. Samples that were in excess of occupational exposure limits (OEL) for respirable fly ash were analysed for toxic metals. Physical observations and a questionnaire were used to determine awareness of employees on the health effects of coal fly ash dust exposure. The results from personal dust sampling ranged from 20.7 to 477.2 mg/m^3 with an average of 101.2 mg/m^3. All the results were above the 5.0 mg/m^3 legal threshhold specified in the applicable South African legislation. Static dust sampling results ranged from 2.2 to 28.7 mg/m^3 with an average of 13.5 mg/m^3. Only 8% of the static dust samples were below the OEL for respirable dust ($\geq$ 5mg/m^3). Results that were obtained from toxic metal analysis were far below the OEL. Good awareness by employees regarding the health effects of exposure to coal fly ash and awareness of respiratory zones was also reflected. Control measures are recommended to reduce the exposure risk to fly ash.

Key words: Workers exposure, coal fly ash, occupational hazards, fabric bag filters.

INTRODUCTION

At a coal-fired electricity generating plant, coal is the main raw material for production and it is expected that employees will be exposed to coal dust (with a probability of developing silicosis because of the prevalence of crystalline silica in coal) at various stages of production. This kind of low grade coal produces a high yield of ash after combustion in the boilers. It is because of this high yield that is, there is a probability of high concentration of toxic metals in the ash that is generated, considering that 15% of ash is produced from combustion of coal (Boswell, 1987). Dermatitis, bronchitis, eye injuries and lung diseases can result from exposure to hazardous dust (Agius, 2001). Fly ash as a waste by-product of the electricity generation process consists mainly of fine particles. The particle sizes of fly ash dust are according to Meij and te Winkel (2000) distributed on the basis of

internationally accepted differentiation between inhalable and respirable fractions in these proportions: 50 μm (55%), 10 μm (20%), 4 μm (5%) and 2.5 μm (1%).

Studies conducted by UNIPEDE (1995) revealed that bulk fly ash has an alpha-quartz content of between 0.1 and 11% in the respirable fraction. The potential for alpha quartz exposure exists if there is high quartz content in the coal feed stock. However, its toxicity is significantly reduced by the high combustion temperature (approximately 1800°C) in modern power stations. This results in the quartz being converted from crystalline to a non-hazardous vitrified form. It is because of the formation of a glassy material that fly ash was considered as being relatively inert and innocuous after combustion Elemental analysis of the fly ash shows that the main components are silicon, aluminium, and calcium (EURELECTRIC, 2000).

Due to its composition and genesis, coal fly ash exhibits pozzolanic properties; it reacts with dissolved calcium hydroxide and water at normal temperature to

---
*Corresponding author. E-mail: engelbrechtjc@tut.ac.za.

form strength developing minerals in a similar manner to cement (EURELECTRIC, 2000). The composition of fly ash is dependent upon the mineral matter present within the coal that is fired in the furnace and the effectiveness of combustion that has taken place (Sear et al., 2004). Cook (1983) specifies that glassy particles in fly ash are probably the most important pozzolanic constituent. The reactions of the glassy particles with lime released during hydration of cement produce secondary cementatious compounds. Cook (1983) further states that there is a general agreement that the finer the fly ash, the higher the pozzolanic activity. In general, the finer ashes react quicker to produce slightly higher early strengths and slightly faster set times. The glass in fly ash is the most reactive component in cementation and an increase in the mineral content would reduce the pozzolanic efficiency of the ash (Lesch, 1987). According to studies undertaken by EURELECTRIC (2000) typical fly ash also contains additional metal oxides such as TiO_2, MgO, K_2O and P_2O_5 at 1 to 3% and Pb_2O_5 at 0.3 to 3%.

Employees at coal-fired power generating plants are generally not exposed to high concentrations of fly ash dust because the ash is enclosed within the production system. However, there are tasks or activities such as maintenance on electrostatic precipitators, hoppers, bag houses and conveyers whereby employees are likely to be exposed to high concentrations of this dust (ESKOM, 1996).

Aim and objectives

The aim of the study was to determine the extent of exposure of workers maintaining fabric filter plants to fly ash and to determine their level of knowledge and awareness on hazards and risks accociated with their excopure to fly ash. The objectives were to:

1. Assess employees' exposure (personal and environmental exposure) to fly ash dust.
2. Compare the sampling results to the legal occupational exposure limits (OELs) for dust.
3. Determine the occupational hazard awareness levels of exposed workers and
4. Recommend measures to control and/or reduce the risk of exposure to fly ash dust.

MATERIALS AND METHODS

The study employed a mixed method approach: 1) Laboratory based analyses, 2) Observations, and 3) Questionnaire for workers. Such an approach allowed for a multi-dimensional analysis of occupational hazards related to exposure bag filters workers to coal fly ash by fabric.

Description of study site

The Duvha Power Station is situated 15 km east of Witbank in the Mpumalanga province of the Republic of South Africa. It is the only Eskom coal-fired electricity generating station that utilises both bag houses and electrostatic precipitators as means of controlling the particulate emissions into the atmosphere. The rest of Eskom's coal-fired electricity generating stations utilises electrostatic precipitators only. Three of the generating units at Duvha Power Station utilise Optipulse pulse-jet fabric filter in the bag houses. In 1993 Duvha Power Station became the first power station in the world to be retrofitted with pulse jet fabric filter plant on three of its six units. The rest of the units function on electrostatic precipitators. These plants contribute largely to the reduction of air pollution by removing 99.99% of the fly ash otherwise released into the air through the station's chimneys. For each corresponding unit there are 4 compartments or cells in a baghouse. There are about 6 724 filter bags in a compartment, each of which is eight meters long, giving a total of 26 896 bags per unit (Duvha Power Station Technical Information, 1996). The bag house as means of controlling emissions is one of the approved, effective and efficient methods in the control of particulate air pollutants. This method, just like any production process requires maintenance whereby the fabric filter bags are replaced with new ones.

The process of replacing bag filters in the bag houses is undertaken alternately in the units. The frequency of the bag replacement is dependent on the performance of the stack emissions. The more dust that is emitted from the stacks, which is indicative of poor performance of the bags, the more frequent this operation will take place. This operation for bag replacement if undertaken is frequently done biannually and this process is carried out manually (Figure 1). Prior to the fabric filter bags' replacement project, the unit or plant is shut down or taken off production cycle for bag house maintenance. This action is undertaken in order to allow for the cooling down of the area and to reduce the risk of exposure to the hazards of flue gases as well as heat. There are 15 employees during a shift that are involved in the task of replacement of bag filters inside a bag house. During the replacement of the fabric filter bags these employees are in direct contact with fly ash dust. The fly ash dust is known to consist mainly of particles that are below the 10 μm size range. This dust fraction is therefore considered to be respirable and will reach the alveoli once it is inhaled, thus could increase the risk of causing pulmonary diseases such as silicosis if it contains crystalline silica (Health and Safety Executive, 1986). This study was initiated because of lack of sufficient scientific knowledge and evidence by employees and employers alike at the power station on the toxicity and health hazards associated with exposure to fly ash dust when replacing bags at baghouse units.

Sampling and analysis

Personal (64) and static (26) samples for respirable fly ash dust were collected between routine maintenance shutdowns of the fabric bag filter plant. The procedure that was followed for the personal and static sampling of respirable fly ash dust in the bag house was a combination of methods health and safety executive (HSE) MDHS 51 (for flow rate calibration of personal samplers), NIOSH 0600 and NIOSH 7300 (for trace metals). The NIOSH 0600 method is for the determination of respirable dust and is applicable for the sampling of any non-volitile respirable dust and is also recommmend for respirable coal dust sampling. The method requires that a polyvinyl chloride (PVC) filter be used for the collection of dust particles. The filters are to be equilibrated and weighed before and after sampling. The method requires in addition that a cyclone be used in collecting the respirable dust fraction. An air flow sampling rate of 1.7 and 2.2 ℓ/min needs to be maintained pending on the type of cyclone that is used. Thirty seven (37) mm diameter, 0,8 μ pore diameter mixed cellulose ester (MCE) filters and support pads were used for the collection of samples. The

Figure 1. Fabric filter bags removed that are ready for disposal.

MCE filters were chosen for their suitability for laboratory analysis of metals by inductively coupled plasma spectoscopy. Personal and static sampling was conducted for the entire shift (8 h) or at a minimum of 80% of the shift-time when bag filters were replaced in the bag house. Where only 80% of the shift time was applicable the unsampled (20%) of exposure concentration remained constant. Throughout the sampling period it was confirmed that the sampling pumps were operational for the entire sampling periods. Where problems existed with the running of pumps, sampling was terminated and then resumed with a new set of sampling equipment (normally during a next work shift). Flow rate calibration was excersised according to MDHS 51. Sampling for quartz were not conducted as the high temperature achieved during the combustion of coal changes the chemical composition and quartz is not formed. Polycyclic aromatic hydrocarbon (PAH) and silica were also not sampled as the study was limited to the workers' fly ash exposure from the combustion process.

The respirable fly ash dust samples with concentrations that were ranging from 40 to 450 mg/m^3 were selected at random, transported and submitted to the laboratory for the determination of identified toxic trace metals. Twenty nine (26) samples were analysed at a laboratory that is accredited by the South African National Accreditation Services (SANAS) for analysis of toxic trace metals. The laboratory analysed the respirable dust fractions for concentrations of toxic trace metals using the inductively coupled plasma (ICP) analytical technique. An aqueous medium was only used where the levels of toxic metals were high on the samples that were dissolved in acid medium.

Observations and worker interviews

Physical observations were conducted and photographs taken before and during the process of replacement of bag filters in bag houses. The purpose of obervations was to establish worker knowledge and practice levels, attitudes and working conditions. In addition, a questionnaire was administered to workers during the same time as sampling took place.

Questionnaires were used to evaluate the level of occupational hygiene knowledge and training of exposed employees regarding the hazards of fly ash dust in order to recommend the development of training guidelines for the control of personal exposure whilst replacing the fabric filter bags in the bag houses. Questionnaires were developed to include questions on training competencies of employees as well as their awareness of the health hazards of fly ash dust. Questionnaires were administered to a representative group of employees that were selected at random to establish the profiles based on the above aims and objectives. It was decided to use a sample of 31 employees from a population of 45 workers engaged in this operation. This number was statistically considered as being representative for the population that is engaged in the replacement of fabric bag filters in the bag houses.

Prior to administering these questionnaires, workers were engaged in a meeting and briefed about the purpose of the study. In this meeting it was identified that some of the workers were incompetent in speaking and writing english and were unable to complete the questionnaires on their own. To ensure consistancy in the data, such employees were assisted by the researcher and the

Table 1. The statistical analysis of fly ash dust concentration results.

Statistical information	Static samples	Personal samples
Sample size	26	64
Range (mg/m^3)	2.2 - 28.7	20.7 - 477.2
Minimum (mg/m^3)	2.2	20.7
Maximum (mg/m^3)	28.7	477.2
Mean (mg/m^3)	13.54	101.2
Standard deviation	7.65	87.8

safety representative of the workers together, in interpreting and completing the questionnaire. To the rest of employees the questionnaires were explained to them and issued for completion. These employees were allowed two days to complete questionnaires in order to allow sufficient time for response. Data collected on questionnaires was analysed using the Epi info statistical software programme (EPI INFO, 2004). The project was approved by the ethical committee of the Institution under which auspices the research was carried out. Interviews were held with an aim to further establish if cases of fly ash dust over exposure were reported to both the medical centre and the safety departments and to determine what precautionary measures are taken in order to alleviate the risk of overexposure to fly ash dust.

Physical observations were undertaken for supporting information that was collected from the questionnaires. A check sheet was used to include all items that were of interest to observe during the replacement of bag filters in bag houses. A digital camera was made available to assist in capturing scenes in the bag houses whilst bag filters were being replaced.

RESULTS

Fly ash dust samples for personal monitoring

The results for personal dust sampling revealed that all the results are at and above the occupational exposure limit for respirable fly ash dust (Table 1). The exposure duration of the individual workers sampled as well as for the static samples was representative of an eight hour shift. These results ranged from 20.7 to 477.2 mg/m^3 with an average of 101.2 mg/m^3. All the results from the personal samples were above the 5.0 mg/m^3 legal threshhold specified in the applicable legislation (SA, 1995).

Fly ash dust samples for static monitoring

The results for static dust sampling (Table 1) revealed that not all the results were below the occupational exposure limit for respirable dust. The results ranged from 2.2 to 28.7 mg/m^3 with an average of 13.5 mg/m^3.

The results for static samples were generally lower than those obtained for the personal samples. The results were in most cases above the occupational exposure limit but with a less magnitude as compared to all the personal samples as shown in Table 1. The dust load of

static samples was significantly lower as compared to the personal samples. This difference can be attributed to the fact that for the personal samples the collection media were located nearer to the point of action during sampling as compared to the static samples that represented a work area. The collection media for static samples were some distance from the point of action because of a lack of access sampling points in the bag house.

Personal samples were representative of all areas where employees were working while replacing the bags. This ranged from three to four workstations where the bag filters were handled. The static samples were located in one area and were not moved as employees were migrating from one workstation to the next. It was observed that while the employees moved from one area to the next, the dust load in an area with less activity was low, and this would therefore affect dust load at a static sample's point as action was minimised in a work environment. For personal samples in all areas, workers generated dust as they were working with the fabric filter bags and therefore the collection media would pick up a load of dust no matter how small the quantity in each area, when an activity is undertaken. The sampling range varied from 20.7 to 448.3 mg/m^3. A standard deviation of 87.8 is not unusual due to the fact that the sampling was not static and workers that carried the sampler moving around were differently exposed and only representative of the specific individual worker.

Toxic metals concentration

For the toxic trace metals most of the results (Table 2) that were obtained were far below the OEL for the respective metals. This was in support of a previous assumption that although fly ash contains trace elements that are toxic, these are present in concentrations that are biologically insignificant (UNIPEDE, 1995). Their concentrations were indeed at trace levels and even at very high atmospheric fly ash concentrations the exposure limits of the respective metals were not exceeded. This study has scientifically confirmed this assumption. The only exception was the lead concentration where the results was in some samples approaching the OEL for this metal. The results of this

Table 2. Statistical analysis for toxic trace metals.

Metal	Average mg/m^3	Minimum mg/m^3	Maximum mg/m^3	Range mg/m^3	OELs (mg/m^3)
Ba	0.04	0.021	0.066	0.021 - 0.066	0.5
Ni	0.02	<0.018	0.03	<0.018 - 0.03	*
Zn	0.024	0.0134	0.055	0.0134 - 0.055	5
Hg	<0.063	<0.007	<0.069	<0.007- <0.069	0.05
Cr	0.013	0.001	0.021	0.001 - 0.021	0.5
Cu	0.0308	0.001	0.08	0.001 - 0.08	0.2
V	<0.0609	<0.07	<0.069	<0.07 - <0.069	0.05
Mn	0.007	0.001	0.029	0.001 - 0.029	5
As	<0.065	<0.063	<0.069	<0.063 - <0.069	0.1
Pb	0.0499	<0.032	0.139	<0.032 - 0.139	0.15

* Not available.

study could therefore not give conclusive information regarding the levels of toxic trace metals in the respirable fly ash dust. Even if the cumulative effect of the toxic chemicals are taken into consideration, no overexposure of workers is expected.

Interview results with safety and health management

The interviews with the Safety and Health Department at the Power Station have revealed that there were no cases or incidents reported of overexposure to fly ash dust at the power station for the past 5 years. It was because of this information that one could not establish exposure trends because of a lack of information. There is no dedicated awareness training regarding the hazards of exposure to fly ash dust. The training that is available is generic and is only given during induction for new and contract workers. There is therefore insufficient awareness and training on the hazards of fly ash dust.

The questionnaires results

The questionnaire results revealed that of all employees that participated in the study, 96.7% agreed that fly ash dust is dangerous and can cause harm to them. 90% indicated that the ash affects them. 45.2% confirmed that they felt uneasy after being exposed to excessive concentration of fly ash, whilst 48.4% said that they never felt uneasy after being exposed to fly ash dust.

Of the employees involved in the replacement of bag filters in the bag house, Figure 2 indicates that 41.9% have been involved in this job for at least 2 years and 32.3% for at least 3 years of their period of employment and have confirmed that they are not working in this area on a daily basis.

Of all the employees that were interviewed, all answered positively that they were given information

about the danger of over exposure to fly ash dust. All of the participants confirmed that the wearing of respiratory protection devices are meant for both employees and management while working in a high-risk area. Employees were able to identify a respirator zone because 87.1% correctly indicated how to describe such an area, while all of them knew how to identify it.

From the sample of employees that participated in the study, all confirmed that they had received training in the wearing of respirators. 61.3% of them confirmed the limitations of the respiratory protective devices that they were using (Figure 3), whilst 19.3% was unsure of the limitations of these devices. 90.4% think that the respiratory protection devices provide sufficient protection. 74.2% confirmed that training on the wearing of respiratory protection is given frequently; daily before starting with work. 68% of employees confirmed the wearing of respiratory protection for all the period that they spend working in the bag house during the replacement of bag filters.

Physical observations while replacing bag filters

When observing employees working in the bag house, it was difficult to establish if a formalised or written safe work procedure was followed. There was no recognisable pattern of how the work is undertaken.

Only one type of a respirator was used during the bag replacement operation. It was therefore not required to compile a list of the respirators that are used at the power station. The employees were using these respiratory protective devices for the duration of operations while working in the bag house during replacing the fabric filter bags. There were however, some respirators that were partly not in good condition. There were those that had soiled cartridges which could ultimately render the device ineffective and insufficient for the intended task. Although these respiratory protective devices were kept in a central

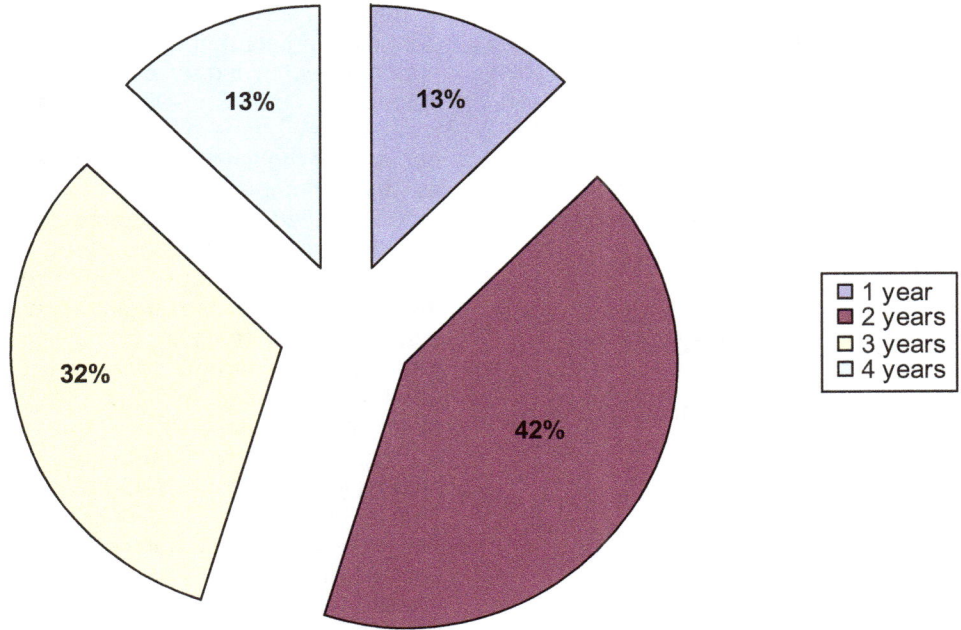

Figure 2. Years of experience working with fly ash.

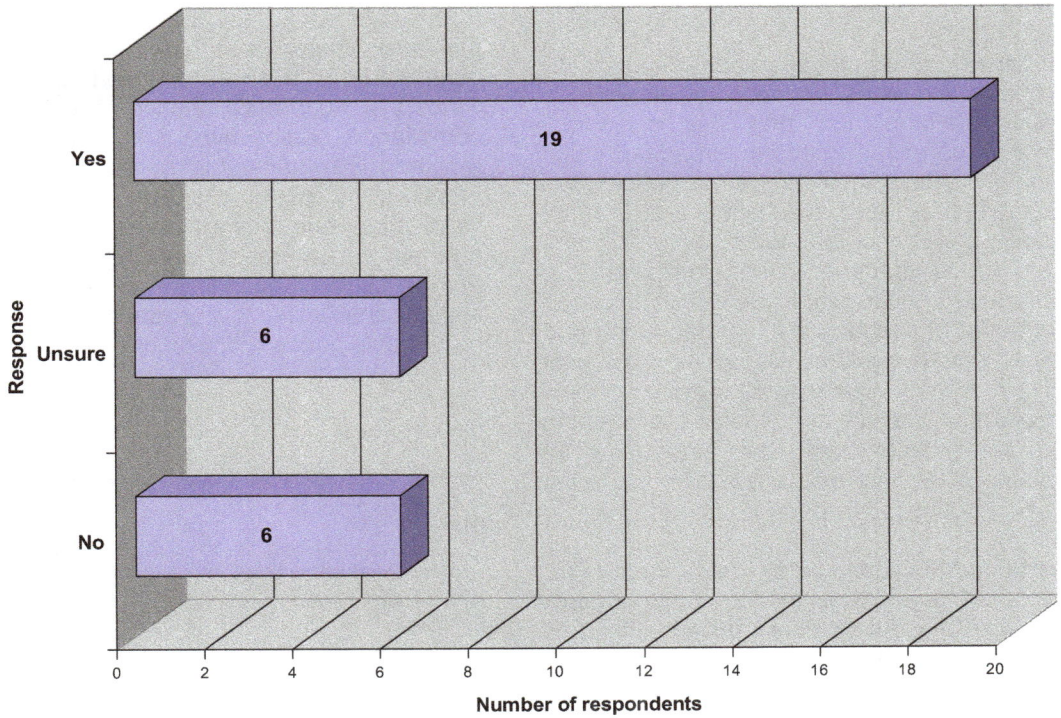

Figure 3. Knowledge on limitations of respirators.

place before and after use, the condition under which they are stored appeared to be inappropriate.

There was a concern regarding personal hygiene because employees were seen leaving/breaking for lunch while fully covered in a heavy load of fly ash dust on their clothing and faces. Attempts were not made for decontamination or showering prior to taking lunch. This action would cause employees to be exposed to fly ash dust through ingestion while on lunch. In addition, this also exposes other employees that are by the nature of their job not working in a dusty area/respirator zone. This practice is prohibited by the Regulations of Hazardous Chemicals Substances (South Africa, 1995).

CONCLUSIONS AND RECOMMENDATIONS

A number of conclusions can be drwan from this study: First, that employees are exposed to high levels of fly ash dust during the replacement of fabric filter bags and that the results for toxic trace metals in respirable fly ash dust were inconclusive as far as over exposure of workers. The hypothesis that employees that are responsible for the replacement of bag filters in the bag houses are exposed to high fly ash dust levels that pose an occupational health risk was partially proven to be true mainly because of the inconclusive results of toxic trace metals that indicated that workers are not over exposed. Cumulative or possible combined effects were verified and results indicated that due to very low concentrations of the toxic metals, overexposure is not evident (Canadian IRSST, 2010).

Second, that training of employees should specifically address fly ash dust exposure because there were no records for such training and that the respiratory programme needs to be reviewed and enforced. Although awareness training is given, there are some deficiencies (for example, ineffective and insufficient respirators, personal hygiene, and decontamination before taking lunch) that need to be tightened up in improving the intervention. This would cause employees to be exposed to fly ash dust through ingestion whilst having lunch and also expose other employers that are by the nature of their not working in a dusty area/respirator zone. It could therefore be concluded that the hypothesis that training guidelines and safety knowledge are inadequate to address the training of workers responsible for replacing the bags in the bag houses, has been tested and proven to be true.

Third, there was no identifiable safe work procedure for the replacement of bag filters in the bag houses. Experience and age of workers could not be linked to causes of overexposure to fly ash dust while replacing fabric filter bags in the bag houses. No medical records of employees' exposure to fly ash dust were available and therefore it was not possible to determine trends of exposure. The records available are for general periodical medicals.

The study, found out that in most instances, personal protective equipment is provided only after all means of environmental controls have been considered. There are

a number of instances where the use of respiratory protective equipment is likely to be the final, but adopted option for achieving acceptable long-term exposures to a substance (BOHS, 1996). This occurs where other control methods are not feasible to be implemented because of the limitations in the work environment as well as the material that is handled. It is therefore recommended that a respiratory protection programme be enhanced in order to reduce the risk of employees' exposed to fly ash dust. This is recommended because there are no feasible engineering control methods that could be instituted for controlling/reducing exposure. A respiratory programme such as NIOSH (1987) should be used as guideline for controlling/reducing the risk of exposure to fly ash dust.

A comprehensive safe work procedure should be developed and employees be trained on its application. This procedure will ensure that all employees work uniformly in the bag houses while replacing bag filters. Just as the Basel Convention EURELECTRIC (2000), whereby coal fly ash is tabulated in lists A and B for hazardous and non-hazardous wastes, this study revealed concentration levels of toxic trace metals below hazardous legal limits. Because of the cumulative nature of some of the chemicals such as mercury and lead, long term term exposure can lead to increased risk of over exposure of workers. These workers are exposed to flyash dust on a daily basis and they all had worked in those conditions for an average of 8 years. Most of them confirmed that they are exposed to high levels of flyash. Since this study only focused on respiratory dust from a compliance perspective, it is recommended that more research is required for the determination of toxic trace metals in fly ash be repeated by expanding the work in more than one power station and to include particle number concentration and size distributions including PM1.0. More emphasis should be placed at those power stations that utilise low grade coal. Future studies of this nature also needs focus on reliability analysis of results.

REFERENCES

Agius R (2001). Occupational and Environmental Lung Disease. United States of America, p. 3.
BOHS (1996). The Manager's Guide to Control of Hazardous Substances. H and H Scientific Consultants Ltd., Leeds, United Kingdom, p. 21.
Boswell JES (1987). The disposal of power station ash by Eskom in South Africa. Proc. International Symposium on Ash – a valuable resource. CSIR. South Africa.
Canadian IRSTT Mixie. Available at: http//www.irsit.gc.ca/files/outils/intertox/jsndx.htm Accessed on 05/02/2010.
Cook JE (1983). Fly ash in concrete – Technicla consideration. Technical Service Division, Gifford-Hill and Co. Dallas, pp. 51-95.
Duvha Power Station Technical Information (1996). Duvha coal-fired giant. Eskom, South Africa. http://www.eskom.co.za/live/content.php?Item_ID=163 (Accessed on 18/02/2010).
EPI INFO (2004). Version 3.01. Statistical software programme of the World Health Organisation, GENEVA, SWITZERLAND.

ESKOM (1996). Duvha power station Technical Information. South Africa, p. 2.

EURELECTRIC (2000). Residues Task Force. Fly Ash from Coal-Fired Power

Health and Safety Executive (1986). Method MDHS 51, Quartz in respirable airborne dusts. HSE, London, United Kingdom.

Lesch W (1987). The mineral and glass content of South African fly ash. National Institute for Occupational Safety and Health. (NIOSH). Cincinnati. OHIO. USA, p. 3.

Meij R, te Winkel H (2003). Health aspects of coal fly ash. Netherlands, p. 29.

MSA Instuments Home page (2008). [On line]. Available at: www.westernsafety.com/msaproducts/msapa26.html. Accessed on 21/04/2008.

NIOSH Publication No. 87-116 (1987). NIOSH Guide to Industrial Respiratory Protection. National Institute Occupational Safety and Health. Cincinnati, Ohio, USA.

Sear LKA, Weatherley AJ, Dawson A (2003). The Environmental impacts of using fly ash – the UK Producers' Perspective. United Kingdom, pp. 1-7.

South Africa (1995). Hazardous Chemical Substances Regulations. Government Notice. R: 1179, 25 August 1995.

UNIPEDE (1995). Permanent Group on Medical Matters. Pulverised Fuel Ash from Coal-Fired Generation. Paris, p. 3.

28

Comparative nephrotoxic effect associated with exposure to diesel and gasoline vapours in rats

F. E. Uboh[1]*, M. I. Akpanabiatu[2], J. I. Ndem[2], Y. Alozie[3] and P. E. Ebong[1]

[1]Department of Biochemistry, College of Medical Sciences, University of Calabar, Calabar, Cross River State, Nigeria.
[2]Department of Biochemistry, College of Medical Sciences, University of Uyo, Uyo, Akwa Ibom State, Nigeria.
[3]Department of Biochemistry, Cross River State University of Technology, Calabar, Nigeria.

Comparative effect of exposure to 20.7 ± 5.8 cm^3 h^{-1}kg^{-1}m^{-3} day^{-1} of diesel and gasoline vapours on the kidney functions was assessed in rats. It was observed that exposure to diesel and gasoline vapours produced a significant increase (P < 0.05) in serum creatinine, urea, BUN, uric acid, glucose and K$^+$; and a significant decrease (P < 0.05) in serum Na$^+$ and Cl$^-$ levels. However, the percentage increase in serum creatinine, urea, BUN, uric acid, glucose, K$^+$; and decrease in serum Na$^+$ and Cl$^-$ levels recorded for the rats exposed to diesel vapour were significantly higher (P < 0.05) compared to the percentages recorded for rats exposed to gasoline vapour. The result of this study indicates that exposure to diesel and gasoline vapours may be a risk factor for nephrotoxicity in rats; and that diesel vapour tends to contain chemical substance(s) that are more nephrotoxic than gasoline vapour.

Key words: Diesel, gasoline, creatinine, urea, electrolytes, nephrotoxicity.

INTRODUCTION

Domestic and industrial use of petroleum, either in its crude or refined forms, has increased tremendously in recent times. Crude petroleum may be refined into such fractions as gasoline, kerosene, diesel, heavy gas oils, lubricating oils, as well as residual and heavy fuels among others (EHC 20, 1982). Diesel, gasoline and kerosene are among the commonly used fractionated products of crude petroleum. These fractions contain aliphatic, aromatic and a variety of other branched saturated and unsaturated hydrocarbons at variable concentrations (EHC 20, 1982; Henderson et al., 1993; Kato et al., 1996; Anderson et al., 1995). The constituents of the vapours from these fractions, to a greater extent, depend on the composition of their liquid forms, which varies with the brand and storage period.

The blend of unleaded gasoline (UG) designated PS-6, API-0I UG and the methyl tertiary butyl ether (MBTE) blended gasoline are among the brands of gasoline commonly used in the United States (Moser et al., 1996). Gasoline, kerosene and diesel are reported to contain predominantly, hydrocarbons with carbon atoms 4 - 10,

11 - 13 and 14 - 18, respectively (EHC 20, 1982). The volatility of these fractions varies with the predominant hydrocarbon species. Unleaded gasoline for instance, is reported to contain about 300 different hydrocarbon species, most of which are highly volatile and may evaporate if left exposed, to constitute ubiquitous chemical pollutants in the environment (Zahlsen and Tri-Tugaswati, 1993). Reports also indicate that API 91-0I UG contains slightly higher percentage of saturated hydrocarbons than PS-6 blend, and that an estimate of 25% or more of the gasoline supplied in the United States in 1995 were supplemented or blended with MBTE (Lorenzetti, 1994). In the course of usage of these products, and other day to day activities, individuals are frequently directly or indirectly exposed to pollutants of petroleum origin in their environments. However, those that are occupationally exposed tend to be at a greater risk of exposure (Smith et al., 1993; Carballo et al., 1995). Human health hazards arising from intermittent, low-dose exposure to petroleum vapours are not quite consistent. The potential harmful effects associated with chronic or sub-chronic inhalation of the petroleum pollutants in the atmosphere constitute the concern of the general public and the scientific community. To identify the potential health risk of chronic exposure to UG, it is

---

*Corresponding author. E-mail: fridayuboh@yahoo.com.

Table 1. Distribution of experimental groups.

Group	No. of rats	Treatment
Control (I)	5	Vapours-free
Gasoline (II)	5	Exposed to Gasoline vapour
Diesel (III)	5	Exposed to Diesel vapours

reported that American Petroleum Institute sponsored a cancer bioassay, in which B6C3F1 mice and F-344 rats were exposed to UG vapour for 6 hrs/day, 5 day/week for 2 years. The results indicated that the carcinogenic effects detected were the induction of male rat kidney tumours and female mouse liver tumours. The kidney tumours were believed to result from the interaction of the metabolites of certain isoparaffinic components of UG with a male rat-specific renal protein, μ-globulin (Chun et al., 1992; Hard et al., 1993). The accumulation of this protein in proximal tubule cells may lead to cytolethality, regenerative cell proliferation and ultimately, renal cancer (Borghoff et al., 1992). UG vapour is also reported to stimulate the growth of diethyl nitrosamine-induced hepatic preneoplastic lesions in mice, and induce an enzyme activity associated with cytochrome $P_{450}2B$ (Standeven et al., 1993; Standeven et al., 1994; Standeven et al., 1993). These reports indicate that mice are more vulnerable to the toxicity effects associated with gasoline vapours inhalation than rats, and that the male rats are affected than the females when exposed to gasoline vapours. In our previous studies, adverse effects of exposure to gasoline and kerosene fumes/vapours on haematological indices, weight changes, liver and reproductive functions in rats were observed and reported (Uboh et al., 2005a, b; Uboh et al., 2007a, b; Uboh et al., 2008a, b). The chemical pollutants from gasoline vapours, like other known xenobiotics, may be metabolically transformed into various metabolites in the body (Hu and Wells, 1994). Some of these metabolites may be very reactive, interacting in various ways with the metabolizing and excreting tissues (mainly the liver and kidneys) to elicit toxic effects (Page and Mehlman, 1989; Nygren et al., 1994). The interaction of these metabolites with the renal tissues may cause cellular injury, hence, damage to the tissues. Once the renal tissues are damaged, the overall functionality of the kidneys may be compromised. The kidney functions may be assessed from the level of some electrolytes (such as K^+ Na^+, Cl^-) and metabolites (such as creatinine, urea and blood urea nitrogen) in the plasma (Nwankwo et al., 2006; Atangwho et al., 2007; Crook, 2007). Renal dysfunction may be caused by several diseased conditions and exposure to certain reactive or toxic metabolites (Crook, 2007; Chatterjea and Shinde, 2002; Jimoh and Odutuga, 2004). Renal dysfunction of any kind affects all parts of the nephron to some extent, although sometimes, either glomerular or tubular dysfunction is predominant. The net effect of renal disease on plasma and urine depends on

the proportion of glomeruli to tubules affected, and on the number of nephrons involved. In this study, comparative changes in some renal function indices, nephrotoxicity, hence associated with exposure of male rats to gasoline and diesel vapours were assessed.

MATERIALS AND METHODS

Animals and animal handling

Fifteen male Wistar albino rats weighing 180 - 200 g were obtained from the animal house of the Department of Biochemistry, University of Calabar, Calabar, Nigeria and used for this study. The animals were allowed one week of acclimatization to laboratory conditions and handling, after which they were distributed, according to weight into three groups as outlined in Table 1. The animals were housed individually in cages with plastic bottom and wire mesh top (North Kent Co. Ltd) and fed with normal rat chow (Guinea Feeds Product) purchased from the High Quality Livestock Feeds stores, Calabar, Nigeria. They were supplied with tap water *ad libitum* throughout the experimental period. The control group (Group I) was maintained in the animal room adequately ventilated under standard conditions (ambient temperature, $28 \pm 2°C$ and relative humidity, 46% with a light/dark cycle of 12/12 h). The test groups (Groups II and III) were kept in the exposure chambers (Vapours cupboards) previously saturated respectively with premium motor spirit (PMS) blend of gasoline and diesel vapours. The liquid gasoline (PMS blend) and diesel were obtained from the Mobil Refueling station, Marian Road, Calabar, Nigeria.

All animal experiments were carried out in accordance with the guidelines of the Institutional Animal Ethics Committee.

Exposure to gasoline and diesel vapours

A modified nose-inhalation exposure method previously described (Uboh et al., 2005a, b; Uboh et al., 2007a, b;), was used in this study. According to this modification, the cages housing the animals in the test groups were placed in respective exposure chambers (2 cages per one chamber) of $2.835 m^3$, each with two open calibrated beakers of $1000 cm^3$ containing $500 cm^3$ of liquid gasoline and diesel, respectively. The gasoline and diesel were allowed to evaporate freely within the respective exposure chambers at ambient humidity and temperature, and all animals in cages were exposed to vapours ($20.7 \pm 5.8 cm^3 h^{-1} Kg^{-1} m^{-3} day^{-1}$) generated from direct evaporation of the liquid gasoline and diesel. The animals were exposed 6 h/day (9.00 a.m - 3.00 p.m), 6 day/week, to vapours for 64 days. At the end of each exposure day, the animals were transferred to gasoline and diesel vapours-free section of the experimental animal house.

During the exposure period, the initial and final volumes of liquid gasoline and diesel were respectively recorded before and after daily exposure. The daily differences in volume were used to estimate relative concentrations of vapours used in this exposure method.

Table 2. Comparative effect of diesel and gasoline vapours on the levels of some serum catabolites commonly used in the assessment of renal functions.

Group	Urea (mg/dl)	BUN (mg/dl)	Uric (mg/dl)	Creatinine (mg/dl)
I	42.10 ± 1.26	19.66 ± 0.59	1.98 ± 0.08	1.18 ± 0.03
II	$51.72 \pm 3.40^{*}$	$24.15 \pm 1.78^{*}$	$2.38 \pm 0.08^{*}$	$2.00 \pm 0.06^{*}$
III	$72.96 \pm 1.86^{*+}$	$34.07 \pm 0.87^{*+}$	$3.73 \pm 0.17^{*+}$	$3.17 \pm 0.24^{*+}$

Value are presented as means ± SD; n = 7; $^{*}P < 0.05$ compared with group I; $^{+}P < 0.05$ compared with group II; Group II; Group I = Control. Group II = Group exposed to gasoline vapours. Group III = Group exposed to diesel vapours.

Collection and handling of blood serum for analyses

Twenty-four hours after last exposure, the animals were anaesthetized with chloroform vapour and dissected. Whole blood from each animal was collected by cardiac puncture into well-labelled non-heparinized sample tubes and allowed to clot for 3 h in iced water. The serum was separated from the clots after centrifuging at 10,000 rpm for 5 min into well-labelled plain sample bottles, and used for assays.

Biochemical assays

Serum urea and blood urea nitrogen: Urea in serum was estimated by the endpoint colorimetric method using Dialab reagent kits (Searcy et al., 1967). In this method, urease enzyme hydrolyses urea to ammonia and carbon dioxide. The ammonia so formed reacts with alkaline hypochloride and sodium salicylate in the presence of sodium nitroprusside to form a coloured chromophore which was measured with DREL 3000 HACH (England) model spectrophotometer.

Serum creatinine

The concentration of serum creatinine was assayed based on the reaction of creatinine with an alkaline solution of sodium pirate to form a red complex (Newman and Price, 1999). The red coloured complex which is proportional to the concentration of creatinine in the sample was measured spectrophotometrically.

Serum glucose

Serum glucose level was estimated, using Dialab reagent kits, by the principle of glucose oxidase reaction (Barham and Trinder, 1972). In this principle, glucose oxidase oxidizes glucose to gluconic acid, and hydrogen peroxide formed as a byproduct. The peroxides whose concentration is in proportion to glucose in sample develops quantifiable colour via 4-aminophenazone in the presence of a peroxidase.

Serum potassium

Potassium in serum was determined by photometric turbidemetric test using TECO analytical reagent kits (Tietz, 1976). Potassium ions in a protein-free alkaline medium react with sodium tetraphenylboron to produce a finely dispersed turbid suspension of potassium tetraphenyboron, whose turbidity is in proportion to the potassium concentration originally in the sample.

Serum sodium

Serum sodium concentration was estimated using Mg-Uranylacetate reaction method described in Dialab diagnostic kits (Trinder, 1957). Sodium in serum is precipitated with Mg-Uranylacetate, the remaining uranyl ions form a yellow-brown complex with thioglycolic acid. The difference between reagent blank analyses is proportional to the sodium chloride.

Serum chloride

Chloride in serum was determined using mercuric thiocyanate reaction method described in Dialab diagnostic kits (Tietz, 1976). Chloride ions in the sample react with mercuric thiocyanate displacing the thiocyanate ions. The displaced thiocyanate ions react with ferric ions producing a coloured complex.

Statistical analysis

All data are expressed as mean ± SEM (that is, standard error of the mean). The results were analyzed by one-way analysis of variance (ANOVA), followed by pair wise comparison between test and control groups using Student's t-test.

Differences between groups were considered significant at $p < 0.05$.

RESULTS

Changes in the levels of serum creatinine, urea, blood urea nitrogen (BUN), uric acid, glucose, as well as sodium, potassium and chloride ions (that is Na^+, K^+ and Cl^- respectively) were used to assess the renal function impairment effect of gasoline and diesel vapours in rats. The results of this study are shown in Tables 2 and 3 as well as Figures 1 and 2.

The results showed that the levels of serum creatinine, urea, BUN and uric acid increased significantly ($P < 0.05$) within and among the groups of rats exposed to gasoline and diesel vapours, compared with the rats in the control group (Table 1). However, the levels of serum creatinine, urea, BUN and uric acid obtained for the group of rats exposed to diesel vapour (3.17 ± 0.24, 72.96 ± 1.86, 34.07 ± 0.87 and 3.73 ± 0.17 mg/dl respectively) were observed to be significantly higher ($P < 0.05$) compared to the levels obtained for the group of rats exposed to

Table 3. Comparative effect of diesel and gasoline vapours on the levels of serum glucose and some electrolytes commonly used in the assessment of renal functions.

Group	Glucose (mg/dl)	Na$^+$ (mEq/l)	K$^+$ (mEq/l)	Cl$^-$ (mEq/l)
I	69.01 ± 3.95	119.20 ± 6.71	4.61 ± 0.16	100.76 ± 0.75
II	110.24 ± 3.36*	91.85 ± 0.65*	5.67 ± 0.09*	97.71 ± 0.40*
III	126.10 ± 5.13^{*+}	74.67 ± 3.53^{*+}	8.03 ± 0.15^{*+}	89.90 ± 1.62^{*+}

Values are presented as means ± SD; n = 7; $^*P < 0.05$ compared with group I; $^+P < 0.05$ compared with group II; Group abbreviations as in table.

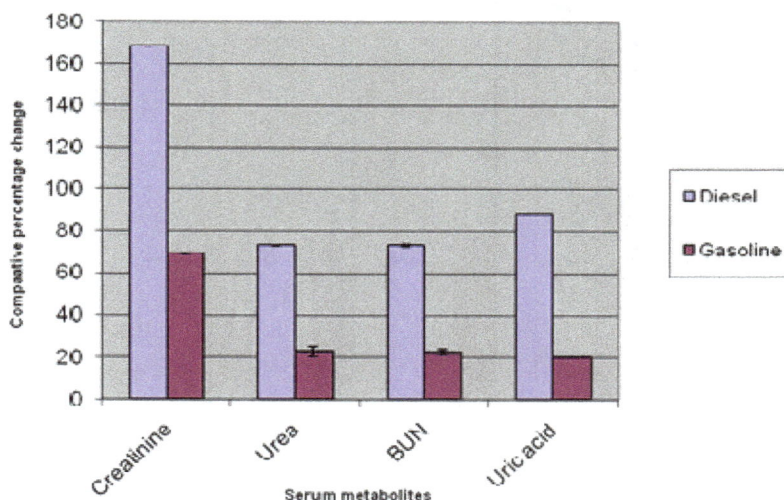

Figure 1. Comparative percentage increase in some serum renal function assessment metabolites in rats exposed to diesel and gasoline vapours.

gasoline vapour (2.00 ± 0.06, 51.72 ± 3.40, 24.15 ± 1.78 and 2.38 ± 0.08 mg/dl respectively). Consequently, the percentage increase in the levels of serum creatinine, urea, BUN and uric acid resulting from exposure to diesel vapour (168.6 ± 0.14, 73.3 ± 1.56, 73.3 ± 0.73, 88.4 ± 0.13% respectively) were significantly higher (P < 0.05) compared respectively with the percentage increase resulting from exposure to gasoline vapour (69.5 ± 0.05, 22.9 ± 2.30, 22.8 ± 1.20 and 20.2 ± 0.08% respectively) (Figure 1).

Moreover, the levels of serum glucose and K$^+$ were increased, while the levels of Na$^+$ and Cl$^-$ were decreased significantly (P < 0.05) sequel to exposure to gasoline and diesel vapours, in comparison with the levels obtained for the control group (Table 2). The results showed that the levels of serum glucose and K$^+$ for the rats exposed to diesel vapour (126.1 0 ± 5.13 mg/dl and 8.03 ± 0.15 mEq/l respectively) were significantly higher (P < 0.05) compared with the respective levels obtained for the rats in the group exposed to gasoline vapour (110.24 ± 3.36 mg/dl and 5.67 ± 0.09 mEq/l respectively); while the levels of serum Na$^+$ and Cl$^-$ obtained for the group exposed to diesel vapour (74.67 ± 3.53 and 89.90

± 1.62 mEq/l respectively) were significantly lower (P < 0.05) compared with the respective levels for the group exposed to gasoline vapour (91.85 ± 0.65 and 97.71 ± 0.40 mEq/l respectively). On the basis of this, the percentage increase in the levels of glucose and K$^+$ and percentage decrease in the levels of serum Na$^+$ and Cl$^-$ for the rats exposed to diesel vapour were observed to be significantly different (P < 0.05) from the respective percentage changes obtained for the rats exposed to gasoline vapour, as illustrated in Figure 2.

The results obtained from this study indicated that exposure to diesel and gasoline vapours may cause impairment of the renal functions, with diesel vapour being more adverse than gasoline. The impairment of the renal function reported in this study implies that diesel vapour contains more nephrotoxic constituents than gasoline vapour.

DISCUSSION

The kidney maintains constant extracellular environment by its involvement in the excretion of such catabolites as urea, creatinine and uric acid; and regulation of water and

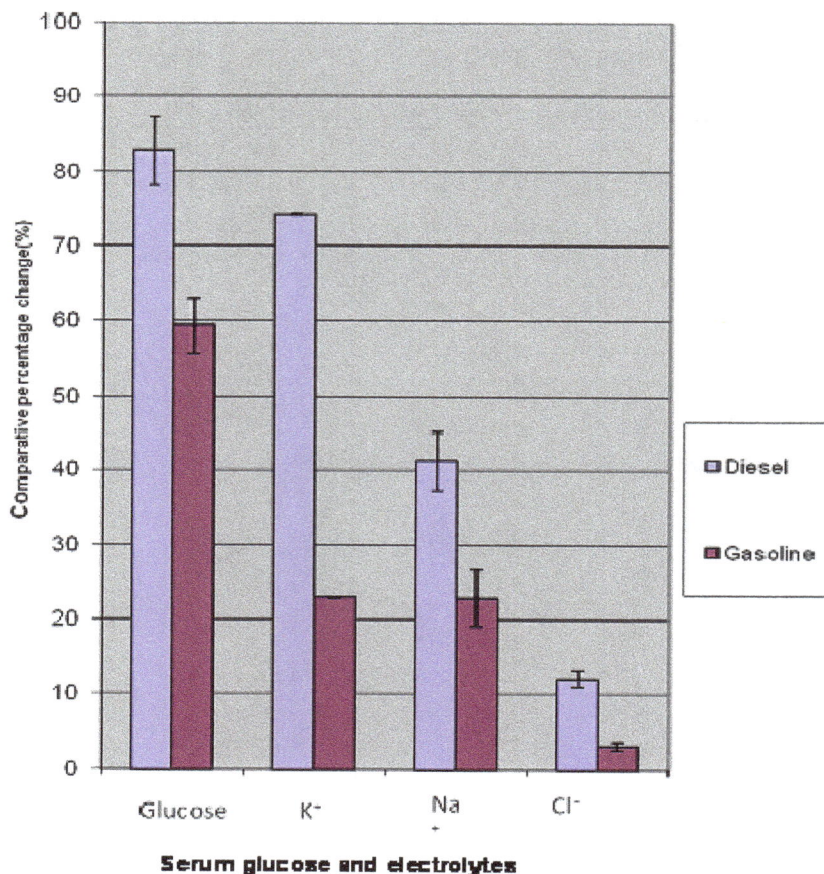

Figure 2. Comparative percentage changes in serum glucose and some electrolytes level in rats exposed to diesel and gasoline vapours.

electrolyte balance. Abnormal concentration of these catabolites and some electrolytes in the plasma or serum is a clear indication of renal function impairment (Nwankwo et al., 2006; Crook, 2007; Gidado et al., 2001; Zanna et al., 2008). Impairment of the renal functions may be caused by exposure to different nephrotoxic substances, in addition to certain diseased conditions. For instance, exposure to lead from automobile exhaust is reported to be a risk factor for nephrotoxicity among traffic policemen (Mortada et al., 2001).

Renal function impairment manifests in a variety of different clinical presentations, some of which may be asymptomatic. The renal function impairment with asymptomatic presentations can only be detected by routine laboratory examinations. Azotaemia, a clinical condition associated with renal function impairment, is one of such presentations that can rightly be detected by laboratory findings. The condition is characterized by elevated levels of serum creatinine, urea and BUN (Cotran et al., 1999). A persistently increased serum creatinine is reported to be one of the risk factors for chronic kidney disease, which may results in renal failure (Mortada et al., 2001; Appel et al., 2003).

In this study, elevated levels of serum creatinine, urea, BUN, uric acid, glucose, and K^+, as well as decreased levels of serum Na^+ and Cl^- are reported for rats exposed to diesel and gasoline vapours. However, the derangement in the levels of these serum parameters recorded for rats exposed to diesel vapour was observed to be greater than that recorded for rats exposed to gasoline vapours.

The results of this study showed that exposure to diesel and gasoline vapours may be a predisposing factor for renal functions impairment in rats. This result agrees with our recent routine laboratory findings that exposure to gasoline vapour may cause elevation of serum urea, BUN and creatinine; an indication of renal function impairment in rats. In our previous studies, we have also observed that exposure to gasoline and/or kerosene vapours causes hepatotoxicity in rats (Uboh et al., 2005a, b; 2007a, b; 2008a, b).

The nephrotoxicity observed in this study suggests the presence of some nephrotoxic chemical substances in diesel and gasoline vapours. For instance Mortada et al. (2001) reported the presence of lead in automobile exhaust as a risk for nephrototoxicity among traffic policemen. While Halder et al. (1985) reported that lead is the component of gasoline responsible for nephrotoxicity observed to be associated with exposure to leaded gasoline. However, the specific chemical constituent(s) and mechanism(s) responsible for the nephrotoxic effect reported in the study is not very clear; but it is believed that the reactive metabolites of the hydrocarbons and other constituents of the vapours must have interacted with the renal tissues to cause derangements in glomerular function. Further study to elucidate the specific nephrotoxic constituent(s) and mechanism(s) responsible for nephrotoxicity reported in this study is in progress.

In conclusion, the result of this study indicates that diesel and gasoline vapours contain chemical constituents whose metabolites may interact with the renal tissues to impair renal functions in rats; and that the nephrotoxicity risk, as judged by changes in serum glucose and electrolytes, associated with diesel vapour is greater than that associated with gasoline vapour. Hence stringent regulation of the amount of the liquid and vapour forms of diesel and gasoline released inadvertently into the environments is highly recommended to the various environmental protection agencies.

REFERENCES

Anderson D, Yu TW, Philips BJ, Schmezer P (1995). An investigation of the DNA-damaging ability of benzene and its metabolites in human lymphocytes using the comest assay. Environ. Mol. Mutat. 26: 305-374.

Appel LJ, Middleton J, Miller ER, Lopkowitz M, Norris K, Agodoa LY, Bakris G, Douglas JG (2003). The Rational and Design of AASK. J. Am. Soc. Nephrol. 14: 166-172.

Atangwho JJ, Ebong PE, Eteng MU, Eyong EU, Obi AU (2007). Effect of Vernonia amygdalina Del Leaf on kidney function of diabetic rats. Int. J. Pharmacol. 3(2): 143-148.

Barham D, Trinder P (1972). An improved colour reagent for the determination of blood glucose by the oxidase system. Analyst 97: 142-145.

Borghoff SF, Youtsey NL, Sweenberg JA (1992). A comparison of European high test gasoline and PS-6 unleaded gasoline in their nephropathy and renal cell proliferation. Toxilet. 63: 21-33.

Carballo MA, Nigro ML, Dicarlo MB, Gasparini S, Campos S, Negri G, Gadano A (1995). Ethylene oxide II: Cytogenic and biochemical studies in persons occupationally exposed. Environ. Mol. Mutat. 25 (25): 81-97.

Chatterjea MN, Shinde R (2002). Renal function Tests. In: Textbook of Medical Biochemistry. 5th Edition. JAYPEE Brothers Med. Publ. Ltd. New Delhi pp. 564-570.

Chun JS, Burleigh-Flayer HD, Kintig WJ (1992). Methyl tertiary butyl ether: vapour inhalation oncogenicity study in F344 rats. Bushy Run Res. Center (BRRC)'s Report; 91NOO13B. BRRC, Export. PA.

Cotran RS, Kumar V, Collins T (1999). The kidney. In: Robbins Pathologic Basis of Disease, 6th Edi WB.Sanders Co. Philadelphia pp. 930-996.

Crook MA (2007). The kidneys. In: Clinical chemistry and metabolic medicine, 7th Edi. Bookpower, Britain pp. 36-57.

EHC 20 (1982). Selected petroleum products, In Environmental Health Criteria 20, United Nations Environment programme. The Intl. Org. and WHO. Geneva pp. 243-246.

Gidado A, Bashirat JY, Gana GM, Ambi AA, Milala Ma, Zanna H (2001). Effects of aqueous extract of the seeds of Datura stramonium on some indices of liver and kidney function in rats. Nig. J. Exp. Appl.

Biol., 2(2): 123-127.

Halder CA, Holdsworth CE, Cocokrell BY, Piccirillo VJ (1985). Hydrocarbon nephropathy in male rats. Identification of the nephrotoxic components of unleaded gasoline. Toxicol. Ind. Health, 1: 67-87.

Hard GC, Rodgers IS, Baetcke KP, Richards WL, McGaughy RE, Valcovic LR (1993). Hazard evaluation of chemicals that cause accumulation of alpha2? - globulin, hyaline droplet nephropathy, and tubule neoplasia in the kidneys of male rats. Environ. Health Perspect., 99: 313 -349.

Henderson RF, Sabourin PJ, Bechtold WE, Steinberg B, Chang IY (1993). Isobutene (2-methylpropene). Toxicol. Appl. Pharmacol., 123: 50-61.

Hu Z, Wells PG (1994). Modulation of benzo (a) pyrene bioactivation by glucuronidation in lymphocytes and hepatic microsomes from rats with a hereditary deficiency in bilirubin UDP-glucuronosyl-transferase. Toxicol. Appl. Pharmacol., 127: 306-313.

Jimoh FO, Odutuga AA (2004). Histological changes of selected rat tissues following ingestion of thermally oxidized groundnut oil. Biokemistri, 16: 1-10.

Kato M, Rocha ML, Carvallio AB, Chares ME, Rana MC, Oliverira FC (1996). Occupational exposure to neurotoxicants; preliminary survey in five industries of Camacari petrochemical complex, Brazil. Environ. Res., 136: 49-56.

Lorenzetti MS (1994). On the road with oxygenates. Chem. Bus. Jan., 15-17.

Mortada WI, Sobh MA, EL-Defrawy MM, Farahat SE (2001). Study of lead exposure from automobile exhaust as a risk for nephrotoxicity among Traffic Policemen. AM J. Nephrol., 21: 274-279.

Moser GJ, Wong BA, Wolf DC, Moss OR, Goldsworthy TL (1996). Comparative Short-term Effects of Methyl Tertiary Butyl Ether and Unleaded Gasoline Vapour in Female B6C3F1 Mice. Fundam. Appl. Toxicol., 31: 173 - 183.

Newman DJ, Price CP (1999). Renal function and Nitrogen Metabolites. CA. Burtis, ER Ashwood (Eds.), Tietz Textbook of clinical chemistry. 3rd Edn, Philadelphia. WB Saunders Co. Pp: 1204.

Nwankwo EA, Nwankwo B, Mubi B (2006). Prevalence of impaired kidney in hospitalized hypertensive patients in Maiduguri. Nig. Internet J. Int. Med., 6 (1).

Nygren J, Cedewal B, Erickson S, Dusinska M, Kolman A (1994). Induction of DNA strand breaks by ethylene oxide in human diploid fibroblasts. Environ. Mol. Mutagen., 24: 161-167.

Page NP, Mehlman M (1989). Health Effects of gasoline refueling vapours and measured exposures at service stations. Toxicol. Ind. Health, 5(5): 869-890.

Searcy RL, Reardon JE, Foreman JA (1967). Urea determination. Am. J. Med. Technol., 33: 15-20.

Smith TJ, Hammond SK, Wond O (1993). Health Effects of gasoline exposure I: Exposure assessment of US distribution workers. Environ. Health Perspect., 101(6): 13-21.

Standeven AM, Blazer T, Goldsworthy TL (1994). Investigation of antiestrogenic properties of unleaded gasoline in female mice. Toxicol. Appl Pharmacol., 127: 233 - 240.

Standeven AM, Goldsworthy TL (1993). Promotion of Preneoplastic Lesions and Induction of CYP2B by Unleaded Gasoline Vapour in Femnale B6C3F1 Mouse Liver. Carcinogenesis, 14: 2137 - 2141.

Standeven AM, Goldsworthy TL (1994). Identification of hepatic mitogenic and cytochrome p-450. Inducing fractions of unleaded gasoline in B6C3F1 mice J. Toxicol Environ. Health, 43: 213 - 224.

Tietz NW (1976). Fundamentals of Clinical Chemistry. Saunders WB company, Philadelphia, PA. Pp: 874-880.

Trinder P (1957). Analyst, 76: 596-600.

Uboh FE, Akpanabiatu MI, Atangwho IJ, Ebong PE, Umoh IB (2007). Effect of gasoline vapours on serum lipid profile and oxidative stress in hepatocyte of male and female rats. Acta Toxicol., 15 (1): 13-18.

Uboh FE, Akpanabiatu MI, Atangwho IJ, Ebong PE, Umoh IB (2008). Effect of vitamin A on weight-loss and heamatotoxicity associated with gasoline vapours exposure in Wistar rats. Int. J. Pharmacol., 4(1): 40-45.

Uboh FE, Akpanabiatu MI, Ebong PE, Eyong EU, Eka OU (2005). Evaluation of toxicological implications of inhalation exposure to kerosene and petrol fumes in rats. Acta Biol Szeged., 49(3-4): 19-22.

Uboh FE, Akpanabiatu MI, Ekaidem IS, Ebong PE, Umoh IB (2007). Effect of inhalation exposure to gasoline fumes on sex hormones profile in Wistar albino rats. Acta Endocrinol. (Buc), 3(1): 23-30.

Uboh FE, Akpanabiatu MI, Eteng MU, Ebong PE, Umoh IB (2008). Toxicological effects of exposure to gasoline vapours in male and female rats. Internet J. Toxicol., 4 (2)

Uboh FE, Ebong PE, Eka OU, Eyong EU, Akpanabiatu MI (2005). Effect of inhalation exposure to kerosene and petrol fumes on some anaemia-diagnostic indices in rats. Global J. Environ. Sci., 3(1): 59-63.

Zahlsen I, Tri-Tugaswati A (1993). Review of air pollution and its health impact in Indonsia. Environ Res., 63: 95-100.

Zanna H, Adeniji S, Shehu BB, Modu S, Ishaq GM (2008);. Effects of aqueous suspension of the root of *Hyphaene thebaica* (L:) mart on some indicators of liver and kidney function in rats. J. Pharmacol Toxicol., 3(4): 330-334.

Permissions

List of Contributors

A. Ologundudu
Department of Biochemistry, Adekunle Ajasin University, Akungba Akoko, Ondo State, Nigeria

A. O. Ologundudu
Department of Biochemistry, Adekunle Ajasin University, Akungba Akoko, Ondo State, Nigeria

O. M. Oluba
Department of Biochemistry, Faculty of Life Sciences, University of Benin, Benin-City, Nigeria

I. O. Omotuyi
Department of Biochemistry, Adekunle Ajasin University, Akungba Akoko, Ondo State, Nigeria

F. O. Obi
Department of Biochemistry, Faculty of Life Sciences, University of Benin, Benin-City, Nigeria

A. A. Onyeaghala
Department of Chemical pathology, University College Hospital, Ibadan. Nigeria

J. I. Anetor
Department of Chemical pathology, University College Hospital, Ibadan. Nigeria

A. Nurudeen
Department of Chemical pathology, University College Hospital, Ibadan. Nigeria

O. E Oyewole
Department of Health Promotion and Education, Faculty of Public Health, University of Ibadan

S. Awe
Department of Biological Sciences, Ajayi Crowther University, P. M. B. 1066, Oyo, Oyo State. Nigeria

E. Tunde Olayinka
Department of Chemical Sciences, Ajayi Crowther University, Oyo, Nigeria

Xudong Xu
Department of General Surgery, Tongji Hospital, Tongji Medical College, Huazhong University of Science and Technology, Wuhan 430030, P.R. China

Quan Sun
Department of General Surgery, Zhongnan Hospital, Wuhan University, Wuhan 430071, P.R. China

Zhisu Liu
Department of General Surgery, Zhongnan Hospital, Wuhan University, Wuhan 430071, P.R. China

Lin Zhang
Department of General Surgery, Tongji Hospital, Tongji Medical College, Huazhong University of Science and Technology, Wuhan 430030, P.R. China

Zhiyong Luo
Department of General Surgery, Tongji Hospital, Tongji Medical College, Huazhong University of Science and Technology, Wuhan 430030, P.R. China

Yun Xia
Department of General Surgery, Tongji Hospital, Tongji Medical College, Huazhong University of Science and Technology, Wuhan 430030, P.R. China

Yaqun Wu
Department of General Surgery, Tongji Hospital, Tongji Medical College, Huazhong University of Science and Technology, Wuhan 430030, P.R. China

Hurtado de Catalfo Graciela
INIBIOLP (Instituto de Investigaciones Bioquímicas de La Plata), CCT La Plata, CONICET-UNLP, Cátedra de Bioquímica y Biología Molecular, Facultad de Ciencias Médicas, Universidad Nacional de La Plata, 60 y 120 (1900) La Plata, Argentina

Astiz Mariana
INIBIOLP (Instituto de Investigaciones Bioquímicas de La Plata), CCT La Plata, CONICET-UNLP, Cátedra de Bioquímica y Biología Molecular, Facultad de Ciencias Médicas, Universidad Nacional de La Plata, 60 y 120 (1900) La Plata, Argentina

Alaniz María J. T. de
INIBIOLP (Instituto de Investigaciones Bioquímicas de La Plata), CCT La Plata, CONICET-UNLP, Cátedra de Bioquímica y Biología Molecular, Facultad de Ciencias Médicas, Universidad Nacional de La Plata, 60 y 120 (1900) La Plata, Argentina

Marra Carlos Alberto
INIBIOLP (Instituto de Investigaciones Bioquímicas de La Plata), CCT La Plata, CONICET-UNLP, Cátedra de Bioquímica y Biología Molecular, Facultad de Ciencias Médicas, Universidad Nacional de La Plata, 60 y 120 (1900) La Plata, Argentina

O. K. Sindiku
Department of Chemistry, University of Ibadan, Oyo State, Nigeria

O. Osibanjo
Department of Chemistry, University of Ibadan, Oyo State, Nigeria

Larry R. Williams
US Army Public Health Command (Provisional) [USAPHC (Prov)], Directorate of Toxicology, Aberdeen Proving Ground, MD, 21010, U.S.A

Cheng J. Cao
US Army Public Health Command (Provisional) [USAPHC (Prov)], Directorate of Toxicology, Aberdeen Proving Ground, MD, 21010, U.S.A

Emily M. Lent
US Army Public Health Command (Provisional) [USAPHC (Prov)], Directorate of Toxicology, Aberdeen Proving Ground, MD, 21010, U.S.A

Lee C. B. Crouse
US Army Public Health Command (Provisional) [USAPHC (Prov)], Directorate of Toxicology, Aberdeen Proving Ground, MD, 21010, U.S.A

Matthew A. Bazar and
US Army Public Health Command (Provisional) [USAPHC (Prov)], Directorate of Toxicology, Aberdeen Proving Ground, MD, 21010, U.S.A

Mark S. Johnson
US Army Public Health Command (Provisional) [USAPHC (Prov)], Directorate of Toxicology, Aberdeen Proving Ground, MD, 21010, U.S.A

Daniel A. Medesani
Department of Biodiversity and Experimental Biology, FCEyN, Pab. II, University of Buenos Aires, Ciudad Universitaria, C1428EHA Buenos Aires, Argentina

Claudio O. Cervino
School of Medicine, University of Morón, Machado 914, (1708) Morón, Pcia, Buenos Aires, Argentina

Martín Ansaldo
Argentine Antartic Institute, Cerrito 1248, 1010 Buenos Aires, Argentina

Enrique M. Rodríguez
Department of Biodiversity and Experimental Biology, FCEyN, Pab. II, University of Buenos Aires, Ciudad Universitaria, C1428EHA Buenos Aires, Argentina

A. B. Medani
Department of Pharmacology and Toxicology, Khartoum College of Medical Sciences, Sudan

S. M. A. El Badwi
Department of Pharmacology and Toxicology, Faculty of Veterinary Medicine, University of Khartoum, Sudan

A. E. Amin
Department of Pharmacology and Toxicology, Faculty of Veterinary Medicine, University of Khartoum, Sudan

Beatriz Bosch
Departamento de Ciencias Naturales, Facultad de Ciencias Exactas Físico-Químicas y Naturales (FCEFQN), Universidad Nacional de Río Cuarto (UNRC), Argentina

Fernando Mañas
Facultad de Agronomía y Veterinaria (FAV) Universidad Nacional de Río Cuarto (UNRC), Argentina
CONICET, Argentina

Nora Gorla
Facultad de Agronomía y Veterinaria (FAV) Universidad Nacional de Río Cuarto (UNRC), Argentina
CONICET, Argentina

Delia Aiassa
Facultad de Agronomía y Veterinaria (FAV) Universidad Nacional de Río Cuarto (UNRC), Argentina
CONICET, Argentina

Abdul Naveed
Department of Zoology, Panchsheel College of Education, Nirmal. A.P., India

C. Janaiah
Department of Zoology, Kakatiya University, Warangl 506 009, India

P. Venkateshwarlu
Department of Zoology, K. D. C. Warangal, Andhra Pradesh, India

Kiyun Park
Department of Fisheries and Ocean Science, Chonnam National University, San 96-1, Dundeok-dong, Yeosu, Jeonnam 550-749, Republic of Korea

Inn-Sil Kwak
Department of Fisheries and Ocean Science, Chonnam National University, San 96-1, Dundeok-dong, Yeosu, Jeonnam 550-749, Republic of Korea

D. O. Nwude
Department of Chemical Sciences, Bells University of Technology, Ota, Ogun State, Nigeria

J. O. Babayemi
Department of Chemical Sciences, Bells University of Technology, Ota, Ogun State, Nigeria

I. O. Abhulimen
Department of Chemical Sciences, Bells University of Technology, Ota, Ogun State, Nigeria

O. S. Adeyemi
Department of Chemical Sciences, Redeemer's University, P. M. B 3005, Redemption City, Mowe – 121001, Nigeria

M. A. Akanji
Department of Biochemistry, University of Ilorin, P. M. B 1515, Ilorin, Nigeria

A. O. Asita
Department of Biology, National University of Lesotho, P. O. Roma 180 Maseru, Lesotho, Southern Africa

E. B. Tanor
Department of Chemistry, National University of Lesotho, P. O. Roma 180 Maseru, Southern Africa

S. Magama
Department of Biology, National University of Lesotho, P. O. Roma 180 Maseru, Lesotho, Southern Africa

N. M Khoabane
Department of Chemistry, National University of Lesotho, P. O. Roma 180 Maseru, Southern Africa

F. C. Onwuka
Department of Biochemistry, Faculty of Basic Medical Sciences, University of Port Harcourt PMB 5323 Port Harcourt, River State, Nigeria

O. Erhabor
Department of Haematology College of Health Sciences, University of Port Harcourt P. M. B. 5323, Port Harcourt, Rivers State, Nigeria

M. U. Eteng
Department of Biochemistry, Faculty of Basic Medical Sciences, University of Calabar, P. M. B. 1115 Calabar, Cross River State, Nigeria

I. B. Umoh
Department of Biochemistry, Faculty of Basic Medical Sciences, University of Calabar, P. M. B. 1115 Calabar, Cross River State, Nigeria

L. C. Kerio
Tea Research Foundation of Kenya, P. O. Box 820, 20200, Kericho, Kenya

J. R. Bend
Department of Pathology, Schulich School of Medicine and Dentistry, University of Western Ontario, 1400 Western Road London, N6G 2V4, Canada

F. N. Wachira
Tea Research Foundation of Kenya, P. O. Box 820, 20200, Kericho, Kenya
Biochemistry Department, Egerton University, P. O. Box 536, Egerton, Kenya

J. K. Wanyoko
Tea Research Foundation of Kenya, P. O. Box 820, 20200, Kericho, Kenya

M. K. Rotich
Chemistry Department, Egerton University, P. O. Box 536, Egerton, Kenya

Zineb Elyoussoufi
Laboratory of Physiology and Molecular Genetics associated with CNRST, Department of Biology, Ain Chock Faculty of Sciences, Hassan II University, Casablanca, Morocco
Laboratory of Experimental Medicine and Biotechnology, Faculty of Medicine and Pharmacy, University Hassan II, Casablanca, Morocco

Norddine Habti
Laboratory of Experimental Medicine and Biotechnology, Faculty of Medicine and Pharmacy, University Hassan II, Casablanca, Morocco

Said Motaouakkil
Laboratory of Experimental Medicine and Biotechnology, Faculty of Medicine and Pharmacy, University Hassan II, Casablanca, Morocco
Medical intensive care unit, Ibn Rochd university hospital, Casablanca, Morocco

Rachida Cadi
Laboratory of Physiology and Molecular Genetics associated with CNRST, Department of Biology, Ain Chock Faculty of Sciences, Hassan II University, Casablanca, Morocco

Abolanle Azeez A. Kayode
Department of Chemical Sciences, Biochemistry Unit, Bells University of Technology, Ota, Ogun State, Nigeria

Joshua Olajiire Babayemi
Department of Chemical Sciences, Industrial Chemistry Unit, Bells University of Technology, Ota, Ogun State, Nigeria

Esther Omugha Abam
Department of Chemical Sciences, Biochemistry Unit, Bells University of Technology, Ota, Ogun State, Nigeria

Omowumi Titilola Kayode
Department of Chemical Sciences, Biochemistry Unit, Bells University of Technology, Ota, Ogun State, Nigeria

P. Dailiah Roopha
Department of Advanced Zoology and Biotechnology, Rani Anna Government College for Women, Tirunelveli-627 008, Tamil Nadu, India

J. Savarimuthu Michael
Department of Advanced Zoology and Biotechnology, Sri Paramakalyani College, Alwarkurichi-627 412, Tamil Nadu, India

C. Padmalatha
Department of Advanced Zoology and Biotechnology, Rani Anna Government College for Women, Tirunelveli-627 008, Tamil Nadu, India

A. J. A. Ranjit Singh
Department of Advanced Zoology and Biotechnology, Sri Paramakalyani College, Alwarkurichi-627 412, Tamil Nadu, India

Esmail Gharedaashi
Department of Fishery, Gorgan University of Agricultural Sciences and Natural Resources, Gorgan, Iran

Hamed Nekoubin
Department of Fishery, Gorgan University of Agricultural Sciences and Natural Resources, Gorgan, Iran

Alireza Asgharimoghadam
Department of Fishery, Gorgan University of Agricultural Sciences and Natural Resources, Gorgan, Iran

Mohammad RezaImanpour
Department of Fishery, Gorgan University of Agricultural Sciences and Natural Resources, Gorgan, Iran

Vahid Taghizade
Department of Fishery, Gorgan University of Agricultural Sciences and Natural Resources, Gorgan, Iran

K. B.Pone
Central Institute of Medicinal and Aromatic Plants, P. O. CIMAP, Kukrail Picnic Spot Road, 226015-Lucknow, Uttar Pradesh, India
Department of Biochemistry, Faculty of Science, University of Dschang. P. O. Box 67 Dschang, Cameroon

P. B. Telefo
Department of Biochemistry, Faculty of Science, University of Dschang. P. O. Box 67 Dschang, Cameroon

I. Gouado
Department of Biochemistry, Faculty of Science, University of Douala, P. O. Box 24157, Douala, Cameroon

F. M. Tchouanguep
Department of Biochemistry, Faculty of Science, University of Dschang. P. O. Box 67 Dschang, Cameroon

P. Dzomba
Chemistry Department, Faculty of Science Education, P Bag 1020 Bindura, Zimbabwe

S. Nyoni
Chemistry Department, Faculty of Science Education, P Bag 1020 Bindura, Zimbabwe

N. Mudavanhu
Chemistry Department, Faculty of Science Education, P Bag 1020 Bindura, Zimbabwe

O.Salawu Emmanuel
Department of Physiology, Ladoke Agential University, Ogbomoso, Nigeria

A. Adeleke Adeolu
Department of Physiology, Ladoke Agential University, Ogbomoso, Nigeria

O. Oyewo Oyebowale
Department of Anatomy, Ladoke Akintola University, Ogbomoso, Nigeria

A. Ashamu Ebenezer
Department of Anatomy, Ladoke Akintola University, Ogbomoso, Nigeria

A. Ishola Olufunto
Department of Physiology, Ladoke Agential University, Ogbomoso, Nigeria

O. Afolabi Ayobami
Department of Physiology, Ladoke Agential University, Ogbomoso, Nigeria

A. Adesanya Taiwo
Department of Biochemistry, Ladoke Akintola University, Ogbomoso, Nigeria

Ngozi E Abu
Department of Botany, University of Nigeria, Nsukka, Nigeria

K. C. Mba
Department of Botany, University of Nigeria, Nsukka, Nigeria

Jacobus Engelbrecht
Department of Environmental Health, Faculty of Science, Tshwane University of Technology, Private Bag X680, Pretoria, 0001, South Africa

Phanuel Tau
Department of Environmental Health, Faculty of Science, Tshwane University of Technology, Private Bag X680, Pretoria, 0001, South Africa

Charles Hongoro
Health Systems Research Unit, South African Medical Research Council, South Africa

F. E. Uboh
Department of Biochemistry, College of Medical Sciences, University of Calabar, Calabar, Cross River State, Nigeria

M. I. Akpanabiatu
Department of Biochemistry, College of Medical Sciences, University of Uyo, Uyo, Akwa Ibom State, Nigeria

J. I. Ndem
Department of Biochemistry, College of Medical Sciences, University of Uyo, Uyo, Akwa Ibom State, Nigeria

Y. Alozie
Department of Biochemistry, Cross River State University of Technology, Calabar, Nigeria

P. E. Ebong
Department of Biochemistry, College of Medical Sciences, University of Calabar, Calabar, Cross River State, Nigeria

www.ingramcontent.com/pod-product-compliance
Lightning Source LLC
Chambersburg PA
CBHW080629200326
41458CB00013B/4567